BUILDING INTERIORS, PLANTS AND AUTOMATION

AUTOMATED, PRECISION, MICRO-IRRIGATION ™ SYSTEMS
A Guide for Architects, Interior Designers, Engineers, Contractors, Interior Landscapers, Building Managers/Developers/Owners, Corporate Office and Facilities Managers

Stuart D. Snyder

Prentice Hall
Englewood Cliffs, New Jersey 07632

Library of Congress Cataloging-in-Publication Data

Snyder, Stuart D.
    Building interiors, plants and automation : automated micro
  -irrigation systems, a guide for architects, interior designers,
  engineers, contractors, interior landscapers ... / Stuart D. Snyder.
        p.   cm.
    Bibliography: p.
    Includes index.
    ISBN 0-13-086224-X
    1. Interior landscaping. 2. Buildings--Irrigation systems-
  -Automation. 3. House plants--Irrigation. I. Title.
  SB419.25.S68 1990
  635.9'82--dc20   89-35723
                                                                  CIP

Editorial/production supervision and
  interior design: Tally Morgan, WordCrafters Editorial Services, Inc.
Cover design: Karen Stephens
Background cover photo: Listenberger Design Associates
Manufacturing buyer: Dave Dickey

 ©1990 by Stuart D. Snyder

All rights reserved. No part of this book may be
reproduced, in any form or by any means,
without permission in writing from the publisher.

Printed in the United States of America
10 9 8 7 6 5 4 3 2 1

ISBN 0-13-086224-X

Prentice-Hall International (UK) Limited, *London*
Prentice-Hall of Australia Pty. Limited, *Sydney*
Prentice-Hall Canada Inc., *Toronto*
Prentice-Hall Hispanoamericana, S.A., *Mexico*
Prentice-Hall of India Private Limited, *New Delhi*
Prentice-Hall of Japan, Inc., *Tokyo*
Simon & Schuster Asia Pte. Ltd., *Singapore*
Editora Prentice-Hall do Brasil, Ltda., *Rio de Janeiro*

DEDICATION

This book is dedicated to John Portman, renowned architect, planner and real estate developer, and to Dr. Billy C. Wolverton, Senior Research Scientist, Head of the Environmental Research Laboratory at NASA's Stennis Space Center, and a leader in the field of space biotechnology.

Mr. Portman, more than any other design professional, has created and inspired the creation of wonderful and verdant building interiors . . . a lover of nature who has deftly combined architectural elements with lavish, natural greenery, achieving an esthetic harmony that touches our spirit.

Dr. Wolverton, another man of vision and sensitivity toward things natural, is furthering the interior landscape movement in another significant way. His spinoff research into the use of ornamental foliage as a means of reducing major air pollutants in our homes, shops and offices promises to improve and prolong our lives in ways not yet fully appreciated.

Both have been an inspiration to this author in developing the concepts described in the following pages.

CONTENTS

ACKNOWLEDGMENTS, xi

Chapter 1
AN OVERVIEW: People, Plants, Buildings and Technology, 1

Chapter 2
USE OF PLANTS IN INTERIOR DECORATING, 5

Overview, 5
Historical Summary, 5
Industry Statistics, 6
Plants and Emotional Health, 8
Plants and Decor, 10
Plants and Leisure Time, 11
Plants as a Natural Element, 12
Humanistic Architecture and Interior Design, 13
Plants in the Workplace, 14
Improved Tropical Plants, 23
Publicity and Education, 23
Development of Interior Plantscapers, 23
Plants and Indoor Air Pollution, 24
Plants and "Healthy" Buildings, 27
Plant-Care Automation, 28

Chapter 3
TECHNOLOGY IN BUILDINGS TODAY, 30

Overview, 30
Building Design and Technology, 30
Buildings as Functional Structures, 31
"Intelligent" Buildings, 31
The Computer-Regulated Structure, 34
Building Intelligence Defined, 35
The Appeal of Intelligent Buildings, 36
The Economics of Intelligent Buildings, 36
High Technology in Housing, 37
Computer Control in Housing, 38
Technology Available for Smart Homes, 38
Project "Smart House", 39
Micro-Irrigation Systems, 39

Chapter 4
THE INDUSTRY THAT BUILDS, DECORATES,
   AND MANAGES BUILDINGS, 40

Overview, 40
Real Estate Developers, 40
Builders, Construction Companies, and General Contractors, 41
Building Managers, 41
Architects, 42
Interior Designers, 43
Engineers, 44
Landscape Architects, 44
Interiorscapers, 45
Interiorscaping Industry Profile, 47
Tropical Foliage Nurseries, 50
Irrigation Contractors, 50

Chapter 5
TYPES OF PLANTS USED FOR INTERIORSCAPING, 52

Overview, 52
Houseplants Versus Interiorscaping Plants, 52
Tropical Foliage Plants, 53
Acclimatization, 53
Nursery Irrigation, 55
Nursery Soil, 56
Plant Sizes Used, 56
Plant Characteristics, 56
Plant Varieties, 57

Chapter 6
PLANT BIOLOGY AND OTHER GROWTH FACTORS, 59

Overview, 59
Plant Growth Cycle, 60
Soil Moisture, 62
Light, 62
Nutrients, 64
Atmospheric Conditions, 65
The Growing Medium, 67
Pests and Diseases, 69
Summary, 69

Chapter 7
INTERIORSCAPE IRRIGATION: MANUAL VERSUS AUTOMATIC, 70

Overview, 70
Manual Plant-Care Techniques, 71
Moisture Meters, 76
Self-Watering Containers, 80
Automated Plant-Care Techniques, 84

Chapter 8
INTERIORSCAPE IRRIGATION: FULLY AUTOMATIC TECHNIQUES, 89

Overview, 89
Sprinkler Systems, 89
Drip/Trickle Systems, 96
Subterranean Systems, 103
Hydroponic Systems, 105
Micro-Irrigation Systems, 107
Uses of Micro-Irrigation Systems, 121

Chapter 9
THE CONCEPTS OF MICRO-IRRIGATION SYSTEMS, 124

Overview, 124
The Pulse-Flow Concept, 125
Moisture Diffusion Patterns, 129
Moisture Diffusion Studies, 130
Mulches and Moisture Retention, 139

Chapter 10
THE EQUIPMENT OF MICRO-IRRIGATION SYSTEMS, 142

Overview, 142
The Basics, 143
Categories of Micro-Irrigation Systems, 143
High-Pressure Systems, 143
Low-Pressure Systems, 146
Sequence of Operation, 147
Control Centers, 148
Timers, 148
Solenoid Valve Controllers, 149
Pump Controllers, 152
Pump/Reservoir Modules, 154
Tubes and Pipes, 156
Fittings, 158
Fitting Specifications, 161
Flow Control Devices, 162
Emitters, 167
Accessories, 169
Packaged Systems, 173

Chapter 11
AUTOMATIC IRRIGATION AS APPLIED TO INTERIORSCAPES, 174

Overview, 174
Commercial Buildings and Facilities, 174
Office Buildings, 176
Corporate Offices, 182
Hotels, 185
Resort Complexes, 186

Restaurants and Lounges, 189
Shopping Malls and Arcades, 191
Retail Establishments, 192
Banks, 193
Clubhouses and Recreation Halls, 194
Hospitals and Medical Offices, 194
Apartment Buildings, 194
Residential Buildings, 195
Single Family Homes, 196
Apartment Suites, 198
Marine Applications, 200

Chapter 12
THE DESIGN AND COSTING OF MICRO-IRRIGATION SYSTEMS, 202

Overview, 202
Defining the Design Problem, 202
The Sample Layout, 203
Technical Considerations, 203
Solenoid Valve and Pump/Reservoir Ratings, 204
Water Flow Variables, 205
Flow Characteristics of Tubing, 205
The Effect of Gravity, 206
The Effect of Tubing Fittings, 206
The Effect of Emitters, 207
The Location of Control Centers, 207
The Routing of Water Distribution Networks, 209
Important Flow Controls, 210
The Designing and Costing of a System Installation, 210
Designing for Commercial Buildings, 213
Example of System Design for an Office Building, 214
Example of System Design for an Office Suite, 217
Example of System Design for a Single Family Home, 223
Computer-Aided Design and Costing, 228

Chapter 13
THE INSTALLATION OF MICRO-IRRIGATION SYSTEMS, 229

Overview, 229
General Layout Considerations, 229
Installing Control Centers, 230
Installing Water Distribution Networks, 232
Routing Tubing Lines for Integral Systems, 232
Routing Tubing in Retrofit Installations, 240
Tubing/Fitting Connections, 246
Installing Irrigation Receptacles and Performing Other Finishing Tasks, 248
Installing Control Center Equipment, 248
Installing Irrigation Receptacles, 249
Installing Irrigation Manifolds, 249
Installing Emitter Tubes, 250
Installing Systems in Planter Boxes, 251
Installing Outdoor Systems, 252
Interfacing with Other Technologies, 256

Chapter 14
STARTING AND OPERATING A NEW INSTALLATION, 260

Overview, 260
Start-Up Procedures, 260
Operating Procedures, 263

Chapter 15
THE FUTURE, 267

Overview, 267
Technology in Commercial Buildings, 267
The Outlook for Smart House Technology, 269
Plants in Buildings, 271
The Outlook for Automated Interior Plant Care, 273
A Typical Day in the Future, 274

APPENDICES

Appendix A: Interiorscaping Plants and Their Preferred Conditions, 277
Appendix B: Tubing and Pipe Fittings Nomenclature, 286
Appendix C: Sources of Supply, 290

BIBLIOGRAPHY, 295

INDEX, 301

ACKNOWLEDGMENTS

The author is indebted to many for their contributions to this book. The librarians at the Florida Atlantic University Library (Boca Raton), the Boca Raton Public Library and Broward County (Fort Lauderdale) and Palm Beach County (West Plam Beach) libraries were exceedingly helpful in searching out books and materials. Their patience and professional response was highly valued, as was the input of Jerry Campbell (U.S. Dept. of Agriculture, Orlando) in helping to define the floriculture industry. Thanks go out to Edmund J. Mazzei (President, Building Owners and Managers Association, International of South Florida), Phil Cialone (President, Tropical Ornamentals) and Joe Cialone (Vice President, Tropical Ornamentals) for their professional reviews and criticisms, which were extremely helpful. To Dr. B.C. (Bill) Wolverton (NASA, Stennis Space Center), Ramesh Krishnaiyer (NASA/STAC, Fort Lauderdale) and Dr. Donald L. Henninger (NASA, Johnson Space Center) for their help in guiding me through a maze of NASA agencies, contacts and documentation. To Paul F. Corey (Editor, Prentice Hall), for his patience and guidance through the unfamiliar world of publication. To Merritt Winsby for the computer assistance he furnished. To Cliff Tager, for making possible the opening of this technology to the world. The contributions of Lillian and Eli Synder are humbly acknowledged for the assistance they gave, in ways that could only encourage and sustain.

My special feelings of appreciation are directed to Harriet Victoria Ayers, without whom this book could not have been written. Her help, understanding and encouragement were profound.

Stuart D. Snyder
Delray Beach, Florida
February 27, 1989

1 AN OVERVIEW: PEOPLE, PLANTS, BUILDINGS, AND TECHNOLOGY

Much has been written about the application of technology to the design and management of commercial and residential buildings. It pervades the literature. Among the concepts common now are CAD (computer-aided design) systems, integral telephone networks, internal fiber-optic data networks, master energy management systems, and computer-controlled fire safety and security systems. The quest for useful, efficient, sophisticated, and visually pleasant buildings stems in part from the need to accommodate the requirements of the computer and communication age in which we live. The future will see more of these electronic marvels integrated into increasingly "intelligent" structures, but there will also be a greater movement to accommodate our human needs. More buildings will be designed with the elements needed to counteract the harmful effects of high technology and our immediate environment. It was probably John Naisbitt of *Megatrends* fame who said it best and most visibly. He has warned of the need in our lives for softening the effects of high tech with what he calls "high-touch"—in other words, the need to rehumanize the workplace, the home, and our after-hours habits and haunts. The new electronic tools at our disposal tend to have a negative effect on our mental and physical well-being. The medical community is concerned that we are inventing ourselves into debilitating emotional problems, which in the long run may decrease the work efficiency that our newer tools were installed to enhance. As key instrumentalists in providing us with high-touch, the architect, interior designer, building developer, real estate executive, and corporate office manager must rethink the importance of softer surroundings. We must be put into contact with elements of decor that help to preserve our links with nature. The impersonal manifestations of our new technology are alien to us and must be tempered with the more familiar sights, sounds, and smells with which our human spirit has become comfortable over the ages. This is part of the psychology behind the rapidly growing use of live tropical plants in our interior surroundings—nature brought indoors. It restores the soul to have plants. People need that like they need fresh water and sunsets and daydreams.

When most of us think about plants, we generally think of gardening. This book is not about gardening, however. It deals with plants as live elements of decor and as important contributors to a healthy interior environment. It is about the new technology that makes it easier and less costly to take care of interior greenery. It is also about the confluence, or coming together, of building and interior design, tenant needs, health concerns, interior horticulture, and technology for state-of-the-art building management.

The emotional benefits of ornamental plants are just starting to be recognized. The past 15 years has seen an unprecedented rise, of boom proportions, in their use. Owners and managers of restaurants, lounges, sports clubhouses, shopping malls, and so on, know that we feel more rested and at peace with the world when surrounded by natural greenery. We are more apt to frequent those places that offer our spirits a rest from the frenetic workplace that occupies much of our lives. Our home environment is becoming a more natural setting for the same reasons. Furthermore, the corporate office manager is beginning to learn that employee efficiency can be helped by sprinkling the office with attractive potted plants and by being located in an office building where the common areas are graced with cool, woodland-like atriums, patios, and rest stations. This important social and economic trend may someday be labeled something like "the greening of American life-style," but as we go through it, the phenomenon is barely noticed, or in many quarters, not even taken seriously. It is certainly not fully understood, either in its motivation or in its implications. One thing we are sure about, however, is that it can't be bad. There are too many recognized benefits attached to the use of live plants to dismiss the subject casually.

Another new facet is developing that is destined to have an even greater influence. NASA environmental research has recently demonstrated the unusually efficient cleansing effects that plants can have on polluted air in building interiors. Many companies and agencies are worried about the air pollution inside energy-efficient buildings—mainly our homes and offices. It is being called "Sick buildings syndrome," as these structures promote human illnesses, some very serious ones. A great deal of work time has been lost, efficiency has declined, and some deaths have occurred because of it. The NASA biotechnology studies are demonstrating the usefulness of live plants in helping to overcome some of these debilitating health problems. All of the factors discussed will happily bring great quantities of live tropical plants into the closer confines of our residential and commercial worlds, serving to enrich our lives.

As with everything else, however, there is a price—the plants must be cared for. Like pets, they bring enjoyment but must be nurtured properly. This can become a burden, particularly in large commercial installations where hundreds or even thousands of plants are involved. It is with these changes in American social patterns, building design, and business priorities in mind that a specialized technology has been developed to reduce the costs and inconveniences of maintaining foliage and flowering plants in all areas of a building's interior. This new technology is called *Micro-Irrigation*™[1] *automated, precision, interior plant-care systems*. Some versions are designed for the house or apartment with only a few potted plants scattered around the furnished living quarters. Now they can be cared for easily and effectively, even if the owners are away on business or vacation. Other versions are designed for the manager of a major office building, who must pay dearly for the services of interior landscape contractors to maintain the greenery in expansive, common areas. And there are hundreds of applications in between the two mentioned here. There is no longer any reason for an installation of ornamental plants, large or small,

---

[1]Micro-Irrigation is a trademark of Aqua/Trends. All subsequent references to Micro-Irrigation in this text refer to this trademark.

indoors or out, to be without the benefit of automated, plant-care technology. The new precision plant-care systems that will be described in detail in this book have been developed specifically for critical indoor applications. They are automated building management tools that help the homeowner and the commercial manager operate and maintain their facilities in ways never before available. They are a new and interesting step forward in the development of "intelligent" or "smart" buildings. The use of innovative technology is a strong trend in the residential markets, as well as the real estate/building industry. Micro-Irrigation automated, precision, interior plant-care systems (hereafter called simply Micro-Irrigation Systems in the interest of brevity) are part of a wave of new products designed to provide us with healthier, more rewarding life-styles, as well as ways to more easily and cost-effectively manage our businesses.

Automated, interior plant care has many implications, having to do not only with horticultural issues but also with building design, interior design, irrigation design, building construction, real estate marketing and management, corporate office management, hospitality industry management, and certain areas of office and marine product design. These facets will be explored in later chapters. The information will be useful to architects, interior designers, engineers, interior landscape contractors, building owners and managers, real estate agents, plumbing contractors, and do-it-yourselfers, as well to those who plan to become specialists in the technology of automated, interior plant care. It is the purpose of this book to provide the reader with an understanding of the implications, as well as a foundation in the techniques, of *auto-interigation* (automatic interior irrigation). We will concentrate on the new precision systems, the technology used in Micro-Irrigation Systems, developed for comprehensive interior coverage. The information contained in this edition is detailed enough for the reader to be able to handle many of the design and installation problems that might be encountered, and will provide the foundation for further study. Another important objective of the book is to induce design, office management, and real estate professionals to take ornamental plants and their needs more seriously. The current state of the industry is such that the needs of interiorscapes are not being taken into consideration with any degree of regularity, understanding, or expertise.

Many of the rules that have become tradition over the years in interior plant care just don't apply when dealing with the technology used in Micro-Irrigation Systems, for it works differently and the mechanisms involved are not the same. Research carried out during the development of these new concepts demonstrated differences in moisture distribution patterns and soil behavior that clearly contradict some of the old "bromides" that have guided indoor plant care. Concepts of interior landscape maintenance are being broadened, and new ideas must be digested by the professionals that deal with it. Because automation in this field is so new, very few know how to use it. There must be some rethinking and relearning in order to cope with these modern methods and to unlock their usefulness. Anyone holding to the opinion that automatic irrigation is harmful to interior plantscapes is grossly misinformed. The hundreds of thousands of plants successfully serviced by these modern techniques over the past decade add strong testimony to that fact. Micro-Irrigation Systems now open new opportunities for complete, fully automated coverage of building interiors, especially in furnished areas where the other technologies cannot be used. It promises to revolutionize the way indoor plants are taken care of.

Some of the new systems are for integration into the structure of a building and must be planned with the same care that goes into the electrical, plumbing, data distribution, and other functional systems. It is not necessary to be an interior landscape specialist or any other type of plant-oriented person in order to design, install, or use these automated service systems—just well informed. Solid knowledge of the subject is important. We feel that has been provided within these pages.

At the time your author conceived the idea that eventually evolved into the technology used in Micro-Irrigation Systems, his interest in decorative plants was simply a personal one. His background in American industry had been technically oriented, in engineering disciplines far removed from horticultural fields of endeavor, but heavily involved in research and new product development. After years of working intimately with the methods and biology of this field, a great deal has been learned and contributed; but the author does not consider himself a plant or interiorscaping specialist. However, one does not have to begin as an expert in a field to seek knowledge, to develop new concepts, or to uncover new truths. Expertise comes with the exploration of areas neglected by others. The author has developed and engineered Micro-Irrigation Systems and other automated technologies for the management of buildings, which most accurately qualifies him as an authority on automated, interior plant-care systems. It is this and other pertinent technology that will be the subjects of discussion in the following pages. Most of the information written about plant biology and care is merely reportage on the part of the author: the passing along of others' expertise. For greater detail about the design of interiorscapes or about the nurturing of ornamental plants in interior settings, one must consult books written by authorities in those aspects of the industry. There are many good ones and the reader is encouraged to delve more deeply into the subject.

This book is the definitive work in the field of automatic interior irrigation (auto-interigation), but we have much to learn beyond its lessons. The discipline is new and is part of the growing list of technological breakthroughs that will make the future a better and more efficient place. It's exciting to be able to make a building do useful work; they are appropriately called "intelligent" or "smart" buildings. Automation is generally recognized as the wave of the future in real estate management, and automated plant care is the newest addition to the building owner and manager's arsenal of cost-effective tools. Those of you who learn the subject well will find a demand for your knowledge. It will create a new and enriching dimension to your career.

Building Interiors, Plants and Automation can be thought of as being divided into nontechnically detailed material and technically detailed material. The former fills the first eight chapters as well as Chapters 11 and 15 and should be of interest to all, but particularly to those in the disciplines of real estate development and management, home building, corporate office management, restaurant, hotel, and other facilities management, and interior design. The more technically detailed material is found in Chapter 8 and beyond, with the exception of Chapters 11 and 15. This material should be of special interest to architects, interior landscape architects, engineers, interiorscapers, plumbing contractors, and specialty contractors wanting to learn about this new discipline. The technically oriented homeowner will also find many sections of this book to be of interest and help.

2 USE OF PLANTS IN INTERIOR DECORATING

Rooms without any living plants often seem sterile and lacking in warmth. On the other hand, rooms containing house plants appear to be full of vitality and a sense of intimacy. The presence of plants, like people, gives meaning and animation to a room full of otherwise inanimate objects.[1]

Overview The natural beauty of decorative tropical foliage graces all aspects of our lives—at home as well as where we work, shop, relax, and play. The reasons are many, touching on life-style, economics, esthetics, and emotion. Psychologists tell us that we have a natural affinity for plants as they are restful, give us a sense of well-being, and evoke visions of exotic places. We are attracted to plants. We love being around them. Some feel these bonds are deeply rooted and stem from our evolution as human beings. Architects and interior designers accent decor with live tropical plants to help soften the straight architectural lines of our walls, floors, ceilings, and furniture. They are able to create a quieting, relaxing atmosphere in our homes. At work, live plants help reduce environmental harshness, add a pleasant note to the workday, and keep us in touch with nature, even in the midst of the electronic and mechanical marvels that surround us. Merchants find customers are more comfortable and receptive in the relaxed atmosphere of decorative greenery. In this chapter, we will deal with factors influencing the strong emergence of interior landscapes, from their most simplistic to their most complex forms.

Historical Summary The use of live plants in building interiors is a rapidly growing and unprecedented trend. There is nothing new about it, using plants indoors that is, but never before has there been so much emphasis on it. As far back as 3,000 years, the Chinese are said to have used plants in containers within their homes.[2] Through the ages, containerized plants have been used indoors in conservatories, greenhouses, and solariums, as well as in other decorative and random ways. It has only been within the past 15 years that live foliage plants have

---

[1]William E. Hague, *The Complete Basic Book of Home Decorating* (Garden City, NY: Doubleday & Co., 1976), p. 493.
[2]George H. Manaker, *Interior Plantscapes* (Englewood Cliffs, NJ: Prentice-Hall, Inc., 1987), p. 15.

seemingly exploded into popularity with American homeowners and design professionals for complementing furnished decor and architectural interiors. It has thus become *an industry*. We have all noted the increasing numbers of stores, shopping malls, lounges, restaurants, banks, clubhouses, and hotels that have become the subjects of interior landscaping. In the home, we have likewise made greater use of flowering and foliage planters, from tiny cactus plants to large trees growing in tubs or planter beds. Live plants are used wherever decor is important. A "green revolution" is in fact taking place.

Industry Statistics The magnitude of this trend is readily apparent when we look at industry figures over the past few years. Professionalism started developing in the interior landscape business about 1970. From that point to 1985, 15 years later, producer shipments of foliage plants for indoor use multiplied by 1600 percent. At $477.0 million, 1988 production was again 1600 percent greater than 1970 figures, down slightly from the 1986 peak of $521.4 million. The construction slowdown is now being felt. While it is true that inflationary factors come into play here and temper growth figures somewhat, the comparison nevertheless shows the trend in terms of boom proportions. To these figures must be added an estimated $45 million in flowering plants sold in hanging baskets and used indoors, plus other categories of potted flowers that find use in interiorscapes. Figure 2.1 illustrates U.S. tropical foliage plant shipments from 1970 to 1986.

Most tropical foliage for interior landscape uses are grown in the subtropical regions of the U.S., the areas that are closest in climate to the natural habitat of the plant species used. Latest statistics show Florida to be the largest producer by far, with 58 percent of the 1988 sales volume of $477.0 million (see Table 2.1). The second largest producer is California, followed by Texas and a host of other states.

The U.S. Department of Agriculture, the agency that compiles these figures, classifies potted, tropical foliage plants as those that are used indoors and on patios. The industry category is called "Floriculture Crops," and the subcategory dealing with indoor plants is entitled "Foliage for Indoor or Patio Use." There has been some uncertainty in the industry and within the USDA statistical services. From 1982 through 1988 (the latest year reported), data is sometimes unavailable (1982 and 1983, for example) or is inconsistent in reporting product categories (1984 through 1986, for example). Some of this had to do with the tightening of federal budgets, which forced the elimination of Floricul-

TABLE 2.1

U.S. Tropical Foliage Plant Shipments for Interior Applications—1988 (Wholesale Value in Millions of Dollars)

State	Value of Shipments
Florida	$279.0
California	92.7
Texas	19.1
Ohio	13.5
Hawaii	12.4
Other	60.3
Total	$477.0

Source: Floriculture Crops, *Annual Summaries,* USDA Crop Reporting Board, Sp. Cr. 6-1 (April, 1988).

Industry Statistics

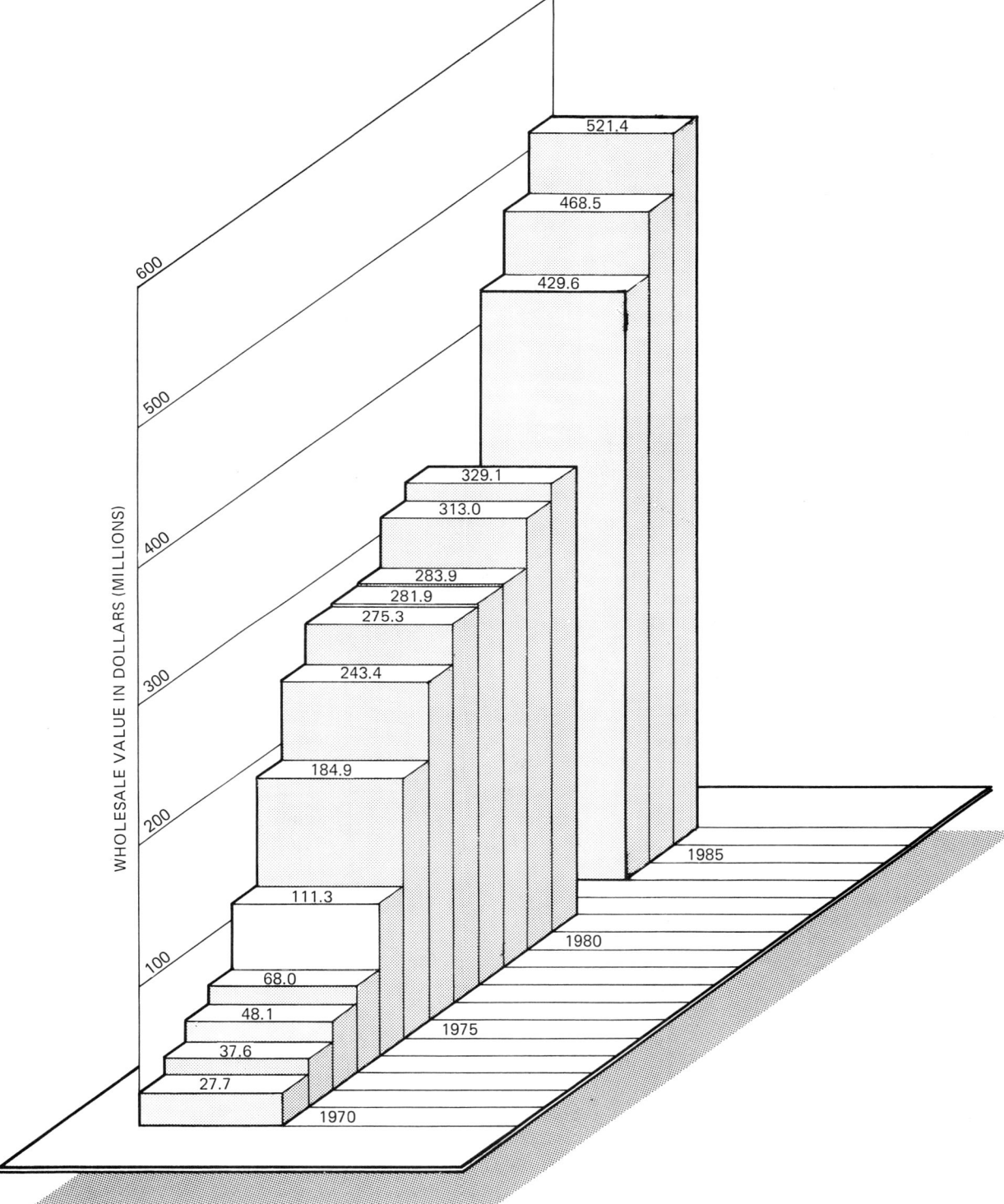

Figure 2.1 U.S. Tropical foliage plant shipments for interior and patio applications—1970 to 1986 (wholesale value in millions of dollars). (Flowers & Foliage Plants, *Production & Sales,* USDA Crop Reporting Board; Horticulture Crops, *Annual Summaries,* USDA Crop Reporting Board; Sp. Cr.)

ture Crop reporting in 1982 and 1983. It took determined industry lobbyists to get the USDA to reconsider and renew statistical coverage in 1984. Other problems were caused by indecision over which product categories should be reported. The USDA now omits flowering potted plants from figures reflecting indoor usage, yet we all know that many flowers are used inside of buildings. The industry and statistical services are still wallowing in the newness of the moment, and it will take a few more years for things to sort out and be cohesive. Meanwhile, the Tropical Foliage Industry continues to grow as the "greening phenomenon" expands.

Plants and Emotional Health

The trends that have been shaping industry growth are expected to continue and most assuredly to strengthen, in spite of temporary slowdowns in commercial construction caused by excesses in current inventory. There are underlying reasons for these forecasts. One has to do with the fact that as valuable as computers and other high tech work tools are to our business economy, they cause some human problems that industrial psychologists, medical doctors, business efficiency experts, and corporate managers are becoming worried about. The human spirit is having trouble adapting to the way our lives are being invaded and manipulated by high tech electronics and space-age procedures, and in various ways we want to fight back. One manifestation of that was the back-to-nature movement of the '60s and '70s.

In his best-selling book *Megatrends,* John Naisbitt proposes that for every technical advance there should be an equal and opposite modifying influence that would tend to rehumanize the world of those affected by the technology, so as to overcome its debilitating effects. He calls these rehumanizing, balancing influences "high-touch." Mr. Naisbitt puts it this way:

> We must learn to balance the material wonders of technology with the spiritual demands of our human nature.
>
> The need for compensatory high-touch is everywhere. The more high-tech in our society, the more we will want to create high-touch environments, with soft edges balancing the hard edges of technology.
>
> Ahead of us for a long period is an emphasis on high-touch and comfort to counterbalance a world going mad with high-technology.[3]

In the Naisbitt concept, high-touch takes many forms: revival of biking or hiking, use of handwritten notes, greater social interaction, handcrafting, and individualistic decorating and gardening among other things. Live tropical plants are one of the strongest influences available for developing a more comfortable and natural environment, the ultimate in high-touch. Mr. Naisbitt is one of the most visible and highly respected of our contemporary seers. His business and social advice is taken very seriously and should have a strong influence on tropical plant use, which is expected to increase considerably in social and commercial value as a result of this movement. Artificial plastic or silk plants will not provide the same *natural* atmosphere. They are just another extension of the high-tech invasion we seek to counterbalance.

The National Aeronautics and Space Administration (NASA) has done extensive research into the use of live plants in spacecraft and space stations to modify the emotional stresses on astronauts. Recognizing the value of natural environments, some space scientists recommend the inclusion of miniature, natural landscapes as microenvironments in cabins and work areas.[4]

[3]John Naisbitt, *Megatrends* (New York: Warner Books, 1982).
[4]Paul N. Klaus, "Decreasing Stress through the Introduction of Microenvironments" (Paper for NASA, Sunnyvale, CA, August, 1987).

Figure 2.2 The glade-like atmosphere of indoor gardens relaxes patrons.

Another author states:

> The green experience is pervasive. Nature matters to people. The examples are endless and involve innumerable nooks and crannies in people's lives. They also provide hints of countless, small changes that can make the human condition just a little more satisfying.[5]

Merchants have found strong commercial value in greening their establishments because the restful environment they create helps to attract customers—we like to be in places like that. As the NASA scientists had noted, it helps to rest the soul. We create comfort and high-touch by bringing a bit of nature indoors, with wallcoverings of grass cloth, wood paneling, stone, and brick veneer. We use pine and floral scents to tease the nostrils and the trickle of a mini-waterfall or fountain to tickle the ear, invading our senses with a hint of the outdoors. Above all else, we surround ourselves with live tropical plants to give us back the feeling of earth, of growing things that evoke the memories of our agrarian past, and perhaps to give us visions of exotic places.

The green color of tropical foliage is another desirable attribute. It is considered a cool color by psychologists, evoking a restful influence. NASA reaches the following conclusions about cool colors in its *Habitability Data Handbook*:

> Dimness, quietness and sedation of the senses in general are associated with the most active effect of cool colors.[6]

---

[5]S. Kaplan and R. Kaplan, *Humanscape: Environments for People* (North Scituate, MA: Duxbury Press, 1978), p. 193.

[6]*Habitability Data Handbook*, Volume 2 (Houston, TX: NASA, Habitability Technology Section, Spacecraft Design Division, July 31, 1971, MSC-03909).

Decorative tropical plants provide these influences with their associated emotional health benefits, a contribution that should not be overlooked.

Plants and Decor Another positive factor in the "greening" of American building interiors is the decorative value of plants in complementing other elements of interior design, forming a symbiotic relationship that helps tie it all together. No longer is line, form, texture, and color limited to furnishings and architectural surface finishes. Interior designers and architects have found that plants serve to unify the design elements, add color, and create interest and softness where none existed otherwise. Several authors put it this way:

> Plants are wonderful accessories. They are the quickest injectors of liveliness and freshness to a room, can add height, variety, colour and drama all at once and are generally worth every penny you pay for them. There is almost no gap in a room that a plant cannot fill and improve, no piece of furniture that cannot be balanced by a spread of foliage. I really would not feel happy in a plantless room.[7]
>
> An interior without living greenery is rather like a meal without salt or spices . . . it may serve its basic purpose, but is dull and uninspiring. Plants bring vitality and visual excitement to a room. They serve as much more than ornaments, thanks to the added dimension brought by their living, growing, ever-changing character and their subtle variations in tone and texture.[8]
>
> Plants are living sculptures. They bring color, texture, and a variety of form into the office. They enliven interior spaces by softening the regular geometry of buildings and furniture. In addition to their visual charms, plants bring some of the intangible delights of the outdoors into the office. They foster interest and provide the satisfaction of nurturing something and watching it grow and flourish. They are a welcome natural element in an artificial environment.[9]

The live presence tends to overcome the stark, monolithic feeling that a concrete, steel, and glass structure can generate. The tie with nature is returned to these construction materials, giving them a sense of compatibility with their new surroundings, as they had before their extraction from the ground. The imagery works as well in residential structures as in commercial ones and provides the finish to a design project that is often overlooked or taken too casually.

Homeowners learned these principles long ago, although they may not have thought about them very seriously. They knew that houseplants dressed the house or apartment, added color, unified decorative themes, and filled bare walls. Women have been particularly appreciative of plants. For ages, the role of homemaker made her the resident decorator. Houseplants were used for decor and to provide a pleasant diversion, as well as a sense of responsibility in caring for the plants. Some people use the term "green pets." For many, prized houseplants are just that. Many emotional attachments are made with the beauty as well as the vulnerability or dependency of these living things—much like animal pets. Other reasons for having plants indoors are said to stem from our primitive need to grow something; prehistoric man had to grow food in order to survive. Psychologists think we still have some of that motivation stirring

---

[7]Mary Gilliatt, *The Complete Book of Home Design* (Boston: Little, Brown and Co., 1984), p. 94.
[8]Michael Wright, ed. *The Complete Indoor Gardener* (New York: Random House, 1979), p. 16.
[9]Judy Graf Klein, *The Office Book* (New York: Facts on File, Inc., 1982), p. 120.

Figure 2.3 Atrium of a modern hotel.

within us. Lacking the room outdoors or the time for full-fledged gardening, we have brought the plants indoors where we can satisfy some of these needs. This is particularly apparent with apartment dwellers.

Plants and Leisure Time

The human need to grow things as a diversion is another factor that has attracted the attention of space researchers on both sides of the Atlantic, as this excerpt from a NASA report shows:

> Since the earliest days at Antarctic research stations, there has been an abiding interest in growing plants. Ryumin (1980) also describes the pleasure that he derived from tending the experimental garden aboard the Soviet Salyut 6 space station. The activity apparently transcended the experimental requirements and the cosmonauts found themselves devoting much of their leisure time to gardening. Since gardening is an activity that provides substantial gratification to many people in both isolated and non-isolated environments,. . . . [10]

The scientific work just cited is merely an extension of our everyday humanity. It is an attempt to make the space worker feel at home in the alien environment of so far away. Gardening, in one form or another, is a preoccupation for most of us because of its salutary value. Its benefits are so well recognized that psychologists use it as therapy to mollify various mental disorders. The Gallup Organization of Princeton, New Jersey, has developed statistics for the U.S. Government as well as for commercial interests. Its latest survey (1987) uncovered the fact that gardening is the number one leisure-time activity of

---

[10]Jack W. Stuster, "Space Station Habitability Recommendations Based on a Systematic Comparative Analysis of Analogous Conditions" (Houston, TX: NASA, 1986), p. 84.

Chap. 2 / Use of Plants in Interior Decorating

Figure 2.4 Plant use in the home (Courtesy of Aqua/Trends.)

American adults.[11] Indoor gardening is a large part of that. Other data generated show that 41 percent of the 88.5 million households in this country (1986) had houseplants. These statistics indicate the magnitude of involvement with live indoor plants in our homes. For some, plants are merely a means of decoration. For others, plants are a means of diversion and personal satisfaction as well.

Plants as a Natural Element Rejection of objects and surroundings that are perceived as being *sterile* or *artificial* or *plastic* has led some to a greater appreciation of living flowers and foliage in interior design. Many of the proponents and activists of the '60s are now architects, interior designers, and landscapers, as well as homeowners, shaping our life-style with their appreciation of nature. Some interior designers caution against use of artificial plants:

> Plants produce fabulous results for rooms decorated on tight budgets. Hang lush plants from the tops of windows and place tall ones in the cor-

[11]*Statistical Abstract of the United States,* 1988.

ners of rooms. Use them to fill in holes near tables or next to lamps. Keep them healthy and bushy by pinching them back all the time. Never, never use plastic plants. If insufficient light is a major problem, use baskets of dried flowers, dried grasses, dried hydrangeas, dried tree branches ... anything at all as long as it's natural and organic. Plastic flowers look unreal and are deadly in any decorative scheme.[12]

Humanistic Architecture and Interior Design

In times past, the mainstream of the design professions had unfortunately ignored many basic human needs and overlooked the elements that could have made buildings comfortable and functional for us. The concentration on design elements for purely esthetic purposes has caused a lot of criticism and self-criticism about their unconcern for the people who had to use their creations. Much of it stems from the old arguments about *form* versus *function*. In his book *Towards a Humane Architecture*, Bruce Allsopp observes:

> People want architecture which is warm and comforting to the senses, architecture which is pleasant to live with, which caters to man as he is and not for man as an abstraction, architecture which is seen to be appropriate to its purpose, bearing in mind the habitual attitudes and responses of people who have been brought up in a living society, not processed in a laboratory. The supreme fallacy of modern architectural thought is that if the architect designs what he knows, by his own introverted standards of pure architecture, to be best, the public ought to grow to like it. Why the hell should they?[13]

This is a scathing criticism of design practice at the time, and hardly an isolated conviction. In their book *Designing for Human Behavior*, the authors chide the profession thusly:

> Architecture has traditionally been the last of the arts to be affected by changes in the structure of society and the cultural environment to which it contributes. The attitudes of architects both toward architecture and building users are still deeply rooted in the humanism of the renaissance rather than in the humanism of the twentieth century. While both imply an interest in classical studies, the modern use of the term humanism includes the practical study of human values and the application of scientific methodology to the problems of mankind. Architects have been extraordinarily reluctant to embrace this new definition.[14]

The issue has also been explored in depth in the work *Architecture for People*[15] as well as in many other corners of design literature.

Interior designers have not escaped criticisms leveled at their discipline either and thus have taken steps to define their purpose. The concept of *interior ergonomics* is one attempt at this. Walter Kleeman, in his book *The Challenge of Interior Design*, explains the importance of a humanistic design philosophy:

> However, the studies cited in this book tell us as designers that there are behavioral implications in every one of our designs. Interior design can

---

[12]Norma Skurka, *The New York Times Book of Interior Design and Decoration* (New York: Quadrangle/The New York Times Book Company, 1976), p. 92.

[13]Bruce Allsopp, *Towards a Humane Architecture* (London: Frederick Muller, Ltd., 1974).

[14]Lang, Burnette, Moleski, and Vachon, *Designing for Human Behavior* (Stroudsburg, PA: Dowden, Hutchinson and Ross, Inc., 1974).

[15]Byron Mikellides, *Architecture for People* (New York: Holt, Rinehart and Winston, 1980).

produce environmental habitability, a very important goal in terms of human well-being. The condition of the ultimate environment, the interiors where we live, love, work and play, cannot help but be one of the most important factors in our survival as rational, functional human beings.[16]

Some of the more enlightened practitioners in the design professions have heeded the criticisms and directives leveled at them and are now humanizing their structures, architecturally as well as in the many aspects of interior design. Use of live tropical plants is one of the important elements being used to achieve the more comfortable and natural interior concepts. Following the lead of such innovators as architect John Portman, greenery is being used in abundance to enrich our lives. However, the positive steps are being taken haltingly and without firm direction. Among the problems that still exist is the fact that lacking the expertise themselves, few practitioners coordinate with horticultural or irrigation specialists; this is yet another indication of the casual attitude taken toward advanced planning in this area. Important decisions must be made in the **early** stages of projects so that the proper plant materials, as well as environmental conditions and technology for their long-term subsistence, can be incorporated into the basic design and specifications. All too often as well, plant supply contracts for interiorscapes are given to, and advice is solicited from, outdoor landscape nurserymen and other professionals specializing in outdoor horticulture. Most of these well-meaning growers and contractors simply do not have enough expertise or the "feel" to offer the proper advice in interior installations. There is simply too much difference between these two branches of ornamental horticulture; conditions and problems are too diverse. Interiorscaping is a field unto itself, with its unique expertise, and must be recognized as such.

Superior tropical plants have been developed by the specialty growers for interior applications, and advanced techniques for their care have been developed by interiorscape contractors and by the industry. Most design professionals are not trained in these matters and tend to minimize their importance. The team, or collaborative, approach (using specialists in these disciplines as part of the project task force) is being misplaced or misused in most modern building projects. This unfortunately costs too many projects their esthetics due to the eventual degradation of plant materials and much greater plant maintenance costs spread over very long periods of time. If property, and facility owners and managers were aware of the enduring costs being built into their real estate because of inexperience or unconcern, much greater attention would be given to these details. Figures 2.5 to 2.8 illustrate the use of plants as an element of building architecture, and Figures 2.9 to 2.14 illustrate the use of plants as an element of interior design.

Plants in the Workplace

Over the past 30 years or so, as more women entered the work force, they also brought to the office their love of plants. This improved a lot of things at work. Management was slow to understand the benefit of plants, both to the female employee and to the office in general. It is still a vague concept with management, but the economic benefits to the company are starting to become appreciated. The same is true of the design community, who had been taking its cue from corporate clients. It wasn't until the 1950s that corporate management and their building designers started to recognize the fallacies in their approach.

---

[16]Walter B. Kleeman, Jr., *The Challenge of Interior Design* (Boston: CBI Publishing Company, 1981).

Plants in the Workplace

Figure 2.5(a) Planters in an office building lobby.

Figure 2.5(b) Planters in an office building lobby.

Chap. 2 / Use of Plants in Interior Decorating

Figure 2.6 Planters in a shopping mall.

Figure 2.7 Planters in a hotel atrium.

Plants in the Workplace

Figure 2.8 Planters in a multipurpose building.

Figure 2.9 Planters in a restaurant dining room.

Chap. 2 / Use of Plants in Interior Decorating

Figure 2.10 Planters in a fast-food restaurant.

Figure 2.11 Planters adorning a hotel lobby.

Figure 2.12 Planters in a lounge area.

Figure 2.13 Freestanding planters in a building lobby.

Figure 2.14 Freestanding planters in a corporate office suite.

Up until that time, the corporate entity and its edifices were most important. Then someone brought to their attention the fact that the people who work in the building were by far their largest expenditure, not the buildings themselves. Over the life of a building, construction, decorating, and operating expenses account for only about 8 percent of the costs; the remaining 92 percent goes to the employees working within the structure, their salaries and wages. It made sense, then, to try to improve employee productivity by making the working environment pleasant and effective. A study was made in 1982 by the Buffalo Organization for Social and Technological Innovation (BOSTI) to determine the importance of the working environment on employee satisfaction and efficiency. It was found to be more important than management had realized.[17] So building and interior designers now take these powerful factors into consideration. It is another aspect of humanizing our buildings. The use of live plants is an important part of this movement.

---

[17] Michael Brill, *The Impact of Office Environment on Productivity and Quality of Working Life* (Buffalo, N.Y.: Buffalo Organization for Social and Technological Innovation, 1982).

The Europeans, particularly in Germany, Sweden, and The Netherlands, have been more advanced in these matters than have U.S. executives. They show more concern and reverence for the worker than their American corporate counterparts and wouldn't tolerate some of the working conditions imposed in the U.S. It was not until recent years that the open-plan office design was used in Europe. They were not used to the concept. A German management consulting/design group, called the Quickborner Team, was commissioned to see how they could adapt open office plans to their own standards. The result was what they called "Burolandschaft", which means office landscaping—the extensive use of live plants to humanize a dehumanizing situation. The plants were densely installed to screen areas, to define circulation routes, and to separate workstations. The visual and audible improvements provided beauty and decor, introduced the salutary benefits of natural elements, gave more privacy to the worker, improved concentration as well as efficiency and the special placement of plants was used to identify status in executive areas. These are the benefits common to all offices, but plants are so important to the German concept that the generic term for their open-plan office layouts became . . . . burolandschaft.[18] Office management research indicates that the use of ornamental plants in the office can improve employee efficiency by 10 to 15 percent aside from providing greater job satisfaction.[19] Figures 2.15 to 2.18 illustrate the use of plants in various office environments.

Figure 2.15 Plants around office workstations.

---

[18]Judy Graf Klein, *The Office Book* (New York: Facts on File, Inc., 1982), p. 36, 40, 120.

[19]Stephen Scrivens, *Interior Planting in Large Buildings* (New York: Halsted Press/John Wiley & Sons, 1980), p. 1.

Figure 2.16 Plants in office conference rooms.

Figure 2.17 Plants in open-plan office spaces.

Figure 2.18 Plants in office reception areas.

Improved Tropical Plants

One of the major influences in the rapid rise of landscaped interiors has been the tropical foliage industry's development of techniques for adapting a large number of plant species for the specialized conditions of indoor cultivation, as well as breeding new strains for this use. The industry has increased its capacity to produce and developed a greater awareness of the needs of the design community as well as the plant specialty shops and the mass merchants who move so much of their product through the consumer markets. Retail sales are estimated at better than $1.3 billion (up from $1.1 billion in 1983). Most of this is through plant specialty shops (60 percent), with the balance through mass merchants.[20]

Publicity and Education

At the same time, there is slowly developing a greater awareness among design professionals, real estate executives, restaurateurs, the public, and others who use plants, on the special requirements of their application indoors. This has been exploited and well covered by newspaper, magazine, and book publishers and through instructional literature distributed by plant stores. This educational process has taken much of the mystery out of interior plant care and has given many people the confidence to use more live greenery.

Development of Interior Plantscapers

Part of this educational process has reached those specialists who have set out to make a living designing interior plantscapes, as well as installing and maintaining them. They have banded together in associations, propagated their own trade magazines, and furthered the process of educating themselves on the technical and business aspects of interior landscaping. They are now called *interiorscapers* or *interior plantscapers* and make up the important corps of profession-

[20]George H. Manaker, *Interior Plantscapes*, (Englewood Cliffs, NJ: Prentice-Hall, Inc., 1987), p. 4.

als who design, install, and service most of the interior landscape installations in commercial and large, residential buildings. Their service has made it possible for the architects and interior designers, as well as the corporate office managers, to use greenery with much more confidence and has boosted plant usage immeasurably.

Most of this has happened only within the past 15 years and has accounted for the growth factors leading to the state of this important social and economic phenomenon as it exists today. New conditions exist that portend equally strong trends in the future.

Plants and Indoor Air Pollution

Something has surfaced recently that is likely to have a profound influence on the use of live foliage plants in building interiors. It is largely the result of NASA research programs into the use of natural biological processes for both wastewater treatment and indoor air pollution abatement. The long-range goal of this research is practical life-support systems for future space stations. A by-product of the work is the conviction that plants can help solve many comfort and health problems in the home and office as well.

Environmentalists and health agencies are worried about the debilitating effects of indoor air pollution, what has become known as BRI (building related illness), particularly in energy-efficient buildings.[21] It is considered by many to be a greater menace to our health than outdoor air pollution and is growing in severity as more buildings are being sealed tight in an effort to conserve energy. Stale, polluted air is constantly recirculated, becoming more polluted all the time from gaseous emissions emanating from combustion, as well as interior building materials, furnishings, and the hundreds of other products that share our space. The problem affects us at home, in the office, virtually everywhere. It is particularly significant because by most estimates, we spend 90 percent of our lives indoors. Some recent comments follow:[22]

> "Indoor air pollution is one of the most important health issues facing the country today," says Sandra Eberle, who runs chemical hazards programs at the U.S. Consumer Product Safety Commission.
>
> Indoor air pollution . . . from radon, smoke, asbestos, formaldehyde and other noxious chemicals and organisms . . . adds billions of dollars to the nation's annual health-care bill and may be a greater menace than the dirtiest air outdoors.
>
> Indoor air pollution is "one of the most significant hidden hazards certainly of this decade . . . if not of this century," warned Mary Ellen Fise of the Consumer Federation of America, a coalition of 200 citizens groups. Similar alarms have been sounded by members of Congress, researchers at leading universities around the world and top U.S. environmental officials.
>
> As the full story slowly emerges, scientists say that indoor air pollution is being seen as one of the major public health threats facing Americans . . . it is becoming the environmental challenge of the 1980s.

---

[21] Peter N. Ylvisaker, "Air Quality: Is It (Wheeze; Cough!) Time to Test?", *Buildings*, May, 1989, p. 62.

[22] Larry Tye, "The Menace Within: Indoor Pollution, A Crisis in the Making," *The Courier-Journal Magazine* (Louisville, KY, September 15, 1985), p. 4.

The Environmental Protection Agency (EPA), The American Cancer Society, and The American Lung Association have all instituted major studies into the insidious maladies caused by building and decorating materials, as well as other common products found in our buildings, which contribute to interior air pollution, sickness, and death. It is well known that respiratory infections, headaches, depression, fatigue, and allergies are triggered, some victims have difficulty concentrating, and more serious illnesses (heart disease and cancer) sometimes result. Combined, they have put millions out of commission at various times, affecting their comfort, their longevity, and their ability to do useful work. It's being called the "sick building syndrome" or the "tight building syndrome," and it is being taken very seriously in all quarters. A comprehensive study by the Walter Reed Army Institute of Research (1988) conclusively linked respiratory illnesses to tight, energy-efficient buildings. There are volumes of additional evidence. An estimated 40 million Americans are functionally impaired by these disorders. People in the prime working age group of 30 to 40 are the most affected. About 150 million days of work are lost each year to asthma and other respiratory diseases. This costs American businesses an estimated $3 billion annually, plus the lost productivity of those working with illnesses. The annual cost to employees is at least $59 billion in lost wages. Total medical costs are estimated at $15 billion a year. Employers pay a large part of the bill. That is an enormous economic toll, and the sick building syndrome is being indicted as a major contributor.[23]

Home and office are under scrutiny, as are other public buildings. The problem is also capable of taking human lives. The Legionaire's disease tragedy in Philadelphia some years ago was an extreme example. Many expensive lawsuits are being leveled against employers and building owners for their responsibility in perpetuating unhealthy work environments. The indoor air pollution problem was brought into sharper focus with the September, 1988, publication of conclusions resulting from a five-year scientific research project by the EPA into the quality of indoor air in U.S. buildings. It was the first large-scale study of its kind here. The project focused on pollutants known as volatile organic compounds (VOCs) in public buildings. It searched out sources and levels of pollution. The findings are too extensive to detail here, but suffice to say that **over 500 different VOCs were found in air samples from some buildings, sometimes at levels 100 times greater than found outdoors** (primarily new buildings). Most of the VOCs are toxic; some are carcinogenic; some are mutagenic. The VOCs include common solvents such as trichloroethylene and tetrachloroethylene; odorous chemicals used as bathroom air fresheners, such as paradichlorobenzene and alpha-pinene (pine scent); and constituents of glues and paints, such as xylene and decane. Other pollutants were also included in the study: particulates, radon, nitrogen dioxide, and formaldehyde. The sources of these noxious gases and particulates in buildings were found to be paints, carpet and molding adhesives, toilet bowl cleaners, cleaning chemicals, carpeting, upholstery and drapery fabrics, particleboard, cabinets, furniture finishes, latex caulking, vinyl and rubber baseboard moldings, telephone and electrical cables, linoleum, pesticides and moth repellents, tobacco smoke, wall paneling, cosmetics, clothing, plastic building trim and pipes, office machines and their

---

[23]"Indoor Air Pollution: Buildings That Make You Ill," Air Quality Communique (American Lung Association of Florida, Summer, 1986); Nancy Anderson Hedberg, "What's All the Fuss About the Sick Building Syndrome?" *Buildings* (February, 1987), p. 58; and John F. Brundage, M.D., Robert McN. Scott, M.D., Wayne M. Lednar, M.D., David W. Smith, M.S., and Richard N. Miller, M.D., "Building-Associated Risk of Febrile Acute Respiratory Diseases in Army Trainees," *Journal of the American Medical Association* (April 8, 1988), p. 2108.

related solvents, paper, as well as many others. Most of these materials are also found in the home, so the warning is universal. A parallel study of residential structures is being done by the EPA. It is called the TEAM (Total Exposure Assessment Methodology) Study. We are surrounded by everyday materials that can do us harm. While we have no cause to panic, we should be very concerned. The combined effect of these different classes of pollutants can cause health problems in varying degrees, depending on a wide range of circumstances. Ongoing studies will clarify many of the open questions.[24] Parallel research is being carried out overseas as well, most notably by the Swedish Council for Building Research.

A number of solutions are being developed to combat the problem. New air-handling techniques are being proposed, as are programs to minimize the offending materials as well as the control of smoking within buildings. The scientists at NASA's Environmental Research Laboratory (Stennis Space Center) are convinced that common houseplants can be the simple, inexpensive salvation in many of these polluted environments. Their research projects in space biotechnology are demonstrating this. Following is an excerpt from a NASA report authored by Dr. B.C. Wolverton in December of 1986. He is the agency's leading researcher in this field and heads the effort. Dr. Wolverton is enthusiastic over the civilian adaptations of his work.

> One exciting area of this research is directed toward use of houseplants for improving the quality of air inside modern office buildings. The photosynthetic process that allows plants to live and grow requires a continuous exchange of gaseous substances between plant leaves and the surrounding atmosphere. The most common gaseous substances exchanged are carbon dioxide, oxygen and water vapor. The plant leaves normally give off water vapors and oxygen and take in carbon dioxide. However it appears that plant leaves can also take in other gaseous substances from the surrounding atmosphere through the tiny openings (stomates) on the leaves. NASA studies with plants have demonstrated the ability of common houseplants such as spider plant, philodendron and golden pothos to reduce the concentrations of indoor air pollutants such as formaldehyde, benzene and carbon monoxide in sealed, experimental chambers.
>
> Because formaldehyde is probably one of the toxic chemicals with the greatest human exposure in the U.S., NASA's initial indoor air pollution studies concentrated on this chemical. . . . There are numerous sources of formaldehyde emissions indoors. Formaldehyde is used in urea-formaldehyde foam insulation, plywood, particleboard, decorative panels, grocery bags, waxed papers, facial tissues, and paper towels. Many common household cleaning agents contain formaldehyde in addition to floor coverings, carpet backings, adhesive binders, fire retardants, and permanent press clothing. Other sources are combustion devices using fuels such as natural gas and kerosene for heating and cooking in addition to smoke from tobacco. . . .
>
> Another, even more promising technique for using plants to remove indoor air pollutants, is to combine the increased treatment capacity of the plant root-microbial-granular activated carbon process with the leaves. . . . This technique has been used very successfully in treating domestic sew-

[24]L.S. Sheldon, R.W. Handy, T.D. Hartwell, R.W. Whitmore, H.S. Zelon, and E.D. Pellizzari (Volume I) and L.S. Sheldon, H.S. Zelon, J. Sickles, C. Eaton, and T.D. Hartwell (Volume II), "Indoor Air Quality in Public Buildings, Volumes I and II (Washington, DC: U.S. Environmental Protection Agency, November 10, 1988).

age and removing toxic chemicals including radioactive elements from soil and wastewater. The process has the potential of rapidly removing relatively large quantities of chemicals and smoke from indoor air and biodegrading these substances. . . . The activated carbon-plant root filter may also be capable of removing radon gas from inside homes. Although studies have not been conducted to date using this process with radon, other radioactive elements have been removed from water and soil using plant roots. Interior landscaping design with live plants is an already rapidly expanding practice for offices, lobbies, reception rooms, hotels, hospitals, restaurants, and many institutions. Since more and more people are spending a greater percentage of their time indoors, it is desirable for psychological as well as physiological reasons to bring some of the natural outdoor environment inside. As more data become available on the ability of foliage plants to improve the quality of air inside buildings, interior landscaping with plants will probably experience an even greater increase in use.[25]

The plant root-microbial-granular activated carbon process mentioned in the report is a technique of drawing environmental air through the plant's growing medium, which is soil and activated charcoal. Pollutants are filtered out and absorbed by the granular charcoal, then degraded by microorganisms in the soil/root mass, and decomposed by the plant after being absorbed into its root system.

Plants for "Healthy" Buildings

Some text, technical tables, and illustrations have been deleted from the NASA excerpt just quoted in the interest of brevity, but the message is clear and the work may have an enduring effect on the way we think about plants, particularly on the way future homes and commercial buildings are designed and used. The effectiveness of plants in cleansing air is startling. In one study detailed in the report "Space Bio-Technology in Housing" by the same NASA group, common spider plants (*Chlorophytum sp.*) were able to reduce the concentration of formaldehyde in an enclosed chamber from 37 parts per million (ppm) to 8 ppm in only six hours.[26] Some plant varieties are more effective than others, but the results are consistent across a broad range of toxic pollutants. Plants create a healthier environment. NASA is establishing experimental houses to demonstrate its theories, and even modest success in these studies can add an important impetus to the use of foliage plants. The Associated Landscape Contractors of America (ALCA) has worked cooperatively with NASA to further research into earthbound applications. The interior design of future space stations is also currently in progress, and live tropical plants are an important part of the plans, not only for air pollution control concepts but also for their recognized contribution to the emotional well-being of the space worker. It is only a matter of time before real estate developers and managers become aware of and realize the commercial opportunity this work represents for them, for the foliage plant industry, and for the interiorscaping business. As information about the health benefits of live plants diffuses through the building management community, ways to more fully utilize these new concepts, in addition to providing greenery as a defensive measure against environmental law suits, will surely be found.

[25]B.C. Wolverton, Ph.D., "Houseplants, Indoor Air Pollutants and Allergic Reactions" (NASA, National Space Technology Laboratory, MS, December, 1986), p. 6.

[26]B.C. Wolverton, Ph.D., "Space Bio-Technology in Housing" (NASA, National Space Technology Laboratory, MS, January 19, 1986), p. 8.

Imagine the competitive advantage of a building project that offers its tenants an interior environment designed for prolific, yet cost-effective use of healthful foliage plants, not only in the common areas but in the tenant suite areas as well, where occupants spend most of their time. The appeal for the concerned corporate employer is obvious, benefiting both the employees as well as the company. Remember, too, that the health of management is also being eroded by the polluted office, and employers will be very sensitive to easy methods of improvement. The promotional advantages for the builder of residential units can be equally as strong. Being able to offer the homeowner a unit specifically designed to promote a healthy interior environment is a competitive factor that should not be overlooked.

Plant-Care Automation

It may have been noticed that we are now dealing not only with design esthetics but with building function as well, which leads us to another related subject that is destined to have a significant effect on the use of foliage plants in building interiors. It is the use of *automation* as a tool to facilitate the care of these multitudes of plants that are to surround us. The idea is not to do away with human plant care, but being so labor-intensive, to use alternate methods of reducing labor content to minimum levels. Manual watering is the single most time-consuming on-site chore, as well as the most demanding one. It is the logical target for automation.

Among the factors retarding the use of more live plants in interior decor are the perceived costs and inconveniences that they might inflict on the owner. Many potential plant users stay away from live specimens for these reasons. They just don't want the trouble or expense of caring for them. So they do without, or they use fewer plants, or they buy artificial plants for decoration. Industry growth figures are a testament to the fact that most people are attracted to live plants. Many object to the artificial or plastic connotations of the silk and plastic simulations. Some realize that silk products deteriorate and fade with time and require frequent cleaning. But the fact remains that more live plants would be used if plant care were easier and less costly. Not that it is difficult, but many people just do not want to be bothered with these chores. There may be the perceptions of not having enough time to look after the plants, or not knowing enough about plant care to do a good job, or not knowing what to do with them when away on business trips or vacation. Many managers of commercial buildings and corporate offices have stayed away because of the cost of maintaining meaningful installations of live plants or perhaps because of the poor quality of service obtained in their occasional tries at it. Now that technology is available to relieve much of the inconvenience, inconsistency, and cost of indoor plant care, many of the objections will disappear. Automated interior plant care promises to do for the inside of buildings what automatic sprinkler systems have done for outdoor real estate landscaping. As time goes on and more buildings are constructed with the new automatic precision plant-care systems integrated into them, this factor will reinforce the general trend in a significant way. It is expected to increase usage by 10 to 30 percent in the not too distant future. Those buildings that offer their tenants and owners a way of reducing foliage plant maintenance costs, as well as making the process more convenient and effective, will be sought out. The greening of our interior spaces and the development of automated systems for its care are taking parallel courses and will result in highly functional buildings harboring beauty, convenience, cost efficiency, and healthful environments.

We have established, then, that live indoor plants are here to stay in our everyday lives and are in fact making quite a social impact. Before we get into the technology of modern plant care, it would be worthwhile to review the state of technology in building/real estate management, the professionals who deal with plants and buildings, and the types of plants used in building interiors as well as their horticultural needs for survival. These factors are all interrelated and will help the reader understand what the new plant-care technology is trying to accomplish as well as the professional disciplines dealing with it. Those discussions will be found in the following chapters.

3 TECHNOLOGY IN BUILDINGS TODAY

Overview In this chapter, we will discuss the state of technology as it relates to building management. This is a relatively new subject. The technical revolution in this field started a mere six years ago and is still vague in many respects. It was not the purpose of the author to detail technologies used in the building/real estate/consumer markets but merely to paint in broad strokes the trends that have developed; for this book is as much about building technology as it is about interior horticultural maintenance. It is important to understand the commercial climate within which automated plant-care technology has evolved and will be used.

As people grapple with their environment and complex business problems, they develop ways of controlling them through procedures, materials, and technology. These will provide answers to the needs of the computer and communications age in which we live, as well as sophisticated ways of modifying our surroundings to bring them more in tune with human requirements. Independent devices and systems have been incorporated into homes and workplaces for many years as built-ins or add-ons (retrofit installations). They have helped alleviate many problems as well as provide for our diversion and comfort. In commercial buildings, security and fire control systems, lighting, heating/air-conditioning controllers and voice, data, and video communication systems were added. Homeowners, meanwhile, were installing security and fire alarm systems, simple intercom/music systems, remote-controlled garage door openers, timed light switches, and central vacuum cleaning systems as part of their quest for a high-tech life-style.

Building Design and Technology The past decade has seen the integration of much of this technology into buildings while under construction or renovation, and new electronic concepts have been developed to further help occupants in their domestic and business endeavors. Buildings have been transformed from their original roles as simple shelters into structures that are becoming technically functional and capable of doing increasingly wondrous things for us, to our amazement and benefit. The design community has learned that esthetics and designed practicality must be combined with technology to integrate all aspects of our daily needs as building users. They are now being addressed by the architect and interior designer as well as the engineer. Designers have become important contributors to the process of incorporating technology into buildings, and they must now plan for "intelligence" at the drawing board in the early stages of a project. Decision

"Intelligent" Buildings

making generally rests with the real estate developer, but recommendations for the incorporation of technology in projects must originate from the design studio as well. That presents a problem, however, because many architects are not properly trained in technical systems. The architectural community has been self-critical about this failing, and much controversy exists among educators as to how they should properly attack the problem. Following are some comments from a recent article of a prestigious architectural magazine:

> The integration of technical knowledge into architectural education is heavily tied to internship and practice itself.
>
> While most of the educators agree that technical education [of architects] is not adequate as taught in most programs today, they also feel that technical education is of higher quality than it was 20 years ago.
>
> Faculty members [of architectural schools] capable of integrating technical subjects and design seem to be a rare commodity.[1]

These comments help to explain the hesitation many architects have in using technical systems in their projects—they are more comfortable with the esthetic design aspects than the technical ones. In these cases, there would be a greater reliance on supplier services and engineering consulting. The team approach is most important here. The problem is even more acute when we get to the interior designer, for the training of these disciplines completely ignores technical integration with design. There are signs that this may be changing, but very slowly. Those interior designers that have a solid feeling for the technical aspects of their projects have advanced through self-education and experience. Unfortunately, most tend to shy away from anything not having to do with design.

Buildings as Functional Structures

Think of a simple building as a monolith. It is not much more than brick, stone, mortar, glass, steel, and wood—without the ability to do anything useful but keep occupants dry and relatively comfortable. Then we add complexity to the shelter: a nervous system in the form of interconnecting wires and cables; a circulatory system in the form of plumbing; a respiratory system in the installation of heating, venting, and air-conditioning; a visual capability in the form of video monitoring systems and infrared detectors; a tactile capacity in the form of contact alarm sensors; muscle power in the form of electromechanical devices that operate garage doors, elevators, entry systems, ceiling fans, central vacuum cleaners, draperies, and louvers; an audio and telecommunication network to give the structure a voice; and then, of course, a brain in the form of computers that centrally control these systems and provide data-processing facilities for building tenants. These systems, with the other available technologies, perform a wide range of tasks. Now we have a "smart" building that can do things. In a wonderful way, the building almost becomes alive. It is now actually able to perform useful work for us, to take on many of the tasks previously performed by human labor, and to excel at things humans cannot hope to accomplish—saving us time and money in the process. This is what building automation is all about.

"Intelligent" Buildings

In many ways, the technology has advanced beyond our willingness or capacity to embrace it, but the "smart" or "intelligent" buildings, as they are being

---

[1]M. Stephanie Stubbs, "Technical Education for Architects," *Architecture* (Washington, DC: The American Institute of Architects, August, 1987).

Figure 3.1 An intelligent office building.

called, are structures that have some of these wondrous, functional systems designed into them to be installed as the building rises or is renovated. Most building systems are expensive, which is causing this phase of the technology revolution to be relegated mainly to Class A commercial projects and upscale residential developments. To insure investment payback, most shared-tenant service buildings are large, with a minimum of 300,000 to 500,000 square feet.[2]

To an important degree, advanced building management concepts were developed in response to the energy crisis of the late '70s. The term *building intelligence* was first used by United Technologies Corporation in the early years of the '80s, and by most definitions intelligent buildings did not exist before 1983.[3] The main thrust of technology that endows real estate with intelligence is voice and data communication networking, security, lighting, and environmental control (HVAC systems), as well as safety systems (fire alarm and controlled emergency lighting), intercommunication (audio), central video sys-

[2]Deborah Dietsch, "Intelligent Building—IT INTO IQ," *Interiors* (April, 1985), p. 116.
[3]Kitty Dawson and Andrew Fineberg, "Building Intelligent Offices," *Venture* (October, 1984), p. 90.

tems, elevator control, other office and building automation, and, of course, tenant-shared data processing (mainframe computers) and telecommunication facilities (PBX, WATS, etc.). None of the current crop of intelligent buildings use all of the options available.

Table 3.1 gives a more complete summary of the technology being incorpo-

TABLE 3.1
Current Concepts of Building Intelligence

Communications

Voice—phones, WATS service, message center, paging
Data—mainframe or minicomputer for mass data storage, data processing, access to internal or external data bases and software (local area networking, etc.)
Switching technology—terminals, PBX (private branch exchange)
Radio and satellite receivers
Teleconferencing
Telex/Twx

Office or Tenant Facility Automation

Electronic mail and message switching
Audio/video systems
Master antenna (MATV)
Window shade/louver adjustment
Other local environmental controls
Computerized restaurant service
Small-zone billing (energy)
Telephone link for energy; other control
Micro-Irrigation, other automated plant-care systems

Building Automation (Building Management)

Direct digital control (decentralized systems control)
Energy management—heating, air-conditioning, and lighting intelligence; timed or sensor monitored, small-zone control, central computer controlled, or remote telephone control capability
Local lighting and appliance control—sensor and timer control
Integrated systems—combined telecommunication and energy management
Elevator control—scheduling, card access, synthesized voice announcements
Computerized building directories
Micro-Irrigation, other automated plant-care systems—for building common areas, centralized control of service in tenant suites

Security Systems

Intrusion detection—sensor and video monitored
Access control—card key systems, central lock control
Audio surveillance—sonic monitors
Video surveillance—TV camera monitors
Central, system monitoring
Automatic police/security agency alert

Fire Protection and Life-Safety Systems

Fire detection and alarm—sensor and video monitors
Emergency communications
Emergency lighting
Smoke evacuation and control—automatic pressure increase on adjacent floors
Emergency elevator control
Automatic fire department alert

Uniform Cabling Plan System

Interconnecting cable network—metallic and fiber-optic
Distributed outlets—for telephones, computers, cable TV, thermostats, lighting, appliances

Sources: Jack Caloz, "Wiring For Intelligence," *Administrative Management* (January, 1987) and Deborah Dietsch, "Picking the Brain of the Intelligent Building," *Interiors* (April, 1985), p. 13.

rated into modern buildings. It may be noticed that some of the devices classified as "automation" by the marketplace are not automatic at all but merely state-of-the-art manually operated systems. Many products are deceptively labeled "automation" or "computer-operated" if they contain solid-state electronics. One does not necessarily follow the other.

The Computer-Regulated Structure

In the most sophisticated intelligent buildings, the computerized nerve center monitors and controls a number of subsystems that establish environmental comfort levels, regulate energy consumption, provide for emergencies (such as fire), and maintain building security. The nerve center can be programmed to anticipate the arrival of the building's work force by adjusting interior temperatures in relation to current outdoor weather conditions. This is done through feedback from strategically placed sensors. If the weather is unusually cold, the computer turns up the building's heating system earlier than usual; if air-conditioning is needed, the computer decides whether it is more economical to bring in outside air or to recirculate inside air.

In addition to receiving data from temperature and humidity sensors, the central computers in many installations can accept data directly from the building's smoke detectors. If a fire breaks out, the detectors immediately alert the computer, which automatically signals the fire department. The computer system might also broadcast evacuation instructions through the public-address system. Concurrently, it might send all elevators down to the first floor and lock their controls so that only the fire department could use them. The more advanced systems would then automatically pump air into the stairwells and the floors above and below the fire to stabilize pressures and help contain the smoke, and would direct the ventilation system to exhaust smoke from affected sectors.

Intelligent buildings have a variety of ways to maintain security. Sensors can detect infrared radiation given off by a person's body heat and, in turn, alert the central computer of a possible security violation. The same types of devices can be used to automatically switch office lighting on or off as employees cross the sensor path. Electronic key cards can limit access to offices or other designated areas. A microprocessor installed in the wall by the entry portal to a secured area will screen the validity of employee key cards as they are inserted into a slot. Unauthorized attempts at entry will be signaled at the central station for investigation by security personnel. In some systems, the card is also capable of adjusting the comfort levels in the offices being entered. If someone came into the building on a weekend to work, for example, the computer would automatically switch the heating or air-conditioning from a weekend level to a workday level.

Computer controllers are the heart of energy management functions. They are usually programmed to monitor the use of electricity and to institute cutbacks during peak demand periods when utility rates are highest. The computer might cause certain nonessential lights to be dimmed or turned off in a given order of priority. It would also turn off nonessential fans, pumps, and water heaters until power conditions were returned to normal.

Telecommunications, data processing, and networking are other examples of computers at work in advanced, functional structures.

In summary, computers are used in modern buildings for reasons of economy, comfort, and convenience, providing building management and tenants with tools of wondrous efficiency.

Building Intelligence Defined

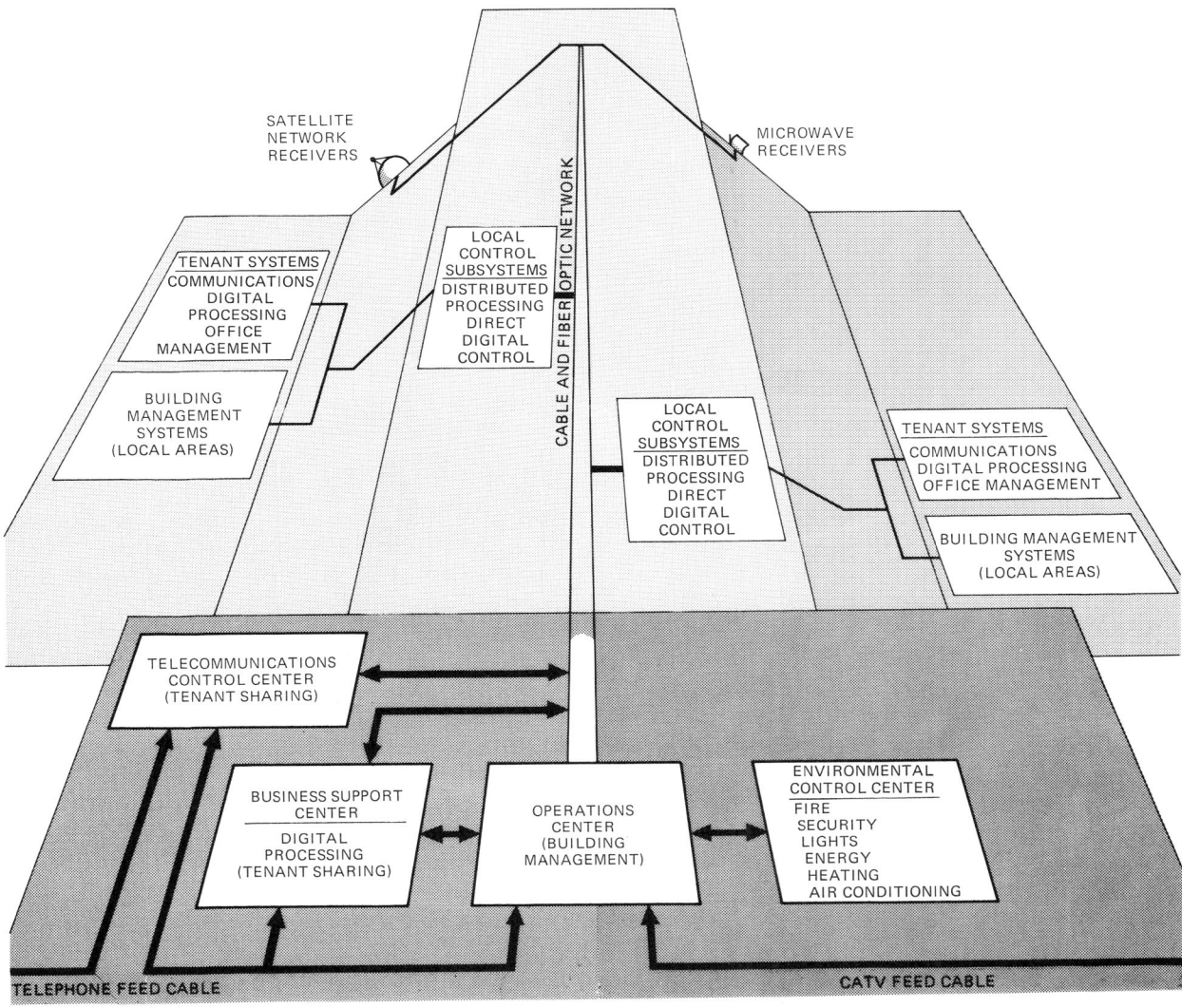

Figure 3.2 Intelligent building systems.

Building Intelligence Defined

There are many definitions for building intelligence, depending on the viewpoint of the user of the term. Things are currently so new that much confusion exists. It will be up to the industry to sort it all out and establish standards. An important step in the right direction has been the formation of two associations addressing themselves to the subject. The Intelligent Buildings Institute (IBI) and the International Intelligent Buildings Association (IIBA), both in Washington, DC, were formed to promote and define intelligent buildings. The simplistic definition of an intelligent building is "a building that uses technological solutions to meet human needs." A set of guidelines on the subject has been written by the IBI and defines an intelligent building as:

> ...one capable, through its structure, systems, services, and management, to be able to adapt to tenant needs and demands both today and in the future, realizing that moves and changes are the most common things in an office occupant's life. The building, and its components, also must

be able to adapt to new technology efficiently and effectively for it to be an intelligent use.[4]

It will be interesting to see if this definition stands the test of time, but for the moment it is the only cohesive factor in a confused industry. The two associations are broadly based and international in scope. They should serve to pull together the scattered factions and interests in the industry, and clarify the situation. As it now stands, some define building intelligence to include mainly those structures that use a sophisticated means for controlling power for heating, cooling, and lighting. For example, having the ability to switch lights on and off as someone enters or exits a room, or for exercising close control over selected areas of a building, so as to be able to offer tenants small-zone billing, keeping individual power costs separate, with the idea of saving money in the process. Other factions believe a building is intelligent only when it gives tenants computerized services, such as central telephone switching (PBX), in-house phone networks, computer networking, in-house mainframe computers for tenant access or in other available forms. These buildings are being fitted for telecommunications hardware, mainframe computers, elevator controllers, "talking" elevators, energy management controllers, fire alarm hardware, satellite dishes for telecommunication and video reception, telecommunication transmitters, cable TV inputs, video- and sensor-response security systems, video building directories, lighting control sensors, and, of course, the bewildering networks of metallic and fiber-optic cables that interconnect the technologies. Another group insists that a building is not intelligent unless it possesses a uniform cabling plan system that interconnects not only current technologies but provides for future ones as well. Yet another school of thought is that only by integrating building management systems with telecommunications and office automation systems (large-scale integration) can a building be made truly intelligent. In such systems, the telephone lines are made a part of the control network.[5] From these examples one can see the dilemma that exists in the industry.

The Appeal of Intelligent Buildings

The appeal that exists for corporate office managers is that they don't have to buy their own systems but are saved the expense by being able to plug into the landlord's services. That makes the *smart*, multitenant building particularly attractive for the small- to medium-sized company. There, they have access to the most up-to-date business and facilities management systems without a heavy capital outlay, as well as the ability to reduce costs by sharing use of the technology. There is also an image-enhancing aura in being quartered in these technical marvels, and there are cost-saving benefits in many of the services being offered. Some buildings have been designed, however, with a heavy dose of *whizbangs*, or devices with the flashing lights, bells, and whistles that don't really mean much. These devices merely produce a sense of awe in the technology itself, and most people have difficulty in comprehending what the devices can really do or not do for them. Sooner or later, the inefficient concepts will be seen for what they really are and will be sorted out of the market.

The Economics of Intelligent Buildings

The cost of incorporating comprehensive high-technology packages into commercial buildings is expensive. Builders and industry consultants figure about

[4]*Intelligent Building Definition Book* (Washington, DC: Intelligent Buildings Institute, 1987).
[5]A.B. Abramson and Clifford Stanley, "Intelligent Building Systems," *Forbes* (June 17, 1985).

4 to 10 percent is added to construction costs,[6] but many in the real estate industry are convinced that sophistication is necessary to attract tenants and provide them with real cost-saving amenities. As one real estate manager put it, "The fact that you can sell some 'sizzle' gives you a higher profile."[7] Many others question the viability of investing in such expensive systems that quickly become obsolete and may not be the commercial magnet that suppliers promote. Most agree, however, that building intelligence is the wave of the future in the real estate industry, but the bulk of the developers are waiting it out, trying to avoid the higher project costs as long as necessary and evaluating the situation as the industry gets deeper into it. Sooner or later, most commercial building projects will incorporate (to some degree) high technology as amenities for tenant use or as building management systems.

Using the stricter definitions of the industry, one source estimates that only about 90 commercial buildings in the U.S. at this point would fit that category. The introduction of high technology into major building projects is going slowly because of cost, rate of return, tenant and developer skepticism, the reluctance of tenants to commit their sensitive business data to a communal reservoir, and the pressures developers feel from the over-produced real estate market. In large intelligent office buildings, the rule of thumb requires about a 60-percent utilization of shared-tenant services before the facility starts making money, and today the break-even point in most of these projects is running three to four years after installation.[8] Many sophisticated buildings, therefore, are having trouble signing up enough tenants to participate, creating a considerable drain on owner resources.

Using loose definitions for the concept of building intelligence, another source estimates that 90 percent of the commercial buildings constructed today incorporate some concepts of intelligence. Much of this technology is being installed into portions of buildings by individual tenants and by building owners in answer to specific needs. Some are part of new construction projects, but much is being added to existing structures in retrofit projects. Therefore, the encroachment of technology into real estate cannot be measured in terms of intelligent buildings alone. Partial building coverage with a high-tech package doesn't make the building smart according to the definition but does endow it with advanced technical features. The point is, in both large and small projects, technology is starting to take hold and have an impact on the construction/real estate industry and on their tenants as well. Technology is the byword and everyone is primed for it.

High Technology In Housing

The use of high-tech amenities in residential building projects is progressing more slowly than in commercial areas. The traditional products of yesterday are still around and have been upgraded in quality. Many homes and apartments are being built with intercom systems, garage door openers, TV cable inputs, electrostatic air filters, solar heating, humidifiers, various versions of security systems and automatic air-conditioning/heating system control, smoke alarms, and central vacuum cleaning systems, as well as automatic light and appliance switching. Compared with the technology of the commercial building

---

[6]Megan Jill Paznik, "Intelligent Buildings Get Smart Enough to Save You a Bundle," *Administrative Management* (January, 1987), p. 25 and Dory Owens, "The Smart Building," *Miami Herald* (August 12, 1985).

[7]Michael Walsh, "Towers with Minds of Their Own," *Time* (June 24, 1985), p. 77.

[8]Megan Jill Paznik, "Intelligent Buildings Get Smart Enough to Save You a Bundle," *Administrative Management* (January, 1987), p. 25.

markets, these are relatively low-cost systems with much less sophistication. Extras or amenities are generally offered as options to the prospective house or apartment buyer. The homes and apartments of the handicapped are now being fitted with remote-controlled and automatic door, window, and drapery/shade openers, as well as appliance control devices. In conventional housing, builders have more tendency to offer upgrades in furnishings and major appliances than technical systems. This is changing, however slowly.

Many mistake the multitude of available entertainment electronics for housing-related technology. The stereo and video systems that flood the market have little to do with facility management as it relates to the household. Only those amenities that deal with residential convenience, comfort, cost-saving, energy efficiency, communication, safety, and security are considered housing-related technologies. This is obviously another area where industry definitions are needed.

Computer Control In Housing

Only in the largest, most expensive residences will one find energy management, security, and fire alarm systems approaching the complexity and sophistication of those being offered to commercial developers. Cost is one reason, of course, but probably more important is the fact that such complexity is simply not required in the home. In recent years, we have seen the computer industry try to convince the homeowner that, among their various applications, personal computers can control home appliances, heating/air-conditioning, and lighting. Other uses they bring to our attention are the computer's ability to help manage home finances and investments, file menus, monitor security and fire alarm systems, and provide educational experiences and relaxation with computer-generated games. These are worthy applications of high technology, but the cost and complexity of operation have kept most homeowners as interested bystanders only. For various reasons, the attempt has failed temporarily, but there will be a time in the not-too-distant future when those applications will be commonplace. The situation is changing as more of us use computers in the office and get a better feel for their operation and capabilities. At the present time, however, relatively few computer-aided residential control systems (as opposed to microprocessor controlled systems) are in use. Two new developments will provide that capability in other ways. A few telephone companies have begun to offer residential energy management control through the phone lines. A computer/controller at the local phone switching centers will activate and deactivate a subscriber's home appliances and HVAC systems. Also, a similar concept has recently been announced that will provide the same type of service through local cable TV companies.

Technology Available for Smart Homes

The main thrust of the technology being developed for housing units is in the field of alarm systems for security and fire detection, power and environmental systems control (HVAC). A wide variety of wired and wireless systems have been made available to the homeowner and building developer and are getting considerable use. In many large residential developments, particularly townhouse and condominium communities, sophisticated central security and fire alarm systems are being installed. These provide for a service company to monitor residential units from a central location. Although used in residential projects, these are more properly considered commercial systems for they are generally sold to the developer or to the community association as a facilities management system. Community satellite dishes fall into a similar category. A relatively new product available to consumers for domestic use is the carrier-frequency energy management and appliance remote-control system that sends control signals through the building's electrical wiring. Originally distributed by BSR, it is currently marketed by X-10 (U.S.A.), Incorporated, and commonly called a *home control system*. It provides a convenient means of automatically dimming and switching lamps and appliances on and off, controlling heating and air-conditioning, interfacing with security systems, and providing a wide variety of remote-control functions, even by telephone from thousands of miles away. These are relatively inexpensive systems; they are flexible, easy to use, and becoming popular in the mass marketplace. Aside from their low cost, the main advantage of the systems is the fact that they do not have to be wired into a home; they are simply plugged in. CEBus is another technology rapidly developing in importance. Sponsored by the Electric Industries Association, it too relies on existing power wiring. These systems will be discussed in greater detail later in the book, for the concepts offers a convenient control element for many automated, plant-care installations.

Project "Smart House"

One of the more interesting projects currently under way promises to have a profound effect on the way we live. It is an attempt by the National Association of Home Builders (NAHB) to drastically change the power and communication systems within the residence. The changes, if implemented in the future, would enable the industry to create very technically sophisticated houses. A research project is under way that is sponsored by the NAHB Research Foundation, a subsidiary of the NAHB. It is called "SMART HOUSE" and focuses primarily on converting U.S. residential buildings to direct current (DC), looped-power systems, with central control for efficiency and flexibility. There are many compelling reasons for such a change, and the effort is being made to initiate a technical revolution in our homes and in the many appliances that service our homes. The implications are far-reaching and will be discussed later in this book when we look into the future.

Many industrial concerns have been watching with interest the development of smart residential and commercial buildings and have seen new opportunities open for them. The results of Project "SMART HOUSE" will determine whether appliance and electronics redesign will be necessary, for the power system of that concept is a radical departure from current technology. System concepts are being developed to address other building owner and tenant needs. Among these are more modest concepts that are less expensive to produce and use. They will permit the real estate industry to incorporate new technology into its projects at a much lower cost and with less risk than when dealing with the complex energy management systems and the other sophisticated building and business management systems. Many of these concepts will produce significant cost savings and would assure a quick return on investment.

Micro-Irrigation Systems

It was within this rapidly developing technical and commercial climate that Micro-Irrigation™ Systems for automated plant care were developed to help in maintaining decorative greenery in building interiors. The strong use of live tropical foliage in all types of interior decorative settings, with the prospect for even greater use in the future, leaves no doubt of the need for technological solutions to cope with the maintenance burden being generated. These systems have been in use in residential and commercial buildings for six years and represent the latest step forward in the development of technology for building intelligence.

4 THE INDUSTRY THAT BUILDS, DECORATES, AND MANAGES BUILDINGS

Overview When decorative plants are used in buildings, a number of disciplines are involved in their planning, selection, installation, and care. This has to do not only with the plants themselves but also with the equipment and methods of maintaining them. Interiorscapes in commercial buildings and in the common areas of large residential buildings are in particular need of the attention of various groups. In this chapter, we will discuss who the industry players are, their responsibilities in the scheme of things, how they interrelate, and how things are done.

Real Estate Developers The decision whether or not to use live plants in the common areas of a building project is a vague one and, according to industry sources, most frequently will emanate from the owner or developer of the building or their agent. Many will have an idea as to the style of architecture, site design, and interior ambiance they prefer, and those ideas will be passed on to the architect for execution. Almost equally important, however, are the projects where esthetic discretion is left with the landscape architect, architect, or interior designer, with the final decisions resting with the developer. Building or real estate developers come in many configurations. They are most often builders by background, in partnership with another company or individual. That partner is usually the financial strength of the development company—a bank, a real estate trust or limited partner, a mortgage or insurance company, an investment banking firm, or possibly a wealthy individual or family. Developers are, for the most part, businesspeople with more inclination for the project costs and commercial potential than for creative or technical detail. Most developers consider it important, however, to be deeply involved in all aspects, for they are the owners of the newly created entity and would have to live with the result, whether it be a housing development, resort, commercial building, or any other type of project. They are the producers of the project. They orchestrate the raising of funding, and the hiring of planners, architects, designers, project and real estate managers, and the general contractor; they also specify many products and services. Following is a quote from a recent issue of *Buildings* magazine:

Where once the design community might have been the primary product specifier, that is no longer true. In virtually every reader survey BUILD-INGS conducts, the question is asked: "Is it important for building ownership and management to play a leading role in specifying products, and if so, why?" Invariably, the response is "Yes," because owners and managers have to live with the product long after everyone else is gone, and experience dictates what works and what doesn't.[1]

They involve themselves in decisions concerning building intelligence and innovative product and service concepts, as well as other things that have a bearing on the vision and commercial viability of the structure they are creating. The technical concepts are considered mainly for the competitive posture they give them and for the corporate image they create. The leading developers will pioneer with innovative technology and design concepts; the rest will merely follow as a means of staying competitive.

Builders, Construction Companies, and General Contractors

Builders, construction companies, and general contractors are the firms that do the actual construction work on a project. In many cases, builders are also the project developers, the ones that initiate a building concept. This is particularly true in the residential and light-commercial segments of the industry. The developer/builder would also be the product specifier or at least the decision maker when it would come to building design, technology, and decor. Construction companies or general contractors are less likely to specify technical products. In many cases, however, the developer leaves these decisions to their discretion or to that of the architect or engineer. Full-service construction companies are more apt to be given this decision-making freedom. General contractors are more in the mold of a construction management company, overseeing the work of dozens of subcontractors, each with their own area of specialty. The final responsibility for the job is theirs, and the main concern, therefore, is construction detail.

Building Managers

Building managers (also referred to as property or asset managers, facilities managers, or real estate managers) are responsible for operating the building or facility after construction, but they are frequently brought into a project during the planning stages to inject their viewpoint. Astute building managers will insist that the architect plan for the installation of advanced systems in order to make their job easier, the operations more cost-effective, and the building more competitive. There is a new breed of building manager called a *Real Property Administrator*. These are highly trained individuals, schooled with college-level courses to a high degree of expertise in their field of management.[2] As real estate management practices become more advanced, so too will the methods and technology they adopt to meet their needs. Automatic interior plant-care systems are one of the newer tools at the managers' disposal, helping to make the building more self-sufficient. These managers know that intelligent buildings are the easiest to lease and maintain. It provides the "sizzle" as the one manager put it, giving the competitive edge in a tight market. As front-line managers, they have the best insight into a building's real needs, living with the project from day-to-day. They understand, for example, that when an architect

---

[1] Craig A. Henrich, "Editorial," *Buildings* (August, 1987), p. 15.
[2] *Information Bulletin*, Arnold, MD: Building Owners and Managers Institute International [BOMI], 1987.

designs a number of planter boxes into a project, he or she is not simply creating a naturally beautiful setting for esthetic appeal, but also an ongoing maintenance cost center. The burden can be minimized by designing them for ease of care. That means practical planter locations and configurations, and the inclusion of up-to-date lighting and irrigation technology to help reduce plant replacements and maintenance labor requirements. The managers generally know that plant-care labor (as well as other maintenance skills) is not a plentiful commodity and is becoming increasingly scarce and costly, and that technology must make up for this deficiency. Those buildings not provided with built-in, state-of-the-art systems will suffer through scores of years with higher operating costs. Until recently, the technology to alleviate some of the plant-related problems had not been available. Now that it is, many building managers will urge the incorporation of this and other high-tech amenities and management systems into their projects.

Architects Architects are the creators of the building or project design and are frequently made responsible for overseeing all the creative and construction aspects of a project. They conceptualize the shape and the size of the structure and specify the structural details (framework, materials, functional systems to be incorporated, etc.). They plan the larger areas of interior plantings if it helps to achieve the ambiance that the developer is seeking. That means incorporating into the plans the larger planter boxes that are part of the permanent building structure (in situ); for example, planter pits in lobbies and promenades, planter boxes in lobbies, restaurants, lounge areas, and atriums. Also within the realm of the architect's responsibility are technical systems used to perform various functions in the building. The irrigation systems used to maintain interiorscapes are part of this, as are the electrical service requirements for planter lighting and automated systems. Some architectural firms are full-service organizations and have an in-house staff with enough expertise to design interiorscapes, specify plant materials, and design irrigation systems. Most architectural firms are small however, and do not have the expertise to detail all of the complex technical and design elements they deal with, and specialty consultants are brought into the project for that purpose. This is called the *team approach*, although it is widely neglected when it comes to specifying plant materials and considering their horticultural needs and the advanced planning required for ongoing maintenance programs (irrigation systems, supplementary lighting, etc.). Ideally, when planning for the installation of an interior plantscape, the collaboration should be between the project architect, a landscape architect with interior experience, an interior designer, an interiorscaper, and an interior irrigation specialist. Each lends his or her own special expertise to the problem. In the real design world, however, seldom are so many disciplines brought together, for practical and/or economic considerations. One of the reasons greenery is taken so lightly is the perception that plantings are among the least expensive elements in the building design. That view is far from the fact, for initial installations can be quite expensive (many hundreds of thousands of dollars) and the ongoing replacement and maintenance costs can be extremely large, especially when we consider that they endure for the life of the building. Consequently, long-term plant care can run into many millions of dollars in major projects. Proper planning can save much of this for the building owners and tenants over extended periods. Automated plant-care systems are capable of paying for themselves in a year or less; savings after that are greater and they increase with time as maintenance wages rise. The informed building owner or manager will make sure that the architect designs a maintenance-effective

Interior Designers

structure, one that is not a maintenance nightmare. Early comprehensive planning is essential. Architects are learning that collaboration with competent specialists can assure design success[3] and are using the team approach with more frequency.

Interior designers are the professionals that conceptualize the design of building interior details, those elements that are not part of the permanent structure of the building. They are concerned with space planning, as well as finishing off the interior decor with esthetic elements, such as furnishings, lighting, architectural woodwork (bars, wall units, facades, etc.), and wallcoverings. They decide on colors, textures of surfacing materials, carpeting, window treatments, partitions and other built-in designs, room layout, furniture, and interior plantscape locations. Most interior designers lack the expertise to design extensive interior landscapes and to specify plant materials or irrigation details, but they can specify the design of planter boxes when part of a partition, as well as lighting for plantscapes. They also decide where in the overall interior design they feel freestanding potted plants should be placed. Most plant-related decisions, however, are left up to the interiorscape contractor, and, unfortunately, too often this is toward the end of a design project almost as an afterthought in many cases. Important elements may, therefore, be ignored or overlooked for lack of time or necessary budget, which could cost the project its esthetic integrity within a short time.

Few interior designers are technically oriented; they are simply not trained in these disciplines. An exception involves the computer and communications-oriented details into which some office design/planning firms have more recently been drawn and which affect client efficiency. Designers must be self-educated in these matters, for the design schools have not included such subjects in their curricula. More recent education in this field is being handled by the architecture schools of universities, resulting in graduates that are much more rounded in their knowledge and better prepared for careers in modern business. These schools are producing professional interior designers, not mere decorators. The better-trained professionals are capable of accepting design responsibilities that go beyond mere esthetics.[4] Creative vision must now be combined with modern practicality to meet more of our human needs. This will be particularly important in the case of the live decorative materials that add to the ambiance designers are creating. The technology is now available to maintain these live decorative materials, and designers can include some of these means in deference to their client's needs. Automated plant-care systems can be designed into cabinets, wall units, partitions, etc., to service the potted greenery in, on, and around them.

The more informed design professionals recognize tropical greenery as one of the important design elements at their disposal and treat it accordingly. In an age when buildings are being put up with stark structural materials and lots of glass, plants have become one of the more important ways of softening their effects, adding interest and color to the interior spaces, providing scale, and tying decorative schemes together with a unifying influence. It is unfortunate that information and training on the subject is lacking in the literature. Few books on interior design cover the subject in detail. There are, of course,

---

[3]Madeleine Parades, "Building Designs Often Overlook Maintenance Needs," *South Florida Business Journal* (May 11, 1987), p. 34 and Jan Kingaard, "The Team Approach to Interior Landscaping," *Western Landscaping News* (October, 1982), p. 32.

[4]Nancy Robinson, "Interior Designers: Shattering the Myths," *Decor* (May, 1987), p. 168.

good references on houseplants, interior landscaping, and interior plant horticulture, but the interrelationships of these with interior design have been somewhat neglected. Design elements, such as window treatments, planter locations, and lighting systems, have an effect on the quality of a planter installation, and the more informed designers consider them while planning.

Engineers Engineering companies are brought into large projects to oversee the technical design and installation work. Generally, they deal with the structural integrity of buildings as well as with plumbing, electrical, heating/air-conditioning, elevator, and other mechanical and electronic systems. Usually, they do the systems design work; sometimes they merely act in an advisory capacity, with specialty contractors, consultants, or equipment suppliers attending to the details. Automated interior irrigation is generally not part of their expertise, mainly because precision systems for interior applications are so new and because the common sprinkler and drip systems have been relegated to the outdoor irrigation contractor. When required, they will coordinate the design work of the interior irrigation specialist for the architect or developer.

Landscape Architects The landscape architect is the professional that creates topographical settings, large and small, using natural and man-made materials interspersed with live plantings to create a specific design effect in an area. Most landscape architects specialize in outdoor settings, but with the popularity of live plantscapes in building interiors has come a widening corps of landscape architects that extend their specialty to interior work as well. Some prefer to concentrate only on interior landscape design projects and have honed their training and experience to that end.[5] It is important that interiorscape projects be designed and managed by professionals with the **proper** training. There is a great deal of difference in environmental and cultural factors in trying to grow plants indoors as opposed to outdoor settings. This factor has been mentioned before and will be stressed repeatedly in this book, for it cannot be overemphasized. Without proper understanding of the special needs of indoor horticulture, the firms responsible for plantings, whether they be landscape architects or interiorscapers, cannot hope to do an acceptable job of planning, designing, installing, or maintaining an interior landscape installation (simple or otherwise). A competent interior landscape architect will have been properly trained to his or her task.

The landscape architect's responsibility is to take the general space plan as developed by the building architect and provide the topographical design, specify the plant materials, and decide on their locations in an overall esthetic plan that provides an interior ambience suitable to the project . . . one that will be both functional and attractive and that fulfills the vision of the building owner. Landscape architects have a wide breadth of knowledge and can deal with the complex issues of large projects and, for that reason, are involved in the early planning stages of sizable buildings. They deal with land contours, soil composition, drainage systems, rocks and boulders, retaining walls, paving materials, walkways and accent lighting, irrigation, waterfalls, and fountains, and, of course, tropical foliage plants.

Plants are selected carefully according to the specific needs of the indoor environment, the anticipated maintenance situation, and the design concept

---

[5]Gregory M. Pierceall, *Interiorscapes: Planning, Graphics, and Design* (Englewood Cliffs, NJ: Prentice-Hall, Inc., 1987).

the landscape architect is trying to achieve. Plant size, shape, and color determine how well a plant might fit into the design scheme from an esthetic standpoint. Other important factors considered are the growth habits and requirements of the plants (both above and below ground) and how well they might adapt to the interior environment being developed. Designers have to look at the amount of available light coming through the windows or skylights and how the quantities and qualities of that light change over the course of the day or, for that matter, over the course of the year as the sun changes its position in the heavens. They question the type of soil available and specify a preferred growing medium. They question the possible effects on the plants of the heating and air-conditioning, and the quality of plant-care services that would be available to maintain the installation. They must consider the drainage of the planter boxes, if these are used, and the temperature levels where the plants would be located. They select plants for hardiness in these environmental conditions and purchase suitable specimens from wholesale nurseries. In many cases, the plantscape design work is done much in advance of the building construction, at a time when it is not possible to accurately determine what the area in question will in fact be like. Many assumptions have to be made under these circumstances. Interior landscape architects frequently provide for the maintenance needs of the greenery by planning the supplementary lighting and irrigation systems, hiring interiorscape maintenance contractors, and collaborating with lighting and interior irrigation specialists.

The interiors of residential projects seldom require the services of landscape architects, except for condominium lobbies and clubhouse interiors in upscale developments. The outdoor environments of communities, however, are extensively planned by landscape architects.

Interiorscapers

The contractor known as an interiorscaper or interior plantscaper is considered by many to be the nucleus of the interior landscape industry. In many projects, particularly small- to medium-sized installations, they are called on to do it all—and most interiorscape projects in this country fit that category. At the low end are the few potted plants a homeowner or office manager may want installed to dress up the home or place of business. At the other end of the spectrum are the major office building, department store, restaurant, and shopping mall projects that require hundreds or possibly thousands of plants to achieve broad area coverage. The interiorscaper would normally be hired for a project on a bid basis to install and/or maintain plant stock. Sometimes, they are also asked to design the plantscape (particularly in small- to medium-sized projects). The large interiorscape companies also do some of the big project landscape design work normally relegated to the landscape architect. Frequently, plants are rented under a lease/maintenance contract, rather than being sold to customers. The average interiorscaper does it all—designs, leases or sells the plants, installs them, and guarantees and takes care of them. Some contractors specialize in only a phase or two of the business, such as maintenance or design only, but most have full-service companies. Many will concentrate on commercial work only, preferring the large projects available—in restaurants, banks, office buildings, and the like. Still others prefer the more personal atmosphere of the residential markets and will sell or lease potted plants to the home or apartment owner and possibly take care of them as well, especially in cases where the client is away a good deal of the time.

Interiorscapers are generally experienced plant people well trained in their specialized field of plant horticulture in indoor environments. They buy stock from wholesale nurseries for installation in their projects. Some of the

Figure 4.1 Interiorscaper servicing a shopping mall.

larger plantscape companies own and operate nurseries, while others maintain some type of holding facility to keep a small stock of plants for current or future contracts or for rejuvenation of sickly specimens. When a project is installed, the interiorscaper can also frequently obtain a separate maintenance contract. These are normally written for one-year periods, but maintenance contracts can be renewed and made to endure for many years, providing the interiorscaper's work has pleased the customer. In actual practice, however, interiorscapers generally find themselves in projects for only a couple of years. Contract maintenance fees can be as low as $25 per month for minor contractors working on very small installations to as much as tens of thousands of dollars billed monthly for major projects.

Some interiorscapers involve themselves in irrigation design and installation, but those are few. Most interiorscapers are still suspicious of automatic plant-care systems. Their main perception is that automation will cause a business loss for them, but unfortunately it sometimes manifests itself with general or unfounded claims that automatic irrigation is bad for interior plantings. This very defensive climate is similar to that in the early years of automation in many of our basic industries. The benefits of automation to the interiorscaper can be considerable, but open-mindedness is coming slowly. Some fear for their industry, with the misperception that automation will replace them. Nothing is further from the truth, however, for no responsible manufacturer of automated plant-care systems would promote products without strong recommendations that use should be only in cooperation with competent, interior plant-care specialists. Most biases are understandably rooted in fear as well as in the unfamiliarity with the techniques of advanced, interior irrigation and how they can be used to the interiorscaper's benefit. Most interiorscapers are not technically oriented, except in botanical matters. This, too, is gradually changing.

Interiorscaping Industry Profile

By and large, the interiorscaping branch of the landscape industry is made up of small companies, mostly family-owned businesses. We see a broad diversity here. Some are part-time practitioners—housewives, college students, and others using their spare time to make a few dollars. The capital investment required to enter this field is very small. Most, however, are full-time businesses manned primarily by owners trained and experienced in the techniques of interior plant horticulture and staffed with hired employees possessing varying degrees of knowledge and experience. The largest interiorscaping business in this country has about $30 million in annual sales volume garnered from dozens of subsidiary interiorscapers, so in terms of corporate size related to U.S. industry in general, none would be considered truly sizable. The second largest interiorscaper does about $7 million in annual volume. Of the top 25 interiorscapers in the industry, the median concern did $3.6 million in 1985.[6] Some interesting statistics about the interior landscaping business came out of a study conducted a few years ago by the magazine *Western Landscaping News*. It was shown to be a small industry but rapidly growing as interior landscaping gains in favor. Seventy-five percent of the respondents had been in the business for fewer than 10 years. It is well known that the industry hardly existed in the early '70s. The pioneers at that time had to learn how to take plant species, which had been nurtured in an outdoor environment for millions of years, and try to grow them indoors under adverse conditions. Everett Conklin is one of those early heroes of the business, and from the efforts of nurserymen/interiorscapers like him came a new area of expertise, a new era of interior plant use, and a new industry. These people did us the favor of bringing to our world beauty and a healthier environment as we had not known before—and that ideal is being perpetuated by the interior landscape industry as it exists today.

The conditions these early interiorscapers found in buildings were so different from outdoor horticulture that plants commonly went into shock and died unless certain techniques were followed. These techniques were developed by trial and error as the interiorscapers went along. Keep in mind that the technology at the time was one steeped in agricultural methodology and houseplant care, and much of it did not easily apply to commercial indoor plant culture. Methods are still being developed, for the industry is young and much has yet to be learned.

Another interesting aspect of the interior landscape industry described by the WLN study was the average gross dollar volume done by interior landscape contractors (interiorscapers) annually. The figure reported for 1982 was $53.1 thousand, with 35 percent of the respondents showing less than $10 thousand and 74 percent doing less than $50 thousand gross dollar volume annually. So

---

[6]Jeff Morey, "1986 Interiorscape Contractor 25," Brantwood Horticultural Research Division, *Interiorscape* (September/October, 1986), p. 41.

as one can see, it is very much an industry of small, family concerns. The interiorscape business is very labor-intensive and requires a great deal of manual work. Lots of digging, climbing on ladders and retaining walls, and hauling around heavy plants and watering paraphernalia is required. For these reasons, the industry is also young in age, with most of the employees being in the statistical age bracket of 18 to 24 years. The industry also hires a great number of part-time employees to help out during peak periods, and virtually all of these employees are young as well. It is also an unfortunate fact (as pointed out by the WLN study) that interiorscape installation and maintenance employees are the lowest paid in the general landscape industry (average hourly wage of $4.29 in 1982). It is not surprising, therefore, that the interiorscape business also has the highest rate of employee turnover.[7] Such high turnover rates are expensive, counterproductive, and very hard on the reputation of interiorscape firms. It is very difficult to keep a quality level of service with new, partially trained maintenance technicians. The industry generally recognizes that it takes about a year to train a new employee to be a fully accomplished manual-watering plant-care technician. There are many difficulties in doing so because of the nature of the job.[8] The industry is doing a fine job in organizing training programs through its trade organizations and such, and it should do much to lift the standards. One factor working against the interiorscaping business, however, is that the 18 to 24 age group upon which they rely so heavily is shrinking, as evidenced by demographic trends. It has been forecast by the U.S. Bureau of the Census that by the year 1995, this critical age group will have dwindled to 9.2 percent of the population from 13.3 percent in 1980 (see Table 4.1). That will come at a time when indoor plant usage will be greater than ever, and the need for trained plant-care technicians acute. The low pay scales and routine, difficult work will make it hard to attract competent employees in a highly competitive job market. As a result, a possible crisis situation is expected to face the industry. New technology, such as the automated plant-care systems now available for interior plantscapes, will help solve many of these personnel problems for the interior landscape industry, as they will reduce the need for extensive training in the critical area of irrigation, reduce significantly the physical labor involved, and permit more efficient utilization of available labor.[9] This subject will be discussed in more detail in the following chapters.

The interiorscape industry is generally under a great handicap. Because of the casual attitude that construction, design, and real estate professionals have about plant use, they frequently wait until the tail end of a project before calling in plant specialists. Much of the time, interior landscape planning is done almost as an afterthought, particularly in small- to medium-sized projects. This is changing, however slowly. Even in large building projects and in upscale residential construction, seldom are plant cultural and maintenance details properly considered at the drawing board. This usually leads to a compromised job, and frequently the installation reflects it, either initially or after a period of time. After all, how, for example, can a plantscape contractor do a proper job of irrigating a bed of plants, by manual means or otherwise, if proper drainage had not been planned for, or hose bibbs not made available at reasonable locations? It becomes worse when the plant specialists get to the site and find that the planter bed has already been back-filled with debris and common

---

[7]"1982 Survey of the Landscape Business," *Western Landscaping News* (November, 1982).

[8]Robert Hyland, "Modern Technology Solves Personnel Problems: Interior Landscape Irrigation," *Western Landscaping News* (February, 1983), p. 28.

[9]Jeff Morey and Stuart D. Snyder, "New Trends for Interiorscapers," *Interiorscape* (May/June, 1984), p. 38.

TABLE 4.1
U.S. Population Changes by Age Group (18 to 24 Years)

Year	United States Population (in Millions)		
	Total U.S. Population	Population Ages 18–24	Percent of Population
1970	205,052	23,714	12.1
1980	227,757	28,492	13.3
1985	239,283	28,741	12.0
1990 (proj.)	249,657	25,794	10.3
1995 (proj.)	257,559	23,702	9.2
2000 (proj.)	267,955	24,601	9.2

Source: *Current Population Reports,* Series P-25, Nos. 870, 952*, 985, U.S. Bureau of the Census.
*Used middle series projections.

soil that does not come close to resembling quality growing media. This is commonly the case. Also, in the latter stages of building construction, the interior lighting will have already been specified, frequently without considering the plantscape's supplementary lighting needs. Perhaps window treatments, which further lower ambient room light, were chosen. Interior irrigation systems are frequently not considered in the early stages, because they too are plant related and get only casual attention. What is not realized is that these systems sometimes require tubing and piping to be incorporated into partitions and concrete slabs during construction, and to do otherwise seriously increases installation costs and can undermine the effectiveness of the system operation, as well as increase maintenance fees the building owner and tenants must pay over many, many years. In some cases, these are minor problems, but in most, they are significant. Leaving plant considerations to chance is a disservice to the building owner, the plantscape designer, and the interior plantscape contractor who would install the plant stock; but most of all, it is a disservice to the interiorscape maintenance contractor who will be charged with taking care of the less-than-ideal installation. They would all be seriously hampered. When the installation starts looking badly, recriminations start. The services in interior landscape projects are so interrelated that many tend to take a defensive attitude, and it usually becomes "the other fellow" who is at fault. In many of these situations, it is simply a lack of proper planning at the design stage that have caused conditions to exist that cannot be overcome by the front-line contractor. Conditions are inadvertently created that are bad for the plants and for those looking after them. The one who generally takes the brunt of the blame is the maintenance contractor.

The author has experienced many bad installations that could have been thriving examples of interior landscapes had the proper elements been included in the project. Many such installations have to do with shopping mall promenades that are being designed with iron-grate-covered planter pits in which large trees grow and with large planter boxes embracing all manner of foliage. In too many of these projects, the planters have not been provided with electrical service for lighting and local irrigation control, nor have provisions been made for a simple water supply line from a central source, with a spigot at the planter for a hose connection. If one were to see what the technician must go through to maintain some of these situations, then the irresponsibility of it all would become apparent. These people are caught up in a bad situation that could have been avoided if the design and specification work had been done more diligently. Many maintenance people (often young women) have to climb tall ladders with heavy watering tanks strapped to their backs in order to take care of planters high up on the walls or in the rafters. A building comes to mind where planter boxes were designed into the edge of a narrow sixth-floor ledge just under the skylight of an atrium court. The trailing plants have to be manually watered twice a week by a plant technician tethered to the end of a safety cable so as not to fall to the atrium floor far below. Automatic precision irrigation networks can easily solve maintenance problems such as these. Sometimes proper planning is waived for economic considerations, but in most cases it is neglected through lack of interest or insight into the future problems that may be created. The recent tendency to use the team approach in large projects is helping to alleviate these problems by bringing a broader base of plant-oriented expertise to the design team. Unfortunately, most projects today involve simpler planning, so undesirable conditions still widely exist.

Tropical Foliage Nurseries

The growers of tropical foliage plants for interior decorative uses are hundreds of wholesale nurseries. In Chapter 2 it was pointed out that most tropical foliage originates in Florida, California, and Texas. The largest supplier of tropical plants to the interiorscape trade does about $7 million a year in gross volume (at wholesale), of which about $4 million goes to indoor installations. Most nurseries supply both exterior and interior plant stock, but concentrate on one or the other. They also concentrate on specific plant varieties. The production of plants for interior applications is specialized and more involved because the plants must be conditioned for low-light environments prior to shipment. There are canopied fields around the country dedicated to conditioning (acclimatizing) plants. They are nurtured under shade cloth of varying densities to get them used to their final destination. Not all nurseries are willing to invest in this extra treatment. The tropical foliage industry has done an excellent job of adapting the hundreds of varieties of plants to the unnatural conditions of building interiors and of catering to the special needs of the interiorscaper and retailer. In many projects, particularly the major ones, interiorscapers make a shopping tour of the nurseries months or even years in advance of delivery to pick out and tag for the installation those plants considered most suitable. The nursery will then acclimatize them gradually to the lighting conditions expected to be found at the final destination.

Irrigation Contractors

Irrigation contractors are primarily local companies that design and install outdoor irrigation systems for homes and recreation areas as well as commercial and institutional buildings. Most installations are relatively simple and therefore don't require much planning. Large public parks, golf courses, residential communities, office parks, and the like, are exceptions. They require extensive coverage and the irrigation systems can get quite complex. These jobs are generally obtained on a bid basis, and frequently a systems maintenance contract is gained for long-term troubleshooting and repair. Few contractors are trained in the specialized field of interior irrigation. It is so new that not many have had the opportunity to install systems in building interiors. Irrigation contractors are asked to bid on installations of drip or sprinkler systems in shopping malls and in the lobbies and other common areas inside of public buildings.

Large planter boxes that require a lot of water and are not in furnished areas of the building can sometimes be automatically irrigated with outdoor-type systems. Precision Micro-Irrigation™ Systems are designed for use in the sensitive furnished areas of the building, such as in the living quarters of a home or apartment, in furnished office suites, or in restaurant dining rooms. The technology as well as the design and installation of these systems is quite different from the garden variety system and requires specialized training and experience. Expertise in outdoor irrigation is of limited value in most interior situations, and now that more comprehensive technology is available, we will find many irrigation contractors learning this specialty to become full-service companies. The development of automated precision systems establishes an entire new area of expertise that will allow other contractors to specialize in interior work, thus broadening considerably the market for irrigation services. This subject will be discussed more fully in later chapters.

5 TYPES OF PLANTS USED FOR INTERIORSCAPING

Overview

In this chapter, we will discuss the types of plants used in decorating the interiors of homes and commercial buildings. It is not our purpose to go into the subject in detail, as that is best left to the experts in the field. There are many excellent books on the subject, and we highly recommend the reader become more knowledgeable through further study. For anyone dealing with the irrigation of interior plants, it is necessary to have some knowledge of the varieties available, how they are used, and what it takes to keep them alive. This chapter will provide an overview of the subject of plant varieties, and the following chapters will discuss the other subjects. Hopefully, further reading will be inspired.

Houseplants versus Interiorscaping Plants

One of the first questions that may come to mind is, "What are the differences between those familiar objects called houseplants and the plants used by professionals for interiorscaping commercial projects?" By and large, there are none. The plant varieties are generally the same, except that larger, more mature plants are usually used in commercial landscaping. There is also a tendency for the homeowner to use more flowering varieties of plants than the professional. There are practical reasons for these differences. Concerning the question of size, the pros learned a long time ago that it is easier and less costly to take care of larger plants. Besides, they look better and provide more apparent value. Plants in larger pots do not dry out as fast, and therefore require less frequent care. While virtually all interiorscapers use potted flowering plants from time to time in their projects, they perceive them to be very short-lived, particularly when compared to potted foliage plants. In a recent survey by *Interiorscape* magazine, it was discovered that 45 percent of the interviewees had to replace flowering plants every two weeks because of their short flowering display, creating high maintenance costs. For this reason, potted flowers find only limited use by the professionals. Another practice is the regular use of cut or dried flowers to liven up small furnished areas.[1]

---

[1] T. A. Prince and T. L. Prince, "How Many Are Saying It with Flowers?" *Interiorscape* (September/October, 1985), p. 46.

Tropical Foliage Plants Most of the species used in interior decoration are classed as tropical foliage plants. They are not special varieties created by nature for our comfort and enjoyment indoors. For most species, their natural habitat is the tropical and subtropical regions of Central and South America, Africa, Asia, the Far East, and the South Pacific. One can visualize them growing in the shade of rain forests and jungles, near rivers and streams, or perhaps beneath a canopy of dense trees, fed from underground rivulets. Still other species call the dry, sandy desert regions home. The conditions may be varied, but the one thing all species have in common is that they had spent a long time getting used to their specific outdoor environment and have adapted accordingly.

Figure 5.1 Bamboo palms in their natural environment.

Acclimatization This quality of adaptation has been used to increase the plants' usefulness, bringing many foliage plants into indoor settings as well as new outdoor environments. Not all species have been able to survive the change, and only those most adaptable are able to be used by the homeowner or interiorscaper. The differences in environmental conditions found outdoors and indoors is great indeed. The light levels found in building interiors are very much lower; this drastically changes the growth habits and biological needs of the plants, mainly in terms of moisture and nutritional requirements. To put the difference in perspective, plants growing outdoors will experience 8,000 to 10,000 footcandles (unit of measurement) of light on a bright summer day. If those plants were then moved indoors, into a home or office for example, they would find typically only 30 to 200 footcandles of light.[2] Little wonder that a plant in those circumstances would experience shock, sometimes with fatal results. A plant that is moved from any environmental situation to another faces possible deterioration, but it is usually temporary. The delicate balances within their systems are easily upset. These stresses are mainly from changes in light intensity and available water, both at the roots and airborne moisture for absorption by the leaves. The movement of the plant could be as slight as repositioning it within a house or office, or it could be a major shift, such as moving it from favorable outdoor growing conditions to unfavorable, alien interior settings. There will be some shock to the plant after any change, but drastic environmental shifts could kill it. Nurserymen have found a way to reduce shock. *Acclimatization* is the adaptation of the plant during growth in the field or greenhouse to conditions closer to those that the plant will face when finally installed indoors. The refined techniques now used to acclimatize plants account in a large part for the success interiorscapers have had in installing and maintaining this beauty indoors. Those nurseries that specialize in producing indoor foliage plants grow them in so-called *shade-houses*, which are simply areas in their outdoor fields that have been fitted with translucent cloth canopies that allow only a portion of the sun's rays through, so the plants beneath are shaded to the degree required by the plant species, its size, and the shipping schedule. Some small houseplants are grown almost exclusively in shade-houses or greenhouses, as

Figure 5.2 Nursery shade-houses.

[2]George H. Manaker, *Interior Plantscapes* (Englewood Cliffs, NJ: Prentice-Hall, Inc, 1987), p. 36.

Figure 5.3 Inside a nursery shade-house. (Courtesy of Brentwood Nursery.)

they are shipped while young. Larger plants are generally started under full sun for the early years of their existence and then gradually grown under deeper and deeper shade conditions, so that by the time they are ready to ship, they are well used to the lighting conditions under which they will be grown in the building interior. As mentioned in Chapter 4, it is common practice in large building projects (when the plant materials have been specified well in advance of the building's completion) for the interiorscaper to visit the various nurseries six to nine months prior to delivery and pick out the particular plants felt to be most suitable for the job. The tagged plants, trees as well as smaller foliage varieties, are then conditioned in dense shade-houses up until the time of shipment. This will ensure that the plants do not encounter undue shock from completely unfamiliar lighting levels. An excellent discussion about the technology of acclimatization can be found in the book *Interior Plantscaping* by Richard L. Gaines (see Bibliography).

Nursery Irrigation Irrigation of the containerized plants is done in the nurseries once or more daily, depending on the plants, the time of year, and other factors. Many water using various types of sprinkler systems. Mist systems are frequently used in greenhouses, mostly on smaller plants. Increasing numbers of growers are using drip and trickle irrigation in the fields and greenhouses, as these systems are more water conservative and efficient. The drip and trickle techniques direct water carefully to the root zones; thus, the water reaches the soil where it will do the plants the most good, helping to reduce fungus and other leaf-borne diseases. Outdoor installation requires much more water than those indoors because of the faster growth of the plants and the harsher environmental conditions under which they find themselves. Outside, watering is done fully and frequently. Wind, bright sunlight, and heat tend to dry the plants quickly, and the lost moisture must be replenished frequently to prevent plant wilt. These watering techniques will be discussed in more detail in a later chapter.

Nursery Soil The soil used to grow tropical foliage varies widely and is difficult to detail. There are a number of soil mixes recommended for potted plants. Most supplying nurseries use the best soil mix for the varieties they grow. Unfortunately however, one will also find large numbers of plants shipped in what is mostly sand. Little humus or loam is to be found in these pots. Such plants are more difficult to maintain because the soil has little ability to hold moisture and nutrients. It takes natural humus and loam or artificial soil additives to provide the needed absorption and retention properties. It is highly recommended that poor soil be replaced by a mix specified for that plant variety. *Use of the proper soil is one of the most important factors in achieving efficient plant irrigation, whether it is by manual means or automatic.* This cannot be stressed strongly enough. The subject will be discussed in more detail in Chapter 6.

Plant Sizes Used Plants for indoor use vary in size from the smallest cactus to the largest palm tree. They are grown at the nursery in plastic containers called *grow pots, grow containers,* or *production containers*. Each plant variety is sold in many different stages of growth, from young to mature, and will therefore be found in various sizes. Plant sizes are usually specified by container volume or diameter and sometimes by plant height. There is a correlation between pot size by diameter and the trade's designations of pot size by volume (see Table 5.1). One will find that some in the industry prefer to size plants by container diameter, others by its volume, and some by plant height.

TABLE 5.1
Summary of Container Sizes by Diameter and Volume

Container Diameter (in inches)	Container Volume (in gallons)
6	1
8	2
10	3
12	4
14	7
17	10
22	20
23	30
30	35
28	45 (is taller than 30 inches)
32	65
38	95
42	100

Adapted from a typical wholesale price list.

Plant Characteristics Plants are chosen for size by the interior landscape designer according to the area they have to fit into, as well as their ultimate size and shape when more mature. After all, it would be a poor-looking plantscape indeed if after a year or two the plants were so tall and ungainly that they presented a "seedy" look.

In addition to size, the designer considers the plant's color, shape, texture, growth habit, normal moisture and lighting requirements, and tolerance to disease, pests, and abuse. These factors are compared with the installation environment and a number of different plant types are chosen to meet the project objectives.

Plant Varieties The varieties of plants chosen for use in interior decorating run into the hundreds. They are specified by using either their common names or their scientific names. For example, the common spider plant is known scientifically as *Chlorophytum comosum*. The Norfolk Island pine tree is known scientifically as *Araucaria heterophylla*. Is it any wonder that the common names are generally the ones used to identify plants? Because some of the names are tongue twisters, the professionals have developed nicknames for them. For example, the *Spathiphyllum wallisii* is known simply as a "Spath," and the *Dracaena deremensis* var. "Janet Craig" is known affectionately as a "Janet Craig." There are a number of plant families that have become very popular with the decorating trades. These include ARALIA, AGLAONEMAS, DIEFFENBACHIA, DRACAENAS, PALMS, SCHEFFLERA, SPATHIPHYLLUM, FICUS, FERNS, BROMELIADS, and PHILODENDRON. Within each of these major classifications there are several, if not many, subclassifications. Table 5.2 contains a list of the foliage plants

TABLE 5.2
Popular Plants for Interior Use

Common Variety	Scientific Name (Common Name)	Common Heights Used
Aglaonema	Abigan	6 to 14 inches
	Commutatum	6 to 14 inches
	(Emerald Beauty)	6 to 14 inches
	(Silver Queen)	6 to 14 inches
Aralia (Polyscias)	Fruticosa (Ming Aralia)	12 inches to 10 feet
	Elegantissima	12 inches to 8 feet
Araucaria	(Norfolk Island Pine)	5 to 15 feet
Caladium	Lindenii	18 to 30 inches
Cissus	Rhoicissus Rhombidea (Ellen Danica)	1 to 2 feet (cascades)
	Rhoicissus Rhomoidea (Grape Ivy)	1 to 2 feet (cascades)
	(Fiona)	1 to 2 feet (cascades)
	Mandaiana	1 to 2 feet (cascades)
	Tetrastigma	1 to 2 feet (cascades)
Croton	(Norma)	8 inches to 8 feet
	(Petra)	8 inches to 8 feet
Cycad	Circinalis (Queen Sago)	30 inches to 10 feet
	Revoluta (King Sago)	30 inches to 10 feet
	Zamia Furfuracea (Cardboard Palm)	20 inches to 4 feet
Dieffenbachia	Camilla (Dumbcane)	12 to 30 inches
	(Tropic Snow)	30 to 42 inches
	(Aurora)	25 to 40 inches
	Perfection Compacta	20 to 36 inches
Dracaena	Deremensis "Janet Craig" (Janet Craig)	18 to 48 inches
	Marginata (Madagascar Dragon Tree)	5 to 20 feet
	Fragrans Cane (Corn or Cane Plant)	4 to 18 feet
	Massangeana Cane (Corn or Cane Plant)	4 to 18 feet
	Reflexa (Pleomele)	36 inches to 16 feet
	Warneckii (Striped Dracaena)	30 to 48 inches
	Yucca elephantipes (Joshua Tree)	12 inches to 8 feet

(continued)

TABLE 5.2 (Continued)

Common Variety	Scientific Name (Common Name)	Common Heights Used
Fern	Bostoniensis (Boston Fern)	6 to 24 inches
	Whitmanii (Feather Fern)	6 to 18 inches
	Platycerium bifurcatum (Staghorn Fern)	6 to 18 inches
Ficus	Benjamina—bush form	2 to 8 feet
	Benjamina—tree form	5 to 24 feet
	Nitida	5 to 24 feet
	Pandurata (Lyrata)	3 to 9 feet
	Elastica Decora (Rubber Plant)	3 to 9 feet
Ivy	Hedera Helix (English Ivy)	1 to 3 feet
	Hendera Canariensis (Canary Island Ivy)	1 to 3 feet
Palm	Areca	5 to 22 feet
	Chamaedorea Erumpens (Bamboo Palm)	3 to 14 feet
	Chamaedorea Sifrizii (Bamboo Palm)	3 to 14 feet
	Caryota Mitis (Fishtail Palm)	3 to 25 feet
	Howeia Forsteriana (Kentia Palm)	3 to 12 feet
	Phoenix Roebelenii (Dwarf Date Palm)	2 to 15 feet
	Beaucarnea Recurvata (Ponytail Palm)	2 to 20 feet
	Rhapis Excelsa (Lady Palm)	3 to 10 feet
Philodendron	(Angel Wing)	1 to 3 feet
	Cordatum	1 to 3 feet
	Domesticum (Elephant's Ear)	1 to 5 feet
	(Emerald Queen)	1 to 3 feet
	(Marble Queen)	1 to 3 feet
	Pertusum (Split Leaf)	1 to 3 feet
	(Pluto)	1 to 3 feet
	Scandens (Sweetheart Plant)	1 to 3 feet
	Selloum (Lacy Tree)	1 to 8 feet
Pothos	(Golden)	1 to 3 feet
	(Large Leaf)	1 to 3 feet
Sansevieria	Laurenti (Snake Plant)	18 to 42 inches
Schefflera	Arboricola (Dwarf Schefflera)	3 to 10 feet
	Brassaia actinophylla (Umbrella Tree)	5 to 22 feet
Spathiphyllum	Deneve #1	18 inches to 3 feet
	Floribunda	18 inches to 3 feet
	Mauna Loa (Peace Lily)	3 to 6 feet
	Tasson	15 to 18 inches
	Variegata "Mini"	15 to 18 inches
	Wallisii (Peace Lily)	15 inches to 3 feet

Sources: Various Wholesale Nursery Catalogs; "Plant Source '85," *Interiorscape* TPIE Show Issue (1985), p. 59; Stephan Scrivens, *Interior Planting in Large Buildings* (New York: Halsted Press/John Wiley & Sons, 1980); George H. Manaker, *Interior Plantscapes* (Englewood Cliffs, NJ: Prentice-Hall, Inc., 1987).

most commonly found in the catalogs of major suppliers to this industry. It is by no means a complete list of the plants used in the interior landscape business, but it is representative of the more popular types. A more complete coverage of the plant types can be found in good books on houseplants and interiorscaping (consult the Bibliography).

6 PLANT BIOLOGY AND OTHER GROWTH FACTORS

Overview Plants are living organisms and as such require certain conditions to keep them alive and sustain their growth. As we discussed earlier, too often the interior landscape plantstock, as well as its important growth, maintenance and environmental needs, is taken for granted by the owner, developer, architect, or interior designer. The designers have more control than they realize over factors affecting the ultimate happiness of plants in their new home. Following are a few examples:

- Placement of planter locations in proper relation to natural lighting.
- Proper design of built-in planter boxes, with adequate drainage provisions.
- Incorporation of effective irrigation systems into planter locations.
- Use of automatic, artificial lighting systems to supplement natural light.
- Assuring the specification and installation of proper soil mixes.
- Proper design of glazing, particularly in large window areas to assure recommended lighting levels can be maintained (plants are frequently "baked" in strong-sunlight locations).
- Location of planters relative to traffic patterns.

To neglect plant needs and other related considerations in the final building design, either by choice or ignorance of the biological principles involved, is a disservice to the project integrity and the financial interests of the building owners. Once the project has been completed, it is frequently too late. Building management is kept wondering why its project looks so bad in spite of the big numbers spent on plant maintenance. Often, this has to do with poor initial planning, which can be avoided. It has been mentioned that learning something about plant growth is an essential part of the educational process for architects, interior designers, and interior irrigation specialists. Anyone dealing with plants must understand in some measure their biological needs. It is certainly not necessary to become an expert on this horticultural subject; but just as the architect and engineer study the stress or light-transmission characteristics of the building materials they deal with, or as the interior designer is informed about the special characteristics of the textile materials, woods, and surface finishes he or she deals with, so too should they have some understanding of this other, more vulnerable design material they use—live plants. The following discussion is meant only to be a summary of the subject, and the reader is encouraged to delve into it more deeply.

Plant Growth Cycle A plant's basic requirements for growth are air, water, nutrients, and a suitable environment that includes sufficient light, proper temperature, and protection of the plant's leaves and roots.

A plant cannot gather its own food, as does an animal. It must absorb the necessary chemical substances from its immediate surroundings, the atmosphere and soil. A plant makes its own food from the carbon dioxide and water it absorbs. With the help of energy from absorbed sunlight, it manufactures sugar, its food, in the leaves. This process is called *photosynthesis*. The broad, flat shape of the leaf is nature's design for an efficient miniature food-manufacturing plant. It provides a large surface area for gasses to diffuse in and out through its many tiny pores and for the absorption of sunlight. During food manufacture, carbon dioxide is absorbed by the leaf from the air and diffuses with water brought into the leaf from the roots by way of the stem. The green substance in the leaves, called *chlorophyll,* when activated by sunlight, is able to chemically generate the sugar food substances. A by-product of this process is the release of oxygen into the air. The sugar is circulated throughout the plant either to be used for plant growth or to be stored as starch. The sugar used for plant sustenance and growth goes through a process called *respiration*. Chemically, it is the oxidation of manufactured food to release its energy and promote cell division, the absorption of minerals, and other necessary biological functions. During respiration, oxygen is absorbed from the air by the leaf to enter into the chemical reaction; a small amount of oxygen is also available from the photosynthetic process. Carbon dioxide is the by-product of respiration and is subsequently released into the atmosphere through the leaves. It is then again available to be absorbed by the plant for further photosynthesis, thus completing the cycle. Water that was brought up from the roots passes out of the leaf through its pores. That process is called *transpiration*. It is a form of sweating, much like animals do. As the water is evaporated away, it has a cooling effect, much like it does on our skin; but it also creates a tiny vacuum or suction throughout the plant, right down to the roots. This generates a "pull" on the moisture in the soil, causing it to be drawn into the root system for circulation through the plant. Water also migrates into the roots by the process of *osmosis*. The absorbed water carries with it dissolved mineral salts and other nutrients that are necessary for the plant's sustenance and growth. High temperatures, a dry atmosphere around the plant, and drafts from open windows or air-conditioning and heating systems will increase the rate of transpiration, releasing more water than normal through the leaves into the atmosphere. This lost water must be replaced by the plant, so it is essential that enough moisture be available in the soil at the root zone for replenishment to take place. In hot weather, therefore, it is necessary to water the plant more frequently. In colder weather or when the ambient humidity is higher, transpiration slows down and less water is released from the leaves. There is less need to replenish the moisture in the soil, so the plant's watering schedule can be reduced. There are other factors affecting moisture requirements and they will soon be dealt with.

Roots need to breathe in order to remain healthy and functional. The spaces between soil particles (pores or interstices) must contain oxygen as well as moisture (see Figure 6.2). That means the soil should not become waterlogged; otherwise, water will fill all of the pores around the roots and drive out the oxygen. The roots will soon drown and rot, causing them to be disfunc-

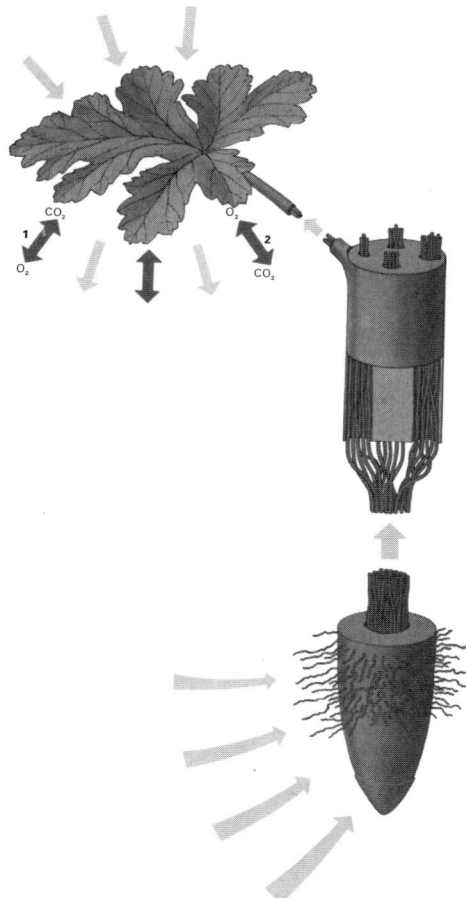

Figure 6.1 The growth cycle of a plant (Courtesy of Richard Lewis from *The Complete Indoor Gardener,* New York: Random House, ©The Original New, Last Partnership, 1979, p. 216.)

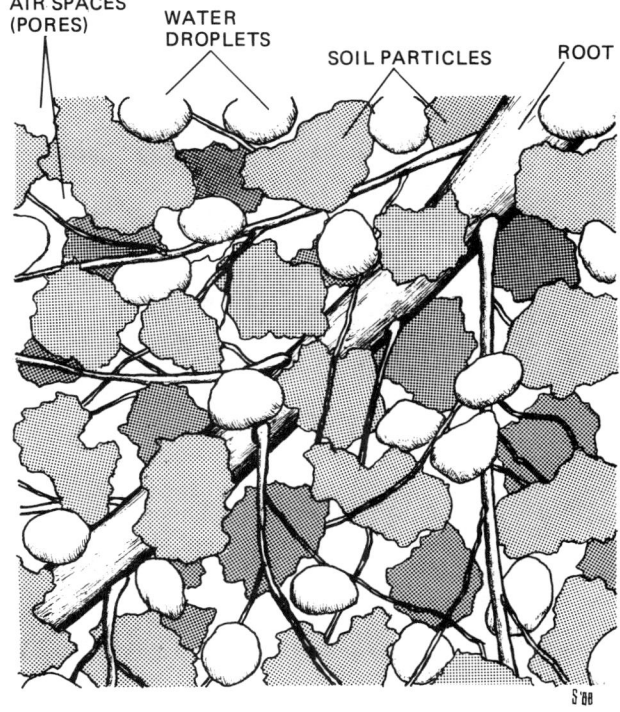

Figure 6.2 Magnified view of the interface between a root system and soil, air, and moisture.

tional. As a result, the plant will die. The opposite happens when water is given off by the leaves faster than the roots can replace it; for example, when the soil dries out. In this case, the plant will wilt because the moisture that is normally in its stem provides much of its turgidity or stiffness.[1]

Nature has provided a balance between the roots, their ability to absorb water and nourishment, and the top growth that the plant system propagates. This balance is commonly known as the *root/shoot ratio*. If there are more roots than the amount of top growth requires, then the top grows out quickly to catch up. This happens when branches are pruned back or broken. In the case of plants in interior environments, much less light acts on the leaves and the photosynthetic and transpirational processes are slowed considerably, with the result that less food is produced and less top growth occurs. This, in turn, reduces the plant's need for a highly developed root system, as moisture and nutritional requirements are kept at a lower level. This is all relative, however, and those indoor plants growing in brighter light do so more vigorously, both at topside and root level. The other side of this story is that if the top growth is much larger in volume than the root system or its ability to gather enough moisture and soluble salts, then the top growth starts to die back (generally dropping leaves and drying up branches) in order to compensate. This happens not only when the root system is not mature enough, but also when the roots become damaged from mechanical mistreatment or from the soil being oversaturated with water, as well as when the soil becomes too dry. That is why proper irrigation is important. It must provide the right amount of moisture at the root level without depleting the soil of its oxygen.

Soil Moisture   Waterborne nutrients are absorbed into the plant system by the roots. Soil moisture, therefore, is one of the key factors in a plant's biological process. The moisture requirement of the various plant varieties used in interiorscaping are commonly found in books on the subject, and there are many. Moisture-related categories are usually simplified as dry, moist, and wet. The largest category by far (the one into which most plants fit) is *moist*. Of the 204 plant varieties listed in Ortho Books' "Mini-Encyclopedia of House Plants," 157 of them carry the watering recommendation of "keep evenly moist."[2] Furthermore, in Manaker's *Interior Plantscapes,* of the 74 plant types delineated by variety and requirements, 60 carry the "keep soil moist" recommendation.[3] Likewise, in the book *Interior Plantscaping* by Gaines, of the 54 plant types delineated by variety and requirements, 49 carry the "keep soil moist" recommendation.[4] Most experts in the field contend that more plants are destroyed by heavy-handed, manual watering than by any other means. Such practices eventually saturate the soil, rotting the plant's root system. The new Micro-Irrigation Systems™ come the closest of all techniques in maintaining even levels of soil moisture. We will discuss this subject in more detail later in relation to interiorscape watering practices.

Light   Light acting on a plant's leaves is a key ingredient in the process of photosynthesis as well as other biological processes, and is therefore necessary for its life,

---

[1]*All About Fertilizers, Soils and Water* (San Francisco: Ortho Books [Chevron Chemical Co.], 1979).

[2]*Houseplants—Indoors/Outdoors* (San Francisco: Ortho Books [Chevron Chemical Co.], 1974).

[3]George H. Manaker, *Interior Plantscapes* (Englewood Cliffs, NJ: Prentice-Hall, Inc., 1987), pp. 233–238.

[4]Richard L. Gaines, *Interior Plantscaping* (New York: Architectural Record Books, 1977), pp. 80–106.

growth, and development. When the subject of light is considered, we must respect it as a complex issue, with interrelated factors of intensity, quality, and duration.

Intensity is the amount of light acting on the plant's leaves, and the photosynthetic process reacts in direct proportion to the intensity of the light; that is, the greater the amount of light, the greater the amount of food manufactured by the plant. Intensity is measured by photoelectric light meters, and the unit of measure is called a *footcandle*. Plants growing outdoors under field conditions will typically experience 10,000 to 16,000 footcandles of sunlight in Florida, Texas, and California, which are the main plant source states. Compare this with the typical light intensity in offices of 30 to 200 footcandles.[5] There is little wonder that the plants must go through a long acclimatization period before they can be used indoors; otherwise, the shock would be too great. Once installed in interior settings, the greenery must be bathed in at least minimum levels of light for them to survive; somewhat greater levels are necessary if plants are to be sustained at their original size and vigor; even greater intensities are necessary if they are to thrive. Minimum interior lighting levels will vary with plant species and how they were raised prior to installation. For example, flowering plants need brighter light than most foliage varieties.

The time in which light acts on the plant is the lighting *duration*. Plants grow better when illumination is sustained for long, continuous periods of time. This is frequently difficult in interior settings, for natural sunlight comes in through windows and skylights at varying intensities as the sun traverses the sky. Consequently, *effective* intensity levels may be available only for very short periods in many interior locations. There is an interrelationship between intensity and duration. The question of how long light of a given intensity must act on a plant each day is complex and varies with a number of factors, but plant species appears to be the predominant factor. The general rule of thumb used by the trade to cover most situations dictates that the plant must have at least 12 hours of exposure to lighting of at least 50 to 100 footcandles in intensity. Most interiorscaping plants will fit into that mold; those classed as high-light-level plants will not.

Moisture requirements are generally specified in relation to lighting conditions, as the two are also strongly interrelated. The more light a plant receives, the more water it needs to sustain its life processes. Any listing of moisture recommendations is contingent on the plant getting the recommended lighting conditions as well.

TABLE 6.1
Plant Categories and Recommended Lighting Levels

Plant Category	Minimum* Light Level	Recommended* Light Level
Low-light	50 fc	75–150 fc
Medium-light	75–100 fc	200+ fc
High-light	200 fc	500 fc
Very high-light	500 fc	1000+ fc

Source: Richard L. Gaines, *Interior Plantscaping* (New York: Architectural Record Books, 1977), p. 70.
*Light levels as measured in footcandles, at 12 hours per day exposure.

---

[5]Richard L. Gaines, *Interior Plantscaping* (New York: Architectural Record Books, 1977), p. 37 and George H. Manaker, *Interior Plantscapes* (Englewood Cliffs, NJ: Prentice-Hall, Inc., 1987).

The *quality* of light is another important factor that must be considered in dealing with plants. Quality has to do with the color (or wavelength) of the radiant light energy that is received by the plant, either from the sun or from artificial light sources. It may be remembered that the radiant energy spectrum contains a narrow band of wavelengths that represent to us visible light. It is within this band that plants also get radiant energy to help their growth. Reddish and bluish visible light is the most effective in the horticultural life processes. Sunlight is rich in these energy wavelengths, but its light quality does change as it passes through the atmosphere on its way to the earth. The least amount of change occurs around high noon, which is why light rays are strongest at that time. The greatest changes occur early and late in the day because the energy must pass through a longer atmospheric barrier before it reaches earth. Therefore, sunlight at the earth's surface is redder during the early morning and late afternoon hours and bluer around midday. The sunlight also takes on some other alterations as it passes into the interior of a building, picking up the color cast of the walls or curtained windows of neighboring buildings, possibly of the painted walls outside its own structure, or of the tinted glass of the building's windows and skylights, as well as of light reflected off surrounding interior surfaces. So the sunlight that finally reaches the interior planters is generally quite different than the sunlight found outside; aside from being much less intense, its quality is not as effective from a plant growth standpoint.

Light quality is an even more important issue when dealing with artificial sources used to supplement natural lighting. Light quality varies with the type of bulb used and the differences can be quite large. Incandescent as well as a wide variety of fluorescent sources are used over planters. Special bulbs are made to stimulate plant growth, but research has shown them to be no more effective than standard lamps and very poor as contributors to general room lighting. They are not used much by designers. The main choice of professionals are fluorescent "cool-white" lamps that have a high blue content, which is good for plant growth. Their other attraction is that they are also good for general room illumination and serve the dual purpose quite well.[6]

Nutrients When we speak about nutrients, we are talking about chemical salts dissolved in soil moisture, which are absorbed into the plant's system and distributed for food production and other biological processes. These chemical salts are referred to as *fertilizers*. Although many different elements are available to the plant, those containing nitrogen, phosphorus, and potassium are considered the *primary nutrients* . . . the most important ones. Smaller amounts of calcium, magnesium, sulphur, iron, and other trace elements are also provided in a well-balanced fertilizer, but the plant species differ in their need of these various components.

Fertilizers can be added to the soil as dry chemicals to be subsequently dissolved by irrigation water, or they can be dissolved prior to watering. Automated irrigation provides the opportunity to water and feed plants through a central system, for fertilizer can be injected into the stream or irrigation water laced with fertilizer can be held in system reservoirs. Slow-release fertilizers are commonly used in interiorscape maintenance. These are dry materials manufactured with a coating so that the chemical nutrients are released into the soil at a very slow rate over a long period of time.

[6]George H. Manaker, *Interior Plantscapes* (Englewood Cliffs, NJ: Prentice-Hall, Inc., 1987), pp. 46–64.

Atmospheric Conditions

Interior plants don't require large doses of nutrients for the same reasons that moisture requirements decrease when they are brought indoors. Very dilute solutions can be fed to plants regularly through automated Micro-Irrigation Systems, or slow-release fertilizers can be used in the soil.

The immediate environment of the plant has much to do with its health. The temperature, circulation, and quality of the air, as well as the local humidity, are some of the important factors. In some circles, this is known as the plant's *micro-climate*, or the close environment as opposed to the larger, more universal environment in which the plant resides.

Because most plants used for interiorscaping are of the tropical variety, they naturally prefer warmer temperatures but are more tolerant of relatively short, cool periods than most people realize. Most plants prefer a temperature range of 65° to 75°F, with slightly lower temperatures at night.[7] Many interior environments are more severe than that, and plants are commonly subjected to baking or near-freezing temperatures. For example, the areas behind large windows bathed in direct sunlight during the midday hours can become quite warm in both summer and winter. Temperatures in these areas can reach well over 100°F. At the other end of the thermal scale, building interiors in cold climates can become fairly frigid at times, particularly next to windows and when the heat is turned down at night or on weekends. Plants have a difficult time coping with these temperature extremes, and they can become damaged, sometimes fatally.

Circulation of air around plants is desirable, so long as it is not excessive.

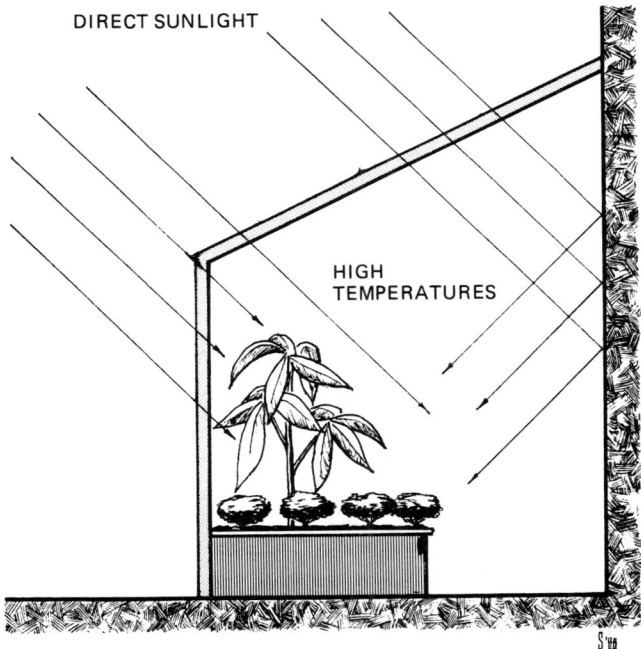

Figure 6.3 High interior temperatures behind large window surfaces exposed to direct sunlight.

---

[7]George H. Manaker, *Interior Plantscapes* (Englewood Cliffs, NJ: Prentice-Hall, Inc., 1987), p. 80.

Too often, fixed planters are designed into buildings, or freestanding planters are placed close to the air-conditioning and heating ducts, where air currents can be quite strong. This can have a drying effect on the plant, as well as raise or lower its temperature dramatically.

Humidity is another important factor. Most plants prefer a relative humidity on the high side and run into problems when the interiors of homes and commercial buildings are dried by non-humidified heating systems. Moisture can be artificially introduced into the air at or near the plants to help them along. Irrigation practices should include dampening the surface soil or mulch around the plants so that evaporation will increase the humidity in the vacinity of the leaves. Automated irrigation systems are quite effective in providing this extra moisture on a continuous basis. Those systems that spray water on the leaves generally should be avoided indoors, however, as this can lead to fungus and other leaf-borne diseases. This problem is particularly troublesome when air circulation around the plants is poor. Plants are frequently grouped close together to promote greater humidity levels; and by placing them near water landscaping elements, like artificial ponds, waterfalls, fountains, and streams, they can benefit from these higher humidity locations.

Air quality is difficult for the designer or builder to control, for it involves a wide variety of polluting sources. The major source of air contamination that could have an adverse effect on interior plant growth is in the air of the neighborhood itself. Heavily industrialized areas carry strong pollutants injurious to humans and vegetation. Construction and decorative materials, as well as activities within a building, add other contaminants. These factors are starting to get more attention now in conjunction with the "sick building syndrome" and indoor air pollution, in general. Most plants have a way of tolerating bad air, but like human reaction, there are frequently serious side effects. The recent studies by NASA researchers also show a wonderful property of air purification

Figure 6.4 Planters around an artificial pool and waterfall.

exhibited by some species. These plants actually absorb and assimilate many of the pollutants that dirty our environment, without harm to the plant. Further research is necessary to more fully define the relationship between tropical foliage plants and air pollution. Meanwhile, suffice it to say that by cleaning up the environment for the benefit of mankind, our foliar companions will also benefit.

The Growing Medium

Plants need something to sink their roots into, providing support as well as a source of nourishment, air, and moisture. That is the function of the growing medium. We don't want to mislead you by calling it soil yet; for although that is the most important type of growing medium, other media are used that cannot be classified as soil. As mentioned previously, **the quality of the growing medium can be as important to the irrigation practices used in an installation as can any other factor,** and one would do well to concentrate on the following discussion.

Soil is a term that cannot be clearly defined. Roughly described, it is the relatively thin layer of loose material around the earth, composed of mineral elements from weathered rocks, dead and living organic matter, air, and water. It is constantly undergoing change from the effects of weather, chemical processes, microbial processes, vegetation, and human activity. There is no precise point at which the loose material becomes soil. As a rule of thumb, however, if it supports vegetation, it is considered soil.[8] Soil used for potting is made up of sand, clay, and silt (the combination is called *loam*), along with other soil modifiers. Natural soil, as it comes from the ground, varies widely in its composition, which is determined by local geology. In its natural state, the soil is seldom well suited for use in potting plants. Without modification, its properties of moisture and air retention, as well as its drainage, pH, and nutrient content are usually not the way they should be. Soils that have a high clay content tend to compact into a heavy mass that has very small pores, excludes air, and does not drain well; it easily becomes oversaturated with water. Coarse, sandy soils, on the other hand, are very open and loose, with large pores in which lots of oxygen can be harbored, but little else. It drains well—too well—and water tends to drain through without much moisture retention. Nutrients are, of course, carried through the soil with the water, and the roots have little chance to benefit from their presence. Sandy soils are, therefore, not considered very fertile. From an irrigation standpoint, predominantly sandy soils can considerably increase the frequency of watering because of their fast-drying characteristics. Soils containing a predominance of silt drain poorly and are also poor growing media.

Loam is a soil composition made up of sand, silt, and clay but, depending on its source, generally has these soil elements in incorrect proportions for a well-balanced growing medium. Natural loams are, therefore, modified with various organic and inorganic materials to change their characteristics into mixtures suitable for the growth of plants in containers, or directly planted in indoor beds. One will find almost as many growing media recipes as there are plant varieties, but the most common revolve around the use of equal amounts of coarse sand, garden loam or topsoil, and sphagnum peat moss. Sand is the ingredient that keeps the mix from packing tightly, and, therefore, promotes better drainage and oxygen retention. Peat moss is an organic ingredient that also promotes drainage and aeration, but its other important contributions stem

[8]*All About Fertilizers, Soils and Water* (San Francisco: Ortho Books [Chevron Chemical Co.], 1979), p. 13.

from its ability to absorb large quantities of water (and waterborne nutrients), thus improving the water-retention and fertility characteristics of the mix. Because of its fibrous nature, it also improves the *wicking* ability of the soil, thus promoting better water distribution in the planter. This is a particularly important property when we deal with Micro-Irrigation Systems, as we shall learn later.

Garden loam and topsoil contain clay and silt particles that also hold water and nutrients. Sand is sometimes replaced in a potting mix with perlite or polystyrene foam beads. Peat moss is frequently replaced with vermiculite, composted leaf mold, or tree bark particles. Other additives, such as charcoal and lime, are used to control soil acidity and toxicity. Bonemeal and other organic materials are used in many cases to improve soil fertility. Proportions of all these elements are changed to accommodate the special needs of some plant varieties. Some of the newer mixes omit the garden soil (loam) content altogether to produce a lightweight medium commonly referred to as *soil-less*. This type of growing medium has advantages to anyone who must grow, handle, or transport the potted plant, although some soil-less media do not provide a sturdy support for the plant, particularly for larger varieties, and toppling sometimes occurs. One might hear about Peat-Lite, U.C. (University of California), and Jiffy-Mix. These are common soil-less media.

Virtually all plants used in interiorscaping are shipped from the wholesale nurseries either balled and burlapped or in containers filled with a growing medium.... Most are shipped in the latter form. It should not be assumed, however, that the medium is of good quality and will promote healthy plant growth. It should be checked. Many shipments contain sandy soil or other substandard growing media with poor moisture retention, that make proper irrigation practices critical. That type of soil is unforgiving, and frequent waterings are a must. Furthermore, it can lead to very heavy maintenance costs if manual irrigation is used. Automated irrigation systems can relieve most of that extra burden, making the best of these bad situations. Fortunately, most growers are responsible firms. It is advisable for the interiorscaper and architect to seek out and deal only with reputable firms. However, if a containerized plant is found to be cursed with poor soil, it is a wise practice to replace it with the proper mix.

Another pitfall is the casual attention that built-in planter beds and pits get during a building's construction. All too often, the soil at the construction site becomes the bedding soil for costly ornamental plants, grown to that stage with great care and at great expense. To sink these beautiful creations into poor subsoil laced with harmful construction debris is an act that could have expensive and long-lasting implications for the building owner and managers. The informed and responsible architect or project overseer will make sure that proper growing media are specified for built-ins, and with his or her horticultural advisers and contractors, make sure these specifications are followed during installation. Once plants are transposed into poor soil locations and begin to take root, it is usually too late to do anything about it from a practical standpoint. The die is then cast for problems over a long period of time.

Acidity of the soil, as measured by its pH, is a factor that the interiorscaper must take into consideration. Most indoor plants prefer soil slightly on the acid side (sour), and the pH must be monitored from time to time to check on the drift that normally occurs. Waterborne minerals accumulate in and on the soil and can reach levels detrimental to the plants. Roots can be chemically burned by high concentrations of mineral salts. Horticulturalists recommend regularly flushing these salts from the container by running large quantities of water through them. In many types of installations, however, this is not practical, and the plants can deteriorate over a period of time from sour soil. Manual irrigation maximizes the buildup of these soluble mineral salts in the soil, for it takes much more water to irrigate by this method. The greater the amount of water passing through the soil (with the exception of actual flushings), the greater the amount of mineral deposits left behind (fertilizer salts as well as unwanted mineral salts). Precision, automatic techniques use much less water in properly irrigating containerized plants, and will therefore deposit harmful salts at a much slower rate. Also, less fertilizer is wasted because water drainoff can be virtually eliminated.

Pests and Diseases   One of the important jobs of the interiorscape maintenance technician is to inspect the foliage for pests and diseases. One might hear about slugs and thrips, mealybugs, scale, and fungus. Any of these or other afflictions can be dangerous to the plant if undetected. There are countermeasures that can usually be taken to eliminate the problems, and these are among the tasks of the maintenance personnel; inspection, diagnosis, and treatment of plants are carried out by the competent technician.

Summary   As can be seen from this discussion of plant growth factors, there are many things that must be considered in placing and maintaining plants in an interior environment. Providing proper moisture is only a single factor, albeit one of the more important ones. There is also a strong interrelationship between many of the growth factors, which must be understood by the designer, the interiorscaper, and the irrigation specialist in order for them to be effective in their work. Overseeing all of this are the real estate owners and managers, who must understand the complexity of the situation and hire only qualified people who are knowledgeable in their specialty. There is a tendency to hire outdoor landscape specialists to maintain interiorscapes . . . even worse, untrained building maintenance personnel are frequently put to that task. Hopefully, this chapter's discussions will promote a greater awareness of the need for **interior** plant maintenance specialists, and industry practices will be refined accordingly.

7 INTERIORSCAPE IRRIGATION: MANUAL VERSUS AUTOMATIC

Overview In previous chapters, we have discussed the strong and increasing use of live potted plants in the interiors of buildings, as well as the developing technologies that transform an inanimate structure into one of functional usefulness that does things for us, making our lives and jobs easier and less costly to perform. The automation of plant care in building interiors is the latest innovation along these lines, creating a new climate in our real estate markets where buildings can be designed and built with prolific use of indoor plants in mind for commercial, economic, and health reasons. From time immemorial, plants have been watered by hand with the exception, of course, of natural rain. With the advent of automatic sprinkler systems, the lawns and gardens around buildings could be cared for more conveniently and economically. Until fairly recently, however, improvements in technique were only a minor part of indoor horticultural maintenance. Now we are closing in on the twenty-first century, and interior plant-care methods are hardly more advanced than they were a hundred years ago. Several years ago, irrigation contractors started installing outdoor technologies in noncritical interior planter locations. Sprinkler systems were first used in promenade areas of shopping malls and in large planter beds of office building lobbies and other common areas. Soon the newer drip or trickle irrigation systems were designed into some of these nonsensitive interior locations. The rest of the building interior was relegated to manual plant care. In most commercial buildings, the majority of decorative plants housed in the structure are in the upper floors, where hundreds or possibly thousands of potted plants grace the corporate office suites, lounges, restaurants, banks, and so on. This foliage receives highly labor-intensive, weekly maintenance, incurring untold millions of dollars in plant-maintenance fees around the country . . . and the world, for that matter. One attempt at reducing labor costs in recent years was the development of self-watering containers. They provided the first real element of easier plant care for the critical furnished areas of homes and commercial buildings and are being used to some extent today by interiorscape maintenance contractors. Their widespread use is being retarded, however, by a number of inherent limitations, leaving almost the entire building interior market open for a more complete, advanced plant-care technology. That technology is now available in automated Micro-Irrigation™ Systems, which will be described in this and the following chapters. These systems promise to revolutionize interior plant care, as did automatic sprinkler systems for exterior landscape care.

Manual Plant-Care Techniques

We have already been reminded that from the beginning of time, decorative plants were watered and otherwise cared for by hand. This is almost universally true of interior plant care, even today in the age of high technology. While outdoor irrigation practices were transformed by automation, indoor technology was virtually nonexistent until several years ago. It can be safely estimated that about 98 percent of all the containerized and direct-bedded plants used indoors are still being fully maintained by manual labor.

Manual plant care takes a number of forms, and watering is only one facet of it. The work is done by *interiorscape maintenance technicians*, who are trained in the horticulture of indoor foliage plants, particularly in the maintenance aspects. Through their considerable labor and knowledge, the greenery is generally taken care of on a weekly schedule, although many companies have lengthened their routines to 10-day cycles. The most time-consuming part of the technicians' work involves the irrigation, or watering, of the plants. Industry sources estimate that 25 to 50 percent of the on-site maintenance time is comprised of watering . . . it is also the most physically demanding part of their work. Irrigation is the main reason for these weekly maintenance visits to a client's location. The other tasks of plant care can be accomplished in two- to four-week cycles, depending on the nature of the installation.[1]

Hand watering is done in most cases by means of a watering can or other container that is filled at the closest faucet, the convenience and proximity of which is dependent on how concerned and diligent the architect was in designing the installation. The water is then transported to the planter location and carried from plant to plant. It is poured onto the soil surface (referred to as *overhead watering*). Water is heavy, with each gallon weighing 8.34 pounds, and the physical strain on field personnel is considerable. They are constantly carrying water-filled containers weighing from 6 to 30 pounds. Frequently, in complex installations, they must climb ladders with the watering equipment or in other ways perform gymnastics around planter beds. Keeping in mind that most interiorscape maintenance work is done by females, the physical burden is considerable. To alleviate some of the strain, devices called *watering machines* have been developed. These are sealed water containers mounted on wheels or casters. Some types use built-in pumps to produce a flow from the container to the planter. Others utilize pressurization of the water vessel to create the flow. In either case, hoses that carry the water to the plant are connected to the container. A convenient hand-operated valve at the end of the hose permits the operator to control the flow. Wands and other handle extensions permit water application beyond arm's length. These machines are used extensively in the trade and have reduced the to-and-fro movement involved in refilling water cans. Watering machines are also very heavy, however. The models commonly available are built to hold between 7 and 44 gallons of water. That represents 367 pounds of water in the larger units, plus the weight of the empty unit, which is usually about 145 pounds. The total is a hefty 512 pounds, which must be wheeled around the client's complex and then reloaded onto a

---

[1]Robert Hyland, "Modern Technology Solves Personnel Problems: Interior Landscape Irrigation," *Western Landscaping News* (February, 1983), p. 28.

Figure 7.1 Typical manually operated watering machines. (Courtesy of Cascade Designs, Inc.)

truck or van, a procedure only to be repeated at the next maintenance stop. The smallest of the units holds 7 gallons (58 pounds of water), plus its empty weight of 38 pounds.[2] This makes for a still considerable 96 pounds when full. Watering machines are a decided help in that they reduce the number of trips to the faucet and the exertion involved; however, much of the physical strain remains.

Properly designed interiorscape installations provide for a source of water and an electrical outlet in close proximity to the planter boxes where possible. The water source is almost invariably a hose bibb connection to the building's cold water line. It should be located no farther than about 40 feet from the most extreme point in the planter: otherwise long, heavy, unwieldly hoses must be employed. Hose bibbs, therefore, are most practically located in the planter boxes or beds, along with an electrical outlet for lighting, water pumping, or general utility use. The unfortunate part of the story is that more often than not, the water and electrical service is not provided in convenient places by the designers, even for large or critical planters that require an unusual amount of irrigation. This magnifies the burden on the plant-care technician, for the time, inconvenience, and physical exertion are increased considerably, as is the cost. The money saved in the elimination of hardware from the original installation is eaten up year after year in higher maintenance expense.

[2]*Aqua-Mate Catalog* (Providence, RI: Aquamatic Systems, Inc., 1985).

Freestanding containerized plants are placed by the interiorscaper or interior designer where they complement the decorating scheme, but their needs, and those of the maintenance technician, are seldom provided for in the building plan. Irrigation of freestanding planters is done almost invariably, therefore, by means of watering cans or machines. In the furnished areas of buildings, such as in private homes, corporate office suites, bank lobbies, hotel suites, lounges, restaurant dining areas, and waiting rooms, great care must be exercised to prevent splashing or drain-off from the bottom of containers. Accidents frequently occur when technicians carelessly wet and soil carpeting, furniture, moldings, and wall coverings. Overhead potted planters are particularly bothersome and difficult to judge, causing frequent drain-off problems. Self-watering containers are used in some of these situations, but they, too, must be refilled manually on a regular, reduced schedule.

The frequency of manual watering of a containerized plant is determined by many factors. The most important of these are the plant's species, its size, its current state of activity or dormancy, the composition of the growing medium, the room temperature, the ambient humidity levels, the air drafts passing the plant, the amount and quality of light it receives, the presence or absence of surface mulch, and the size, shape, and composition of the container. That's a complex menu of conditions, and it points up the variation from plant to plant that would normally be encountered. Common manual irrigation practices call for an indoor plant to dry out for a period of time to allow oxygen into the root zone; then the soil is thoroughly soaked until drain-off occurs from the bottom of the container. The soil at this point is saturated with water, displacing the life-giving oxygen. Well-drained soil will pass most of the water through the container quickly. Sandy soil will pass it through too quickly, and heavy clay soil will pass it through too slowly. Soil that has been dried out too much will have shrunk from the walls of the container and the water application will drain down this space quickly, without much benefit to the plant. In these cases, drain-off will also be excessive. For these and other reasons, the instructions for watering given to the homeowner are not usually the way the professionals would do it, as those techniques would not be commercially feasible. Firstly, the drainage underneath planter boxes and pits is always suspect and usually inadequate. One cannot, therefore, pour gobs of water into a plant to soak it thoroughly. Any drain-off would most likely ruin something, not to mention make a mess. This situation also exists with freestanding and hanging container plants in offices, homes, and restaurants. The maintenance technician is not able to move most of them to a faucet because of the size of the plants, their weight, and numbers. Keep in mind that the ornamental plants used in commercial interiorscapes are generally larger than those used by the homeowner. Medium-sized plants in 14-inch pots (7-gallon size) weigh about 50 pounds. Larger plants weigh 225 to 275 pounds (in 30-gallon, 23-inch containers) to over 1,000 pounds (in 95-gallon, 38-inch containers).[3] The commonly suggested watering frequency is also impractical for the interiorscaper, who cannot be available to service plants when each reaches its proper time for irrigation. Instead, clients must be serviced on a regular, frequent schedule, regardless of the condition of the plants. Weekly maintenance cycles are most frequently used so as to avoid excessive water stress.

When the maintenance technicians arrive on the scene each week, they find plants in various conditions of moisture need, and simple, quick testing is required. There are a number of techniques used to judge the condition of soil

---

[3]JoAnn Johnston, "Specimen Containers and Specimen Weights," *Interiorscape* (May/June, 1986), p. 14.

moisture in each plant. Some use a finger to feel for moisture; others a stick probed into the soil; and many use a moisture meter. These simple tests give them an indication as to how much of a watering is required at that time. There is judgment involved, and because of the nature of the job, it is not an easy matter to train personnel. It takes a good deal of time for a trainee to get proficient at the task, as there is always the tendency to underwater or overwater. Industry managers concede that it usually takes the best part of a year to fully train a maintenance technician in the art of manual irrigation.[4] That is a long, expensive road littered with plant replacements and frustrated, discontented clients. Company profits are at stake as well. The large turnover that interiorscapers commonly experience requires constant hiring and training. Maintenance labor is a major problem for them, and the irrigation aspects of the job are the principal part of it.

There are many other facets of interiorscape maintenance besides irrigation. One of the most important is inspection of the plants for pests and diseases. This means getting close to the plant and carefully examining the stems, the soil, and the leaves for signs of improper lighting, nutrition, watering, insect infestations, and disease. Leaves are inspected top and bottom. This inspection must realistically be done about every two to four weeks to detect any changes in the plant's condition. It is not an easy task, for the kneeling and bending is physically demanding; it is also hard on the eyes, particularly in poorly lit areas. If any signs of these problems are detected, the technician must take quick corrective action to prevent further decay of the plant's health. In the cases of insect infestations or disease, this generally involves the application of an insecticide or herbicide by either spraying or wiping it on. Occasionally, the plant-care chemicals are applied systemically; that is, carried through the root system by the irrigation water.

Another of the tasks relegated to the maintenance technician is that of fertilizing the plants. This is done manually by dissolving a small amount of soluble plant food in the irrigation water and applying the diluted solution during the weekly maintenance visits. Another way of doing it is by means of the newer slow-release fertilizers that are mixed into the potting soil. By means of water hydrolysis or soil bacterial action, the fertilizers slowly release and diffuse their nourishing chemistry into the soil over long periods of time. This means that they don't have to be replenished often, and fertilization thus becomes an easy task for the interiorscaper. There are a number of forms of these slow-release nutrients, the most common being coated globules, tablets, and briquettes. Because of the much slower metabolic rate of plants grown indoors, fertilization rates are greatly reduced; consequently, this is not one of the high-frequency tasks of the technician.

Another maintenance aspect is the rotation of the plants, in which they are left in position but gradually rotated over a period of time so that all the leaves are given even exposure to the lighting. This helps the plants grow fuller and more symmetrical in shape.

During the course of the maintenance visits, it is also necessary to clean and sometimes polish the plant leaves. Cleaning is done by wiping off dust and other particles from the leaves. This is done not just for esthetics but so that the pores of the leaves are not clogged with dirt. Leaf polishing is done by applying a chemical for that purpose. Some of the chemicals tend to clog the pores, and many interiorscapers do not believe in the practice for that reason.

[4]Robert Hyland, "Modern Technology Solves Personnel Problems: Interior Landscape Irrigation," *Western Landscaping News* (February, 1983), p. 28.

Manual Plant-Care Techniques

Figure 7.2 Interiorscape maintenance technician at work.

Plant replacement must be done periodically whenever the plants appear wan or have become diseased. Sometimes, the move is just temporary in order to rejuvenate the plant in a better growing environment. In order to present a fresh, healthy appearance in a commercial interiorscape, plant replacement is frequent and common, particularly when flowering varieties are involved.

As one can see, the manual plant-care process requires a solid knowledge of the subject, good eyes, and a lot of stamina. The number of decisions that must be made by the maintenance technician each day is very high. Each plant is scrutinized individually, and decisions are made as to its general condition, moisture requirements, type of corrective actions needed, and so on. So, in addition to the physical burden of the job, the mental strain is also a factor of worker discontent. Any improvement in plant-care techniques that can relieve some of these burdens and stresses from the technician should be seriously considered by the interiorscape contractor. Changes in irrigation methods can be the most effective, particularly the adoption of automatic systems.

Moisture Meters Because of its ease of use, many interiorscapers have adopted the moisture meter as their main device for judging the amount of moisture in the planter soil, indicating the plant's need for water at the time of measurement. These are precision instruments that depend on a moisture probe to generate a small electrical current when placed in the damp soil. The amount of current generated is in proportion to the wetness or dryness of the soil and is read from a meter attached to the probe. The greater the degree of soil moisture, the higher the meter reading. There are a number of brands and designs of moisture meters on the market, most of which are very inexpensive. Many have the meter and probe as an integral one-piece unit. Others are made in a two-piece configuration, in which they are separated but connected by a wire. The latter type is more convenient, as the meter can be held in one hand where it can be easily seen while the probe can be inserted into the dark recesses of the planter. For this reason, the two-piece style has become more popular with the professionals.

Moisture meters are used as purchased, and the readings are recognized as relative indications of soil moisture levels, accurate enough for most plant-care tasks. Because of the differences in design from one brand to another, the readings cannot be compared except in broad terms. The low numbers on the

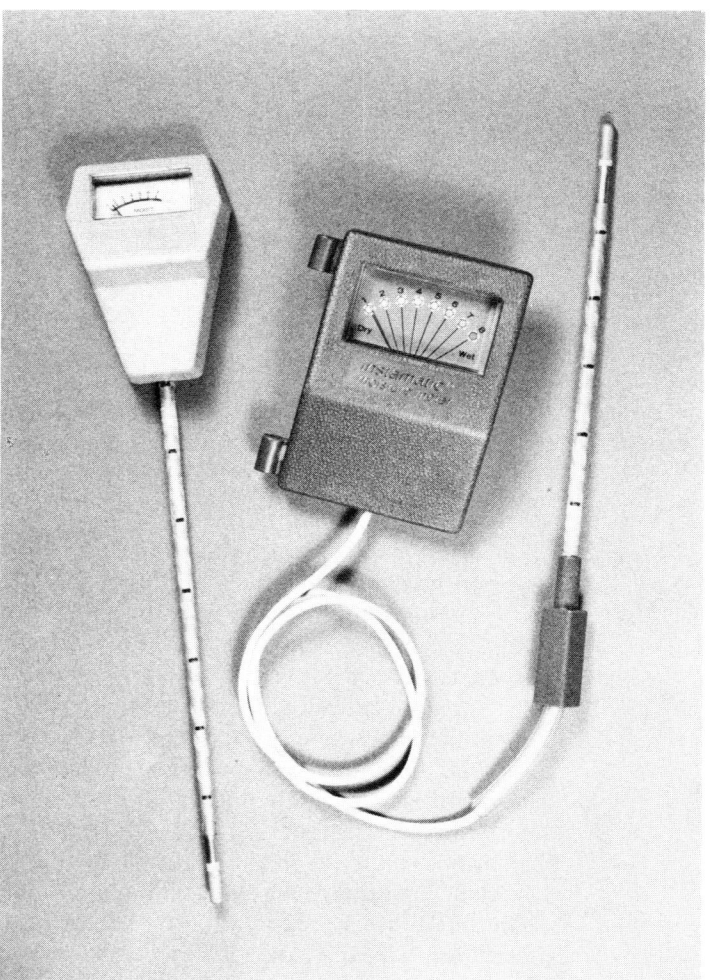

Figure 7.3 Two types of moisture meters.

meter represent dry soil readings, while the higher numbers indicate that the soil is damp or wet. There is usually little attempt to calibrate the meters, as that type of accuracy is of lesser interest to the maintenance technician. Most technicians learn after a while that mineral salt content in the soil has a large bearing on the accuracy of the readings. As salts build up in the soil over a period of time, they increase the value of moisture readings. Soil that is on the dry side but that has a high salt content can read the same as salt-free damp soil. These differences are recognized by the experienced maintenance technician and taken into consideration. The meters can also tell the technician when the soil should be leached (rinsed) of these salts, and corrective action should be taken at that point.

During the course of developing the technology for precision Micro-Irrigation Systems, we found it necessary to make detailed studies on the diffusion patterns of moisture in plant containers so as to measure the effectiveness of these new methods. Accurate moisture readings were needed for comparison purposes. Conventional moisture meters were used for practical reasons, but they had to be modified because their inherent design led to some inaccuracies. The meters were good enough for use in field maintenance but not quite good enough for more scientific applications. The Instamatic Moisture Meter (two-piece configuration) was chosen for these studies, as informal surveys indicated that it was probably the most popular meter used by professionals. Before the moisture diffusion studies were started, a series of experiments were run to determine the idiosyncrasies of the instrument. It was first noticed that moisture readings change, or drift, over a period of time, and a study was done to determine the extent of that change. The drift was studied over a period of 8 minutes at various locations in the planter and at two depth levels (2 and 4 inches below the surface). As can be seen in the summary graph in Figure 7.4, the drift in readings extends over the full 8-minute period but is fairly stable in the period 30 seconds to 2 minutes after insertion of the probe. It was concluded, therefore, that taking readings 30 seconds after the probe has been inserted into the soil will give reasonably consistent results.

It was next noticed that readings were highly dependent on the depth of the probe in the soil in a way that was not altogether expected. Even in saturated soil, where the moisture content was fairly uniform, the readings changed. The readings could vary by as much as several points in uniformly wet soil. This led us to consider that another factor had a bearing on the results. This phenomenon was explored by using the meter in plain water. No meter reading was obtained by inserting the probe in distilled or deionized water, so it was confirmed that mineral salts or other electrolytes must be present in the medium being measured. In plain tap water, the readings more than doubled from insertion at a depth of 1/2 inch to insertion at a depth of 5 inches. Water presents maximum wetness; so if these readings had been taken in soil, there would have been serious variations not having to do with moisture content at the various depths. The readings in water increased with the depth, independent of the moisture level; so with an unmodified meter, no matter what the moisture content of the soil, one would expect the readings to increase with depth. This, of course, is an inaccuracy. The tests were duplicated using additions of common salt (sodium chloride) in the water as an electrolyte. They confirmed the original results, only at higher readings (also confirming the effect of mineral salts on the readings). These results are illustrated in Figure 7.5(a). It can be seen that the meter needle registered nothing (1.0) until the probe penetrated the water deeply enough for its insulator to be below the surface, and then the readings rose rapidly toward their maximum . . . it took 4 inches of depth, however, to reach that maximum. It must be realized that in

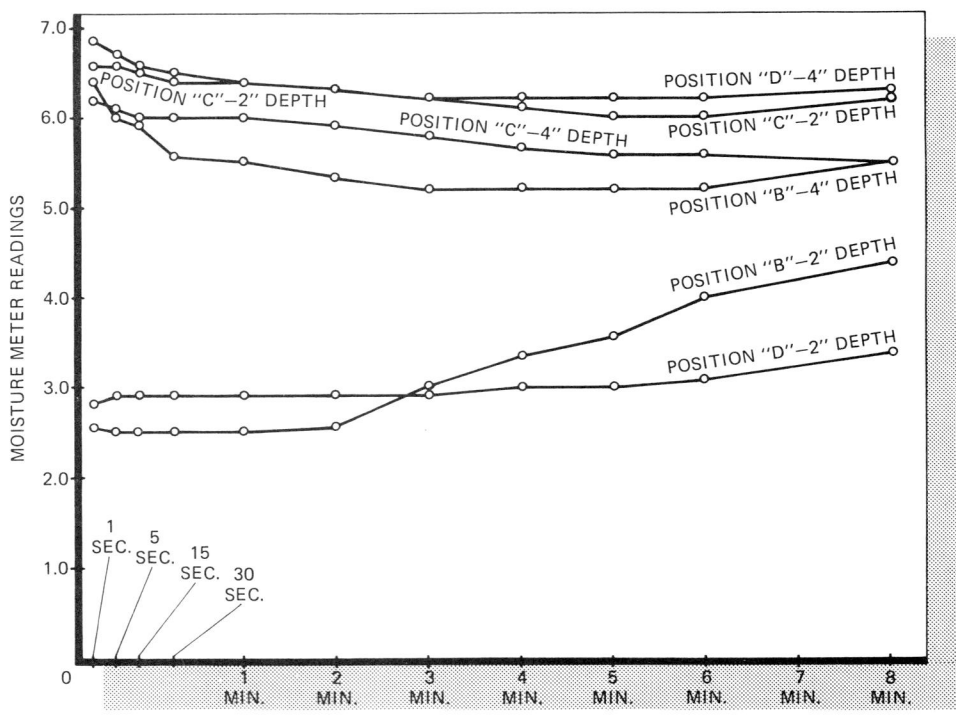

Figure 7.4 Instamatic Moisture Meter (modified) stabilization studies. (Courtesy of Aqua/Trends.)

water, maximum readings must be reached almost immediately for depth measurements in any medium to be meaningful.

Because of these inaccuracies, it became necessary to modify the moisture probe for the purposes of Aqua/Trends' studies. A probe is basically made up of three parts: two metallic electrodes separated from each other by an insulator (see Figure 7.6).

Electrical energy to activate the meter is not generated until damp soil bridges the insulator that separates the electrodes, creating an electrical path between the two conducting elements. This is illustrated in Figure 7.7. Previous experiments demonstrated that inaccuracies are created when the long metallic sleeve is left fully exposed.

As a solution to the problem, the probe section above the insulator (long metallic sleeve) was wrapped with an insulating tape to reduce its effective surface area. An area roughly equivalent to that of the probe tip (below the insulator) was left exposed. This is illustrated in Figure 7.8. The insulating wrap can be Mylar-based Scotch tape; white electronic shrink-tubing can also be used. Vinyl electrical tape was found to shrink too much (unwanted contraction). After the insulating wrap has been stabilized, depth marks are painted at one-inch intervals measured from the probe's insulator. The resultant moisture meter is much more depth sensitive.

Testing with this modified instrument was more predictable. At a depth of one inch in tap water, the reading was the same as for all other depth readings. Refer to Figure 7.5(b) for a comparison between the wrapped and unwrapped moisture meter probes in plain tap water as well as in saline solutions.

The next step was to calibrate the modified moisture meter in soil of various levels of dampness. Perfectly dry soil was gradually wetted at regular incre-

Moisture Meters

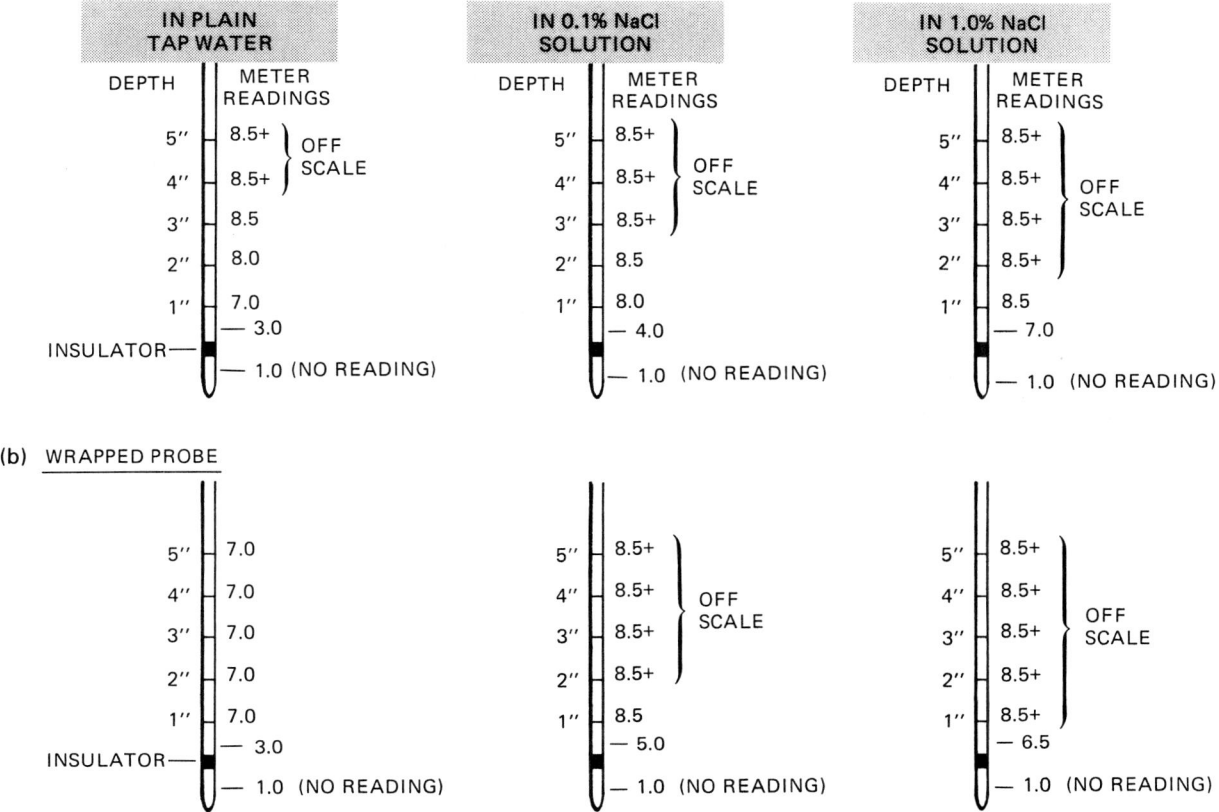

Figure 7.5 Testing of the Instamatic Moisture Meter through meter readings in tap water and saline solutions. (Courtesy of Aqua/Trends.)

ments, with moisture meter readings taken at each stage and at three depth levels. The results are shown in Table 7.1. Measurements show the meter to be skewed toward the damp side. The slight variations in readings at different depths was presumably caused by the settling of water to lower soil levels. It was difficult to maintain a homogeneous moisture diffusion in the confined, test soil sample over the time period used.

The bottom line to these studies is that for everyday plant maintenance, conventional, off-the-shelf moisture meters will do an adequate job if the technician considers their idiosyncrasies. These meters can, however, be easily modified for greater consistency and accuracy.

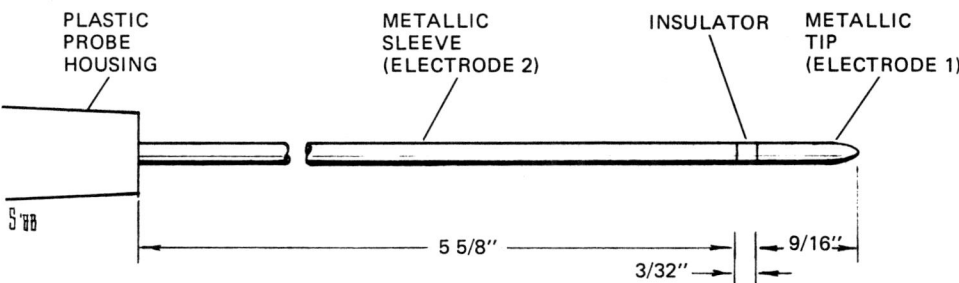

Figure 7.6 Probe detail of an Instamatic Moisture Meter.

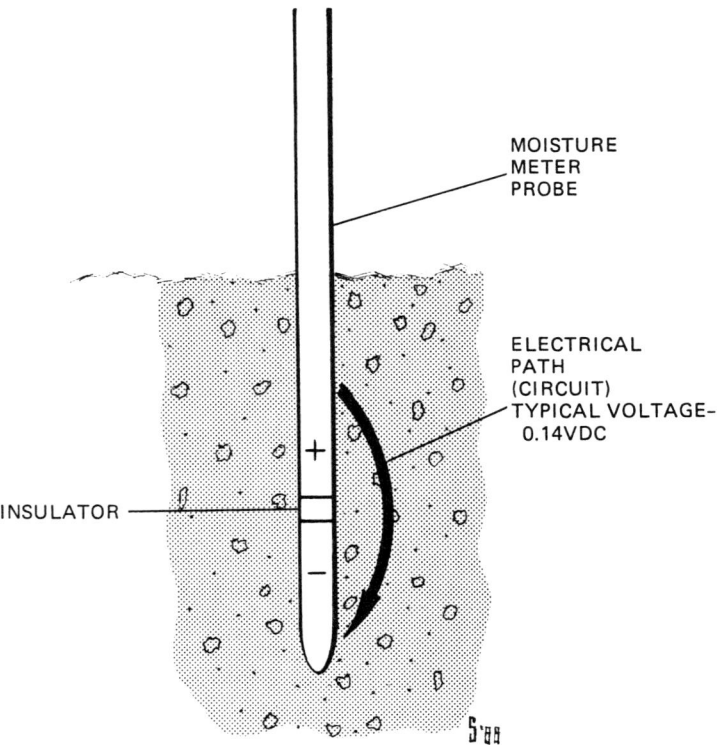

Figure 7.7 Operation of moisture meter probes.

Self-Watering Containers

One of the more recent attempts to reduce the labor content of interior landscape maintenance was the introduction of the self-watering container. It is the first easy-care maintenance system developed specifically for interior foliage plant installations. Originating in Europe where they have gained more popularity than in this country, these devices are variations on the old household technique of using a braided or woven fabric wick to transfer water from a tray or other source through the bottom of the container to the underside of the root ball. The *wicking* (movement of water upwards through the narrow pores of the fibrous wick and then the subsurface soil) is by means of naturally occuring capillary action. The technique is called *subirrigation* by some, because the moisture diffusion is upward rather than downward as in overhead watering.

Self-watering containers are frequently referred to as being *automatic* devices or systems. This is a misnomer, for the technology is limited in its longevity before human attention is required. As unattended irrigation systems go, self-watering containers are not in the same class as fully-automated drip/trickle or precision Micro-Irrigation Systems.

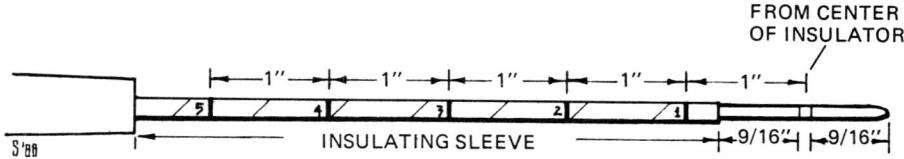

Figure 7.8 Modified probe of the Instamatic Moisture Meter. (Courtesy of Aqua/Trends.)

TABLE 7.1
Calibration of Modified Instamatic Moisture Meter

Soil Moisture Content**	Moisture Meter* Readings		
	At 1" Deep	At 2" Deep	At 3" Deep
Dry soil	1.0	1.0	1.0
Dry salt	1.0	1.0	1.0
10% salt	1.0	1.0	1.0
15% salt	2.0	2.5	4.0
20% salt	6.5	6.5	7.0
25% salt	7.0	7.0	7.5
30% salt	7.5	7.5	7.5
35% salt	7.5	7.5	7.5
40% salt	7.5	8.0	8.5
45% salt	8.5	8.5	8.5

*Instamatic Moisture Meter.
**Percentages by weight.

Self-watering containers are made as either double-walled containers or partitioned containers, having a planter section into which the foliage is directly planted and a water-reservoir section with an air space between the two sections. They are made of sturdy plastic, and the outer shell is molded in a selection of decorative colors. Woven or braided wicks made from nondegradable fibrous materials (such as nylon or fiberglass) are interfaced between the reservoir and the soil sections so as to transfer water between the two. The water reserve is mostly under the planter, but in at least two designs, it also wraps around the sides of the container. A filler hole or tube is provided for convenient refilling. In some designs, the wick is wrapped around the filler tube, which then performs a dual function. Gauges and sight glasses are provided in most models to indicate the level of water remaining in the reservoir.

At least two planter designs use a wickless system featuring a sensor that is inserted into the soil to detect the moisture level. The soil is in limited direct contact with the water reserve. The sensor is a porous ceramic element that allows air to penetrate the tube to which it is attached whenever the soil becomes dry. This has the effect of decreasing the vacuum in the reservoir cham-

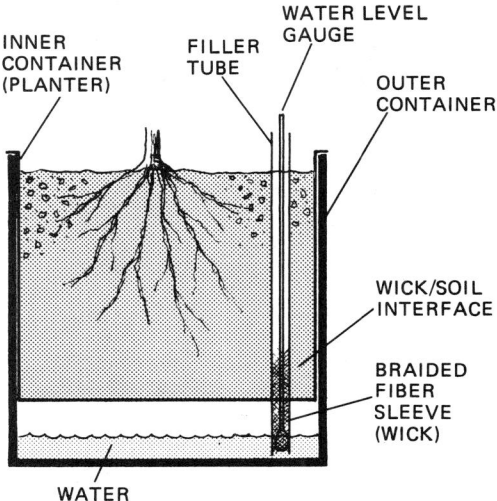

Figure 7.9 Typical wick-type self-watering container.

ber, exerting a slight air pressure against the water surface. That, in turn, renews the capillary transfer of water through the soil mass at the bottom of the container (see Figure 7.10).

Another type uses elongated plastic reservoirs that are buried in planter areas under direct-planted foliage. Moisture diffuses up from the reservoirs to the root levels by capillary action. A filler tube rises to the surface. These independent reservoirs can be interconnected with tubes to make an extensive underground network.

Among the main benefits of using self-watering containers, the foremost is their ability to reduce the labor content of the plant-care cycle. Most of these units are able to sustain a continuous watering period of from two to eight weeks. This is dependent on the size of the reservoir, the size and species of the plant, the number of plants coexisting in the same container, and the rate of transpiration of the plants as determined by their growth cycle and local environmental conditions. The composition and texture of the growing medium also has a bearing. As the plant grows larger, the reservoir must be refilled more frequently. This ability to reduce the number of manual waterings can potentially save a considerable amount of time and cost. Maintenance cycles can be reduced from once a week to every two to four weeks. Interiorscape maintenance contractors can utilize labor more efficiently, realize lower labor turnover by increasing employee satisfaction, reduce training costs, increase customer satisfaction, and reduce plant replacement costs. Furthermore, the maintenance technicians can concentrate on the other aspects of plant care and be able to do them more effectively. As a result, there is usually more consistency in the quality of plant care, particularly when a novice is at work or during periods of technician change, vacation periods, or building shutdown. The plants benefit from more consistent watering as the containers are meant to maintain a fairly even moisture level, thus reducing water stress. Fertilization can be done by dissolving nutrients in the irrigation water. Aeration of the root system is more efficient, because the soil is not compacted as it is with overhead watering. As you can see, much can be said for using this newer technique to reduce the amount of time and effort devoted to plantscape irrigation.[5]

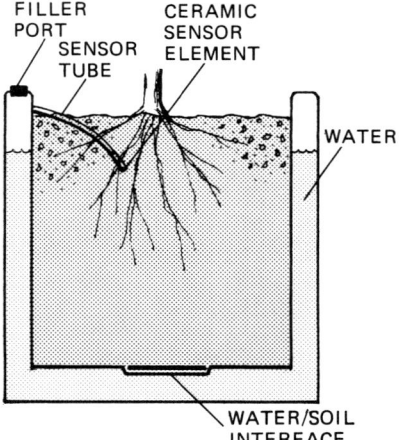

Figure 7.10 Typical sensor-type self-watering container.

[5]Robert Hyland, "Modern Technology Solves Personnel Problems: Interior Landscape Irrigation," *Western Landscaping News* (February, 1983), p. 28.

There is another side of the coin, however. Interiorscape maintenance contractors do not use self-watering containers in large numbers; in other words, the use of these devices is limited. There are a number of deficiencies in the concept that keep it from gaining greater market penetration. One has to do with the decorative aspects of the containers. They are plastic, and molded in a wide range of colors to meet a variety of decorating schemes, but this variety is generally not sufficient for the decorator and the plastic is considered inelegant by some. Then there is the problem of color obsolescence when the decorating scheme changes in a few years. The planter color is fixed, as is its shape. They tend to be larger and bulkier than plain containers, because of their integral water reservoir. The interiorscaper objects to the fact that in most types, foliage must be grown directly in the self-watering planter and cannot be kept in the nursery container, as is the preferred technique. Thus, plant rotation and replacement are made more difficult. Most of the drier-soil plants cannot be used with these systems, as they would be overwatered. Contractors also feel that the purchase price of self-watering containers is prohibitive, especially when large numbers of plants are involved. Containers must be replaced with larger ones when plants outgrow them. Another objection is that capillarity draws water up only so far, leaving the surface layers of the planter fairly dry and confusing maintenance personnel. Another objection is that water-filled, self-watering containers are heavy and cannot be easily used in macrame plant hangers (for overhead installations). The buried reservoir systems have also caused problems when sturdy roots grow down to the system level and engulf the reservoirs and tubing, causing breaks, leaks, and difficulties in removing plants. Others contend that the maintenance time savings are really not there because it takes so long for the technician to refill the reservoirs, and the frequency of filling increases as the plant grows.[6]

It must be remembered that for all intents and purposes, manual watering chores are not eliminated with these systems—for refilling reservoirs is indirectly a manual watering chore. They are simply deferred and reduced in frequency. The biggest disadvantage mentioned by maintenance contractors has to do with the accuracy of irrigation. They feel that the watering control varies with too many conditions: the soil composition, the mineral content, and the size of the plant. They feel they can't count on the planter to do what it was installed to do and is, therefore, too risky for commercial use. As mineral salts build up on the underside of the root ball, they restrict proper capillary action, slowing down moisture diffusion. This has been mentioned as a particularly difficult problem with regard to those self-watering containers that use moisture-sensing devices. The porous ceramic element in the sensor tends to clog with soil particles and mineral salt buildup after a period of time, and this changes the monitoring characteristics of the system, reducing its accuracy. There is a decided drift in soil moisture level during this period, which is caused by sensor malfunction.

One of the unspoken factors tending to retard greater use of self-watering containers is the concern by interiorscapers that self-watering technology will replace them or in other ways reduce their ability to make a living. Part of this is not without foundation, for some clients who install self-watering containers will try to do it on their own. Many have the uninformed perception that watering is the only real concern in plant care, and if it is being done for them the easy way, all they must then do is peek in every once in a while. Many others hire amateur "green thumbs" to do the supplementary maintenance work and

---

[6]George H. Manaker, *Interior Plantscapes* (Englewood Cliffs, NJ: Prentice-Hall, Inc., 1987), p. 119.

risk the quality of overall care given to the installation, for these people seldom have the training or motivation of professionals. Some will hire contractors that specialize in outdoor lawn-and-garden maintenance services to look in on their ornamental interior plants. As mentioned previously, that would be a risky course of action because outdoor specialties are so different from interior ones. The net result is that some business is lost in these misguided situations.

The other part of the interiorscapers' concern is that they must reduce the maintenance charge to the client on contracts involving self-watering containers, as their labor content is reduced. The client expects some savings from the self-care features of the system, and contractors are frequently reluctant to offer them. Some of this resistance is justified. As just mentioned, real savings through use of self-watering containers are sometimes minimal or possibly not there. Contractors will frequently avoid being put in that position by simply not using the self-watering containers. They also feel that if they have to reduce the bottom line of the maintenance contract or leasing/maintenance contract, they are losing business. In situations where real labor savings are gained, however, this perception is usually false. For example, whenever the contractor is able to reduce the maintenance cycle from once a week to once every two or possibly three weeks (which can be safely done in many installations), the labor savings and enhanced utilization of labor are significant. Many more accounts can now be covered by the same field staff. Significant savings can also accrue from greatly reduced plant replacement and training costs. Some of these savings should be passed on to the customer. In doing so, it makes the interiorscaper more competitive. But the fact remains that psychological resistance by the interiorscaper has a negative effect on the use of newer technologies—this one and others.

Notwithstanding these difficulties, self-watering containers do present some real advantages to the contractor and to the homeowner. They will, therefore, have a place in the industry and will continue to be used in limited applications. They present a viable supplement to other technologies and can be used to advantage where competing systems are not feasible.

Automated Plant-Care Techniques

The use of *fully automated* plant-care techniques is the attempt by the industry to reduce interiorscape maintenance labor content to minimum levels and, at the same time, provide the highest efficiency. The impetus for this has come from the commercial real estate industry. It has been looking for ways to manage buildings more conveniently and cost-effectively on the inside as well as outside. As solutions develop in the marketplace, it appears that fully automatic interior systems will provide the basic technology for the building structure, to be supplemented wherever necessary by other concepts, such as self-watering containers.

Automatic sprinkler systems have been used for many years to take care of the exterior landscape—lawns, trees, and shrubs. These systems have saved real estate managers enormous sums of money in landscaping maintenance costs; consequently, ways of doing the same in their property's interiors are constantly being sought. The priorities here are dependent on whether the interiorscape is a major feature of the building, such as in shopping malls and lobbies of Class A office buildings, or merely a minor decorative feature. Large common-area planters generally mean significant maintenance costs. By and large, there has been no similar concern for building tenants—the corporate offices, restaurants, etc. They also have plant-care expenses and problems. Until recently, however, viable plant-care technology was not available as an amenity—thus, the issue was ignored.

A few years ago, irrigation contractors were asked to install their systems in large planter boxes and beds in the lobbies of office buildings and the promenades of shopping malls. The only systems available to them were the sprinkler and drip/trickle irrigation technologies developed for outdoor landscaping maintenance and for agricultural production. It has taken a while for the industry to refine its thinking and techniques to be able to cope with the more delicate situations indoors. For the most part, the industry has not yet caught up with the demands of this new market. Concept and equipment limitations have hampered efforts to satisfy a broad range of interior plantscape needs. Because sprinkler and drip/trickle irrigation technologies were originally developed for agriculture and outdoor landscape use, they are predicated on the use of large amounts of water spread over large areas and distributed for long periods of time. The outdoor soil dries out rapidly from sun, wind, and plant absorption. Thus, it takes a hefty system to keep up with the aqueous demands outdoors. On the other hand, indoor plantscape requirements are very different. Much smaller quantities of water are distributed, with more preciseness and with a greater degree of safety. This is particularly true in the furnished areas of buildings where outdoor irrigation technology is not sophisticated enough to handle these more refined demands. Irrigation contractors shudder when they think of their sprinkler or drip systems being applied in someone's living or family room, or in the lushly furnished offices of an executive suite. It is simply not practical to consider in most instances. For that reason, the technology used in Micro-Irrigation Systems was developed to successfully cope with *all* automatic irrigation situations indoors. It is the first fully automatic system developed specifically for interior applications.

There are several features common to all automatic irrigation systems. First of all, there must be a water source. This is sometimes a well, pond, or stream from which water is pumped. Interior installations are more often sourced from city water supplies. When easy access to city water mains is not available, holding tanks or reservoirs from which the water is pumped are used. Then there must be pressure to generate flow from the source. Pumps and city water pressure provide this. Next, there must be a network of tubing or pipes through which the water flows from the source to the plant locations. In the water distribution network, there must be flow and sometimes pressure-control devices that vary the flow rate and direction in order to meet installation needs. Finally, in automatic systems, there must be timing devices to regulate the period between watering cycles and the length of time that flow is permitted. These timing devices are also controllers, for they activate and deactivate the pumps and solenoid valves at properly timed intervals.

There are five irrigation technologies that are used in building interiors and that can be *fully* automated: sprinkler systems, drip/trickle systems, subterranean systems, hydroponic systems, and Micro-Irrigation Systems. All but Micro-Irrigation and hydroponics systems started outdoors as part of agricultural irrigation. The various types of irrigation technologies used in building interiors are represented in Figure 7.11.

Automated interior plant care is in its very early stages. The field of *automatic interior irrigation* is in a rapid state of flux. We now refer to this new area of expertise as "auto-interigation." Now that automated Micro-Irrigation Systems are available for total building coverage, the vistas of application have multiplied many times over. Their versatility will provide the impetus for developers, as well as corporate office and building managers, to use them in their

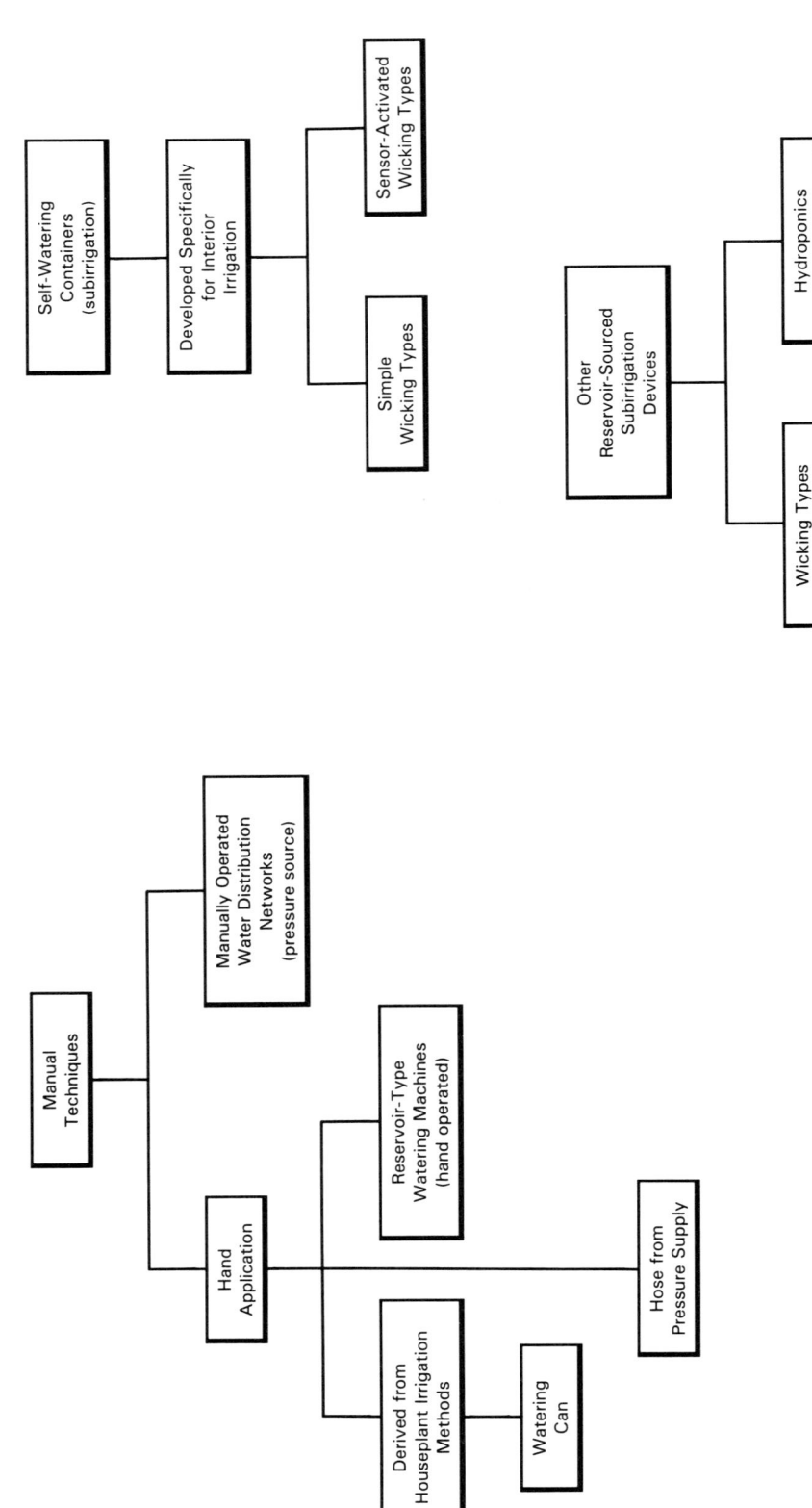

Figure 7.11 Types of irrigation used in building interiors.

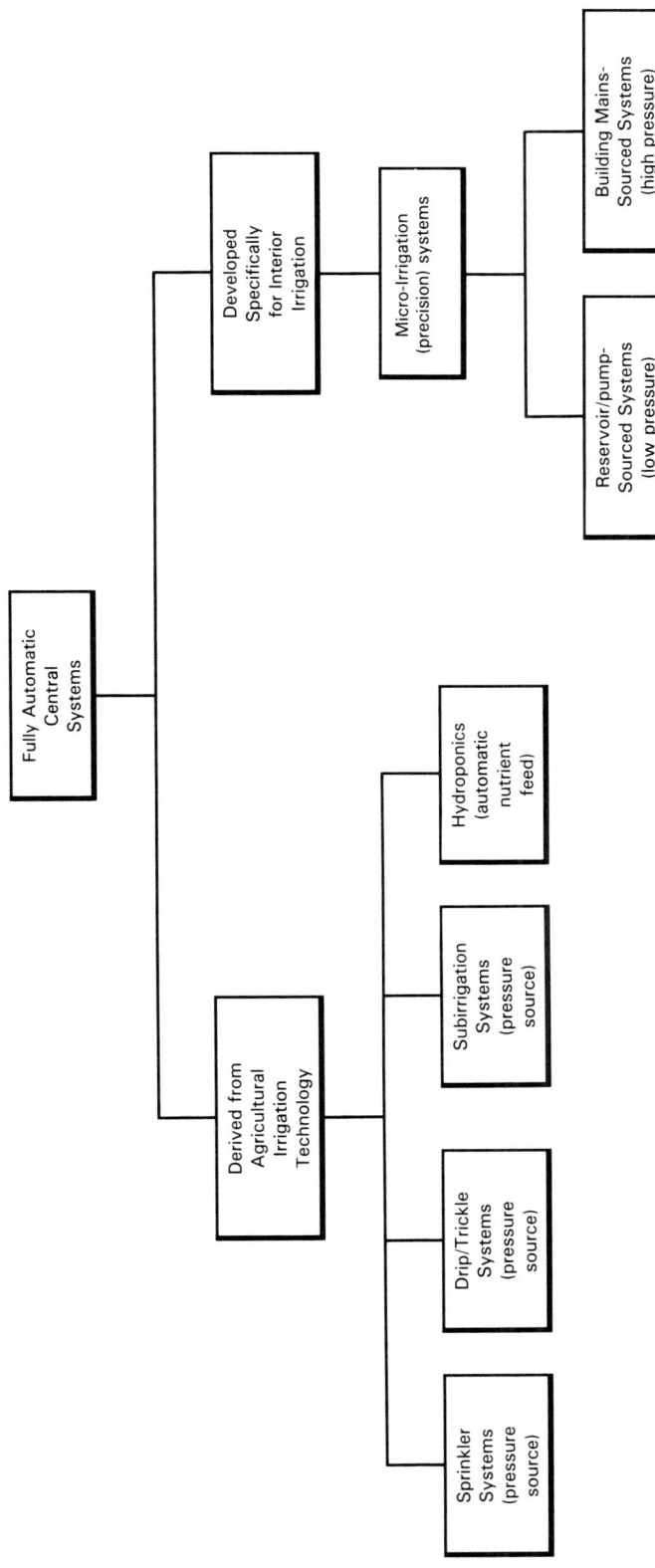

Figure 7.11 Types of irrigation used in building interiors (*continued*).

major projects. Large restaurant and hotel chains can reduce interior plant maintenance costs by tens and hundreds of thousands of dollars per year in their many locations. Builders of housing units have a new high-tech amenity available to increase the functionality of their products. Smart homes are more salable. The many social and commercial implications of these developments will present themselves more clearly in the future. Fully automated irrigation systems will be discussed in greater detail in Chapter 8.

8 INTERIORSCAPE IRRIGATION: FULLY AUTOMATIC TECHNIQUES

Overview Having quickly described the general characteristics of automatic systems in Chapter 7, we will now delve into some of the details that differentiate them, so that a clear understanding can be gained as to their uses, advantages, and disadvantages. The fully automatic irrigation technologies that have been used inside of buildings are (1) sprinkler systems, (2) drip/trickle systems, (3) subterranean systems, (4) hydroponic systems, and (5) precision Micro-Irrigation™ Systems. These technologies as a group comprise the field of auto-interigation (automatic interior irrigation). Each system will be discussed in some depth in this and the following chapters.

Sprinkler Systems This is the most important of the various systems involving fully automatic irrigation. It is the one first thought of when the subject is mentioned, and the one most commonly used outdoors. Sprinklers were developed many years ago for field irrigation to aid in agricultural production. It is probable that most field systems are still operated manually, but recent years have seen the slow shift to automatic control. Because of the long irrigation cycles used by farmers, automation is not as important a feature with sprinkler systems as with the other applications. Sprinkler systems, however, were soon found to be invaluable in maintaining landscapes as well, particularly on golf courses, college campuses, and around commercial buildings. These applications were more suitable for automatic control, as many zones were usually involved and operation was frequently scheduled for more than once per day. Soon, homeowners were offered the benefits of automatic sprinkling for their properties, using the same technology as developed for heavy-duty landscape maintenance jobs. In some areas of the country today, it is rare to find a home being constructed without a lawn-and-garden sprinkler system.

About a dozen years ago, irrigation contractors started to install sprinkler systems in building interiors as many architects, real estate owners, and managers began experimenting with them as building management aids. Prior to that, they had been used in rare project applications, but the recent years have seen interest grow. The applications chosen for sprinkler systems were almost invariably for noncritical areas in shopping malls, office buildings, and institutional structures. Built-in planter boxes and pits in building promenades, as well as tiled or paved lobbies and atriums, were the main areas of concern for the designers and managers. These areas cost the most to maintain and presented other problems as well. Although they generally contained many hundreds of plants, these areas were not as demanding from an applications standpoint. Because the sprinkler systems broadcast relatively large amounts of water over large surfaces and because they were not considered reliably safe, those planters far from the critical furnished sections of the buildings were a reasonable target for automated irrigation. In most cases, the early installations were offshoots of the outdoor landscaping sprinkler system . . . meaning they were separate branches of the same system, devoted specifically to watering the interior zone. These interior branches were usually set up to have an independent flow duration and flow rate control, and in some cases, zone pressure was regulated as well. Automation of these more complex sprinkler systems made sense, as manual control required greater labor demands and a special diligence on the part of maintenance employees. In more recent times, independent systems have been designed for the interiors of large commercial buildings where the added cost and control could be justified. Residential indoor installations have been used over the years, mainly in upscale housing and almost invariably in coarse planter areas, like atriums and large planter boxes and beds, safely away from the furnished living quarters.

Sprinkler systems as used indoors are normally sourced from city water mains, but on occasion well water is used. A filter is generally used at the source to ensure that the small orifices throughout the system do not become clogged by dirt. Those installations that are automated have an electromechanical or a solid-state timer/controller, which switches on the system at predetermined times and for predetermined watering periods. The timer/controller operates a pump or solenoid valve at the appropriate times to achieve that result. In more complex installations, various zones (or stations as they are sometimes called) of the building are established as branches to a common system. The controller permits the flow to each zone in sequence by activating remote, zone-controlled solenoid valves (either electrically or hydraulically). Because of the lesser water flow requirements of interior installations, flow and pressure control devices are frequently used to reduce either one or the other hydraulic element. The water distribution network is a series of interconnected pipes and tubes leading from the water source to the locations of the planters. Pipe sizes in indoor applications can generally be kept fairly small (1/2 to 1–1/2 inches in diameter) and are normally copper, iron, PVC, or polyethylene plastic (called "poly" tubing). Sprinkler heads and other emitting devices are connected into the piping at appropriate locations and spray or dribble water at and around the plants. Sprinkler systems are normally operated in the pressure range of 45 to 80 psi (pounds per square inch) outdoors and about 25 to 55 psi indoors. Higher pressures promote greater flow rates, which are generally not required indoors.

There is an entire catalog of different types of emitters available for sprinkler systems, with literally scores of different configurations. *Emitters* are those small devices installed at appropriate locations that spew water in some manner toward the plants. The emitters most commonly used in sprinkler systems are, of course, sprinkler heads that spray water in a predetermined pattern toward the plants at and over the foliage, stems, branches, and surface of the ground. The spray patterns are generally full circle, half-circle, quarter-circle, and strip (a very narrow stream). These patterns are illustrated in Figure 8.2. Other types of heads used in these systems do not spray at all. Shrub bubbler

Sprinkler Systems

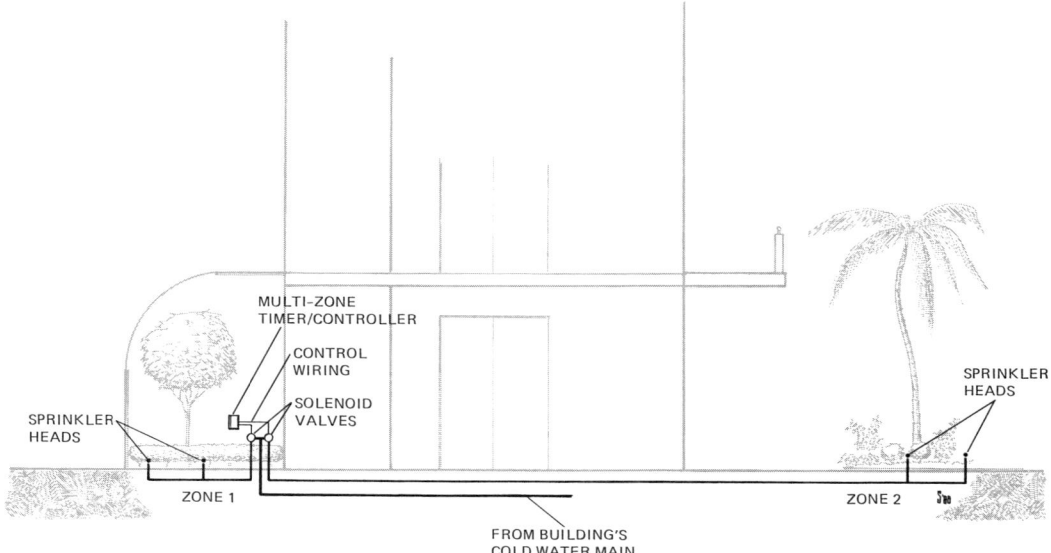

Figure 8.1 Typical multizoned interior sprinkler system.

heads simply dribble water onto the ground in the immediate area of the head. As a result, the water distribution pattern is tight, and capillary action in the soil is relied upon for moisture diffusion to other parts of the planter. If the bubbler heads are grouped close enough together, then fairly even coverage of an area can be obtained.

Sprinkler heads and other emitters are positioned in the installation to cover the area in need of irrigation, and the types used are selected according to what is to be accomplished at their location. Sprinkler heads are meant primarily for broad-area coverage to irrigate crops and turf. Some sprinkler heads

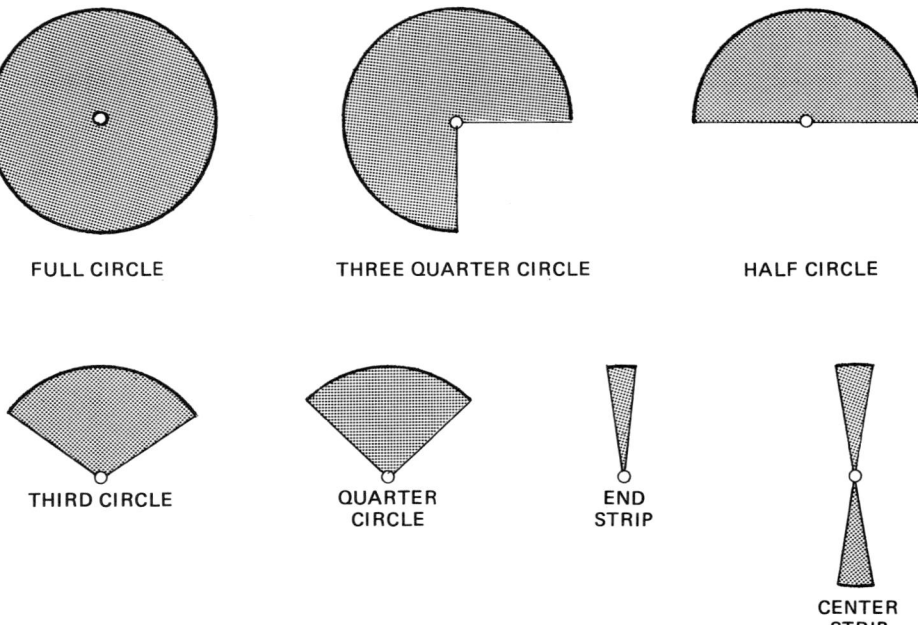

Figure 8.2 Sprinkler spray patterns.

are capable of spraying water great distances (50 or 60 feet) to get maximum coverage. Indoor applications of these systems are much different. Seldom will an interiorscape planter be so large in area as to require broad-area sprinkling. Sprinkler heads can be used in planter boxes and pits (1) if their flow rates are reduced to manageable levels by adjusting the head itself, (2) by lowering the system water pressure, (3) by selecting heads with low flow rates, or (4) by using all of these options. Many interiorscapers don't like to work with sprinkler systems because of the poor control over spray patterns. Mischievous tampering with the spray heads by children and vandals frequently causes hazardous conditions, as floors become slippery when wet, particularly the tile and marble floors commonly found in shopping malls and office building lobbies (see Figure 8.3). Some building managers will not permit sprinkler systems in their facilities for those reasons.

The length of time a sprinkler system is permitted to operate indoors varies with a number of factors, but the real determinants are the plant's moisture requirements at the time of irrigation, the soil composition, and the type of time switch used for the installation. Sprinkling is very inefficient and indiscriminate as it wets everything in its path, including most of the ground in the planted area. Much of the moisture evaporates from the spray before ever reaching the ground. Furthermore, water absorption by the soil is slow in penetrating the depths. Water from sprinkler irrigation tends to move laterally in the soil and is slow to sink, being held up against gravity by soil capillarity (see Figure 8.4 and Table 8.1).[1]

Figure 8.4 and Table 8.1 clearly demonstrate the long periods of time it takes for sprinkler watering to penetrate down to the ornamental plant root masses. With turf grasses, this is not a problem, for the roots are only a few inches below the surface. But upright plants must be sprayed for extended periods in order to achieve the proper irrigation.

Timers are used to determine the interval between sprinkler operations and the duration of watering (the "on" time). They use 24-hour or 7-day clock cycles. Timers (also called *time switches* or *controllers*) for sprinkler applications are usually coarse incremented, meaning their on–off cycles are set in incre-

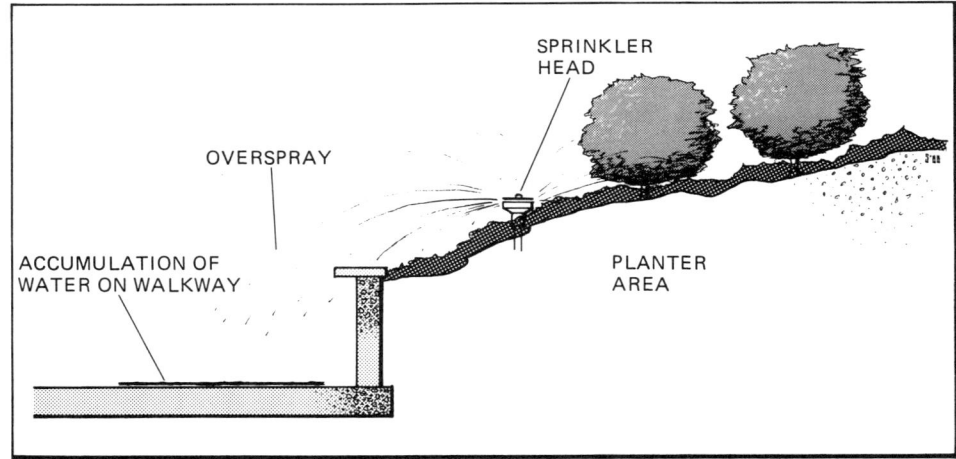

Figure 8.3 A misadjusted indoor sprinkler creates hazards in public areas by overspraying onto walkways.

[1]Jack Kramer, *Drip System Watering* (New York: W. W. Norton & Co., 1980), pp. 16, 17.

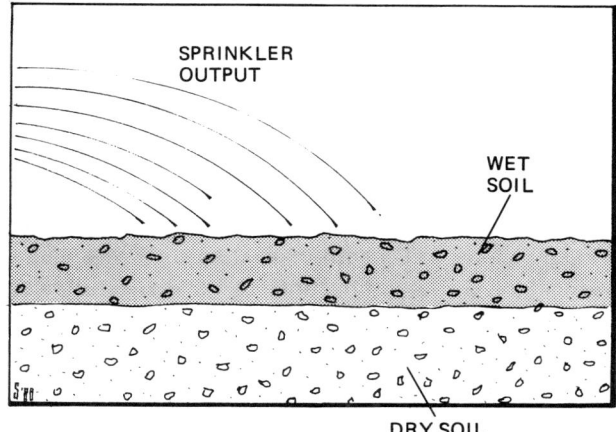

Figure 8.4 Typical water distribution with sprinkler and other spray irrigation.

ments of 3 to 15 minutes. The least amount of time in which the system can be effectively operated is important in interior work, for one of the dangers of any type of irrigation here is the tendency to overwater. The time switches made for commercial irrigation systems are of the electromechanical, electronic, and combination (hybrid) types. The more expensive timing devices permit sprinkler operation as short as 1 minute in duration; however, most are designed with a minimum "on" period of 3 to 15 minutes. The tendency is to operate indoor sprinkler installations for longer durations than necessary—a habit which is being carried over from outdoor work.

Relatively long irrigation cycles are generally necessary, however, because sprinkler systems are the least efficient of all automatic irrigation methods. They broadcast water over a larger area than required, wasting much in the process. There is really no need to irrigate the large quantity of support soil found in planter boxes between the plants. It absorbs much of the water unnecessarily. The root zones are the only important ones. Where water conservation is critical, the long irrigation cycle represents an added burden. Another reason for long irrigation periods with sprinkler systems has to do with the compacting of the surface soil over a period of time as it is moistened by the water spray. The compacted soil takes longer to penetrate, so more water has to be applied in order to ensure proper wetting at the root zones. The problem is further compounded in planters filled with improperly formulated soil mixes. Heavy clay soil is difficult to penetrate, and the surface of fine-textured sandy soils is frequently very difficult to wet out when dry and, thus, difficult to penetrate. This is an unexpected condition, for one would expect that the "open" nature of sandy soil would easily accept water (Note: coarse sand does accept water

TABLE 8.1

Sprinkler Irrigation Time Required for Water to Penetrate Various Soil Types

Soil Depth	Coarse Sand	Sandy Loam	Clay Loam
12 inches	15 min.	30 min.	60 min.
24 inches	—	60 min.	—
30 inches	40 min.	—	—
48 inches	60 min.	—	—

easily). It dries out rapidly and surface tensions must be overcome before the sand can be rewetted. That takes time and frequently a lot of water. Water tends to run off across the surface. One would be surprised at how dry sandy soil can be under the surface layers even after a rainstorm. The top inch or so might be wet, but it can be completely dry below. Using more closely spaced irrigation cycles helps to keep the surface soil from drying out, but that can easily lead to saturation indoors unless cycles are kept short and flow rates low. The use of mulch also helps to keep surface soil damp but many materials, such as cypress bark and peat moss, absorb a great deal of the irrigation water, keeping it from the root zone. More water is usually needed to overcome this effect, so one can see the dilemma that is encountered when trying to water soil that is difficult to wet out. That is particularly true of sprinkler systems using spray heads. The more concentrated application of water from bubbler heads that are close to the root zones can prevent much of this.

Some professionals strongly recommend against the use of sprinkler systems for interiorscapes. The objections have to do with the fact that spraying the leaves of foliage plants often leads to fungus diseases and the ultimate demise of the plants. Interior plant installations are particularly vulnerable to such health problems. There are several contributing factors. Overwatering with an automatic sprinkler system is easy to do. The water is sprayed over a broad area and for a relatively long period of time. Much of it lays on the soil or mulch surface and becomes a continuous source of evaporation, enveloping the plant leaves in a highly humid atmosphere. Add to that the water that is sprayed directly on the leaves by the irrigation system. Now introduce the factor of low air circulation in some planter locations, creating a stagnant, overly moist micro-climate. Fungus diseases breed easily under such conditions, and many professionals have faced the problem often enough to want to avoid use of automatic sprinklers. Some misinformed practitioners unfortunately tend to categorize all automated systems in the same negative way without taking their differences into consideration. For instance, drip and Micro-Irrigation System technologies are more precise in their water placement and seldom foster such problems.

With the proper design and installation, however, sprinkler systems can prevent some of these plant diseases. If possible, the spray from sprinkler heads should be directed under the leaf canopy of the plants, so only the soil is wetted. If the sprinkler head has a flow adjustment, it should be turned down so that less water issues during each watering cycle. That, of course, will reduce the coverage of each head and more closely spaced heads will be necessary. Shortening the irrigation time where possible will also help. Sometimes bubbler heads can be used near each plant so as to dribble water onto the ground above the root zones, rather than spraying it. There are basically two types of bubbler heads, one that emits a gentle stream of water (generally multidirectional) and the other that trickles water down the riser on which the head is mounted (see Figure 8.5).

Because of superior flow control, bubbler heads can overcome many of the plant disease problems common to sprinkler systems and should be used as frequently as possible indoors. Bubblers designed for very low flow rates are available, down to a flow of 0.25 gallon per minute. That's pretty stingy for sprinkler technology, which more commonly operates at 1 to 8 gallons per minute per head.[2] By operating a sprinkler system equipped with bubbler heads for short periods, much better control is provided and, if used properly, can irrigate

---

[2]*Irrigation Products Catalog* (Riverside, CA: The Toro Company, Irrigation Division, 1987) and *Turf Irrigation Equipment Catalog,* (Glendora, CA: Rain Bird Sales, Inc., Turf Division, 1987).

Sprinkler Systems

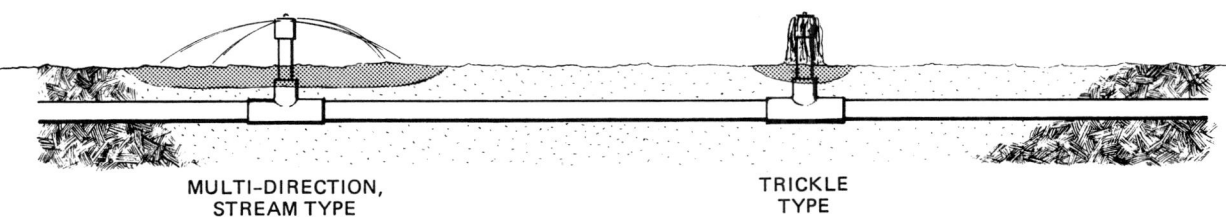

Figure 8.5 Types of bubbler heads.

planter boxes with a reasonable degree of assurance that the bed will not be saturated. If necessary, multiple cycles can be used daily.

Proper drainage of the planter boxes is supercritical when automated sprinkler systems are used. Because of the poor control over moisture distribution, soil pockets can easily become oversaturated, promoting heavy drainage from the planter subsurface. This is more acute when plants are left in their nursery containers, which is the preferred practice by interiorscapers. The design of planter boxes is too often flawed, creating a potentially messy and frequently damaging situation. The preferred planter box design calls for the installation of a large-diameter perforated pipe under the planter box to carry drainage away from the area. The pipe is surrounded by a few inches of coarse crushed rock to complete the drainage bed. A woven soil mat (generally made of polypropylene) is laid over the crushed rock, and the support soil or growing medium filled in above it (see Figure 8.6). Too often, all that is provided in a planter box to catch drainage is a heavy plastic liner (see Figure 8.7). That can work in some situations when one is confident of the irrigation system's flow rate and control, but it generally does not with sprinkler systems.

One can see how difficult it would be to provide proper drainage for freestanding potted plants if we were to try to irrigate them with an automatic sprinkler system. Not only would water placement be a problem, but the flow rates are so great that the pots would soon be flooded. Saucers placed under the containers would not be adequate to handle the drain-off. In furnished areas

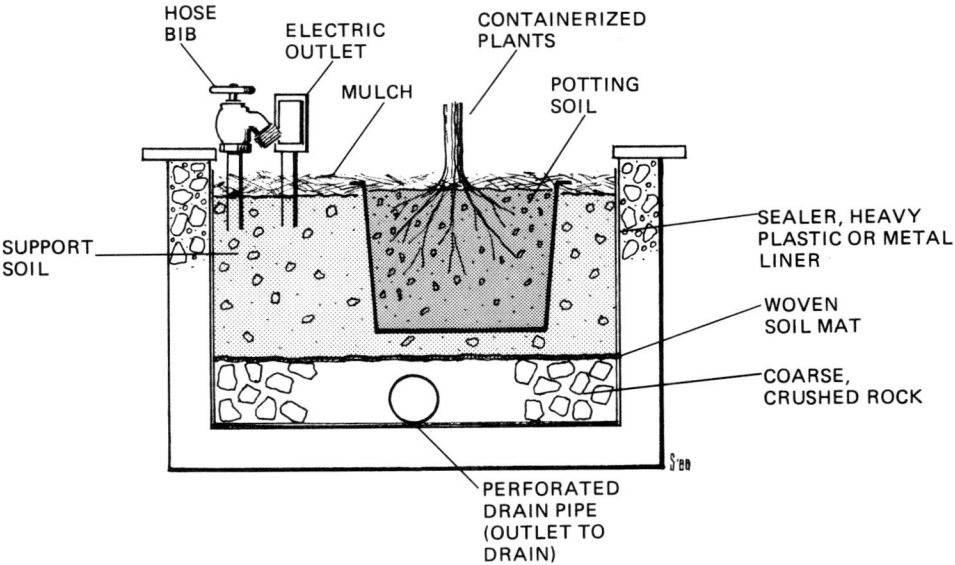

Figure 8.6 Recommended design for built-in planter boxes.

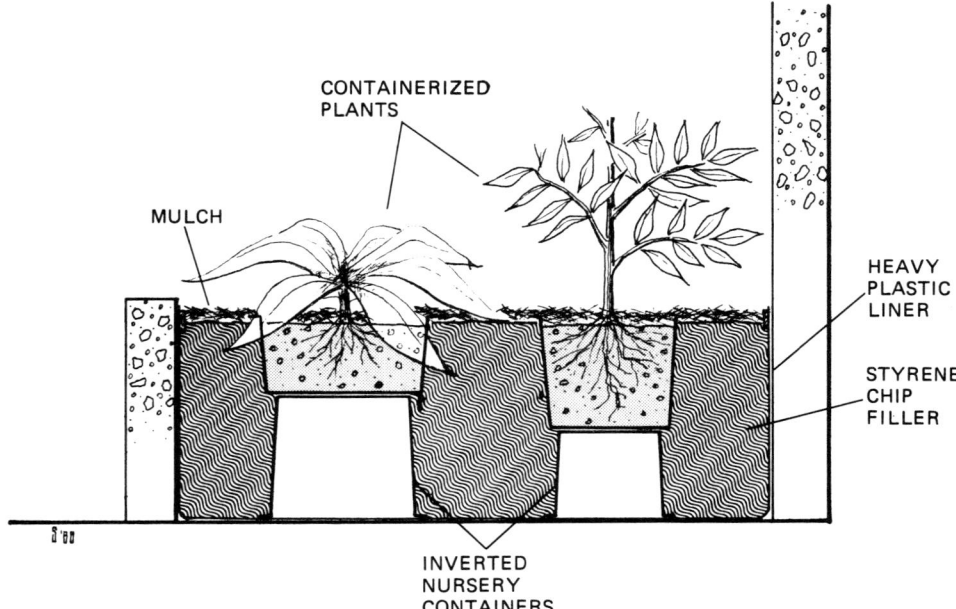

Figure 8.7 Plastic-lined planter box.

of a building, that can be serious. Those problems, coupled with the inappropriate nature of the equipment, make irrigation contractors shudder when confronted with the idea of trying to water freestanding potted plants in furnished areas with a sprinkler system. Particularly distressing is the prospect of a leak in the piping, fittings, or emitters with water spewing out for the many long minutes the system is on. With inherently high flow rates, that can be disastrous in the critical areas of houses, office suites, stores, or restaurants.

In summary, sprinkler systems are the coarsest form of automatic irrigation and must be used with caution in interior settings. They are mostly suitable for open areas in building lobbies, promenades, and atriums, where planter boxes and pits are safely removed from sensitive, furnished portions of the building. Sprinkler equipment is too unrefined, water distribution too voluminous and uncontrolled, and the duration of operation too long for widespread use in building interiors. For those reasons, it will remain the major method of irrigation in outdoor landscape and agricultural applications but will see decreasing usage indoors as more appropriate technology gains favor.

Drip/Trickle Systems Another common irrigation technology that is frequently automated is the drip system (sometimes called trickle irrigation). It, too, was developed for outdoor plantings in agricultural production; later, it found applications in greenhouse horticulture, and more recently in building interiors. Because of the inefficiencies inherent in sprinkler irrigation, better methods were sought by farmers and nurserymen to reduce water utilization. Sprinklers spread water wantonly over a broad area, wetting everything in their path; this is necessary when watering turf, but very wasteful when crops and ornamental plants are being grown. Large amounts of water evaporate even before reaching the ground, and most of it (40 to 60 percent) is wasted wetting out the surface soil. The surface soil becomes compacted from the heavy applications of water, further decreasing efficiency. When water conservation became imperative in many areas of the world, sprinkler irrigation was found too expensive to use.

The idea of dripping water into the soil around plants is generally credited to the Germans, who around 1860 buried clay pipes with loose joints. Water was permitted to drip from these openings and saturate the soil around crops. In the mid-1930s, Australian farmers overcame a drought situation by drilling holes in galvanized pipes in order to irrigate their peach orchards. The slow drip of the system efficiently wetted the ground around the precious trees. At about the same time, an Israeli engineer developed a way of slowly applying water to plants under low pressure by passing it through a tiny coil to extend its path yet allowing the water to issue from a relatively large opening. The method discovered then was the beginning of modern drip irrigation technology. It has become an important alternative means of irrigating plants.[3] The technology has been refined many times and in many ways since, but the concept remains the same. It is based on the highly controlled flow of water applied directly to the root zone of each plant by means of specially designed emitters. These small devices reduce the system's pressure by passing water through a relatively long and complex internal labyrinth. Because of the long flow path, internal friction, and turbulence, the water meekly trickles out of the emitter at a very slow rate. Other concepts are used as well, but the labyrinth type predominates. Although drip systems represent a type of overhead watering, one can see how different they are from sprinkler irrigation. Water is presented to the plant in a highly localized manner at the surface above the root zone, relying on gravity and capillary action for the moisture to diffuse throughout the root zone (see Figure 8.8). For this reason, the drip system's effectiveness is somewhat dependent on the soil texture and composition. Proper use of drip irrigation keeps the soil in an evenly moist condition, preventing wide fluctuations from overly wet or dry. When these systems are automated, they can provide a slow, regular irrigation regimen, very well suited to the needs of ornamental plants as well as crops. Less water is used, providing irrigation with high efficiency.

Within the past twenty years, irrigation contractors have installed drip systems in greenhouse nurseries, where they assist in containerized crop pro-

Figure 8.8 Moisture diffusion from a drip emitter in good soil.

---

[3]Jack Kramer, *Drip System Watering* (New York: W. W. Norton & Co., 1980), p. 36 and *Basic Drip Irrigation Design for Landscaping* (Riverside, CA: The Toro Company, Irrigation Division, 1985), p. 1.

duction. More recently, installations have been made in interior landscaping settings, exclusively in the common areas of buildings—the noncritical sections that had previously been the exclusive domain of modified sprinkler systems. These built-in planter boxes and pits are far enough away from furnished areas not to be worrisome.

Emitters are plugged into a network of water distribution tubing that is generally polyethylene plastic (poly tubing), interconnected with control valves. In automated systems sourced by a city water supply, the control valves are electrically or hydraulically operated solenoid valves. They are connected to sprinkler system timer/controllers, which turn them on at preprogrammed times and for predetermined durations (see Figure 8.9). Because of the very slow flow rates, operating times are extended; they are sometimes longer than sprinkler system irrigation periods.

The key elements in drip technology are the emitters. They come in many different types and flow rates and are very effective for what they do. Most emitter designs rely on a relatively long and complex path for the water to squeeze through on its way from the water supply tubing, which is under modest pressure, to the plant. This labyrinth reduces water pressure to manageable levels and drops flow rates considerably. These systems are designed primarily for outdoor uses that must cover large farm areas with many topographical variations. Water pressures change over these different grade levels, as does emitter output. Uphill flows can reduce emitter input pressures and flow rates considerably. Downhill flows can conversely increase pressures and flow rates. Installations with long and complex pipe and tubing layouts are also subject to severe pressure losses because of surface friction along the tubing walls and through valves and fittings. Emitters at the outer extremities experience less input pressure than those closer to the water source. Systems engineers have gotten around these problems by designing pressure-compensating emitters and intermediate flow devices that automatically adjust for variations in input

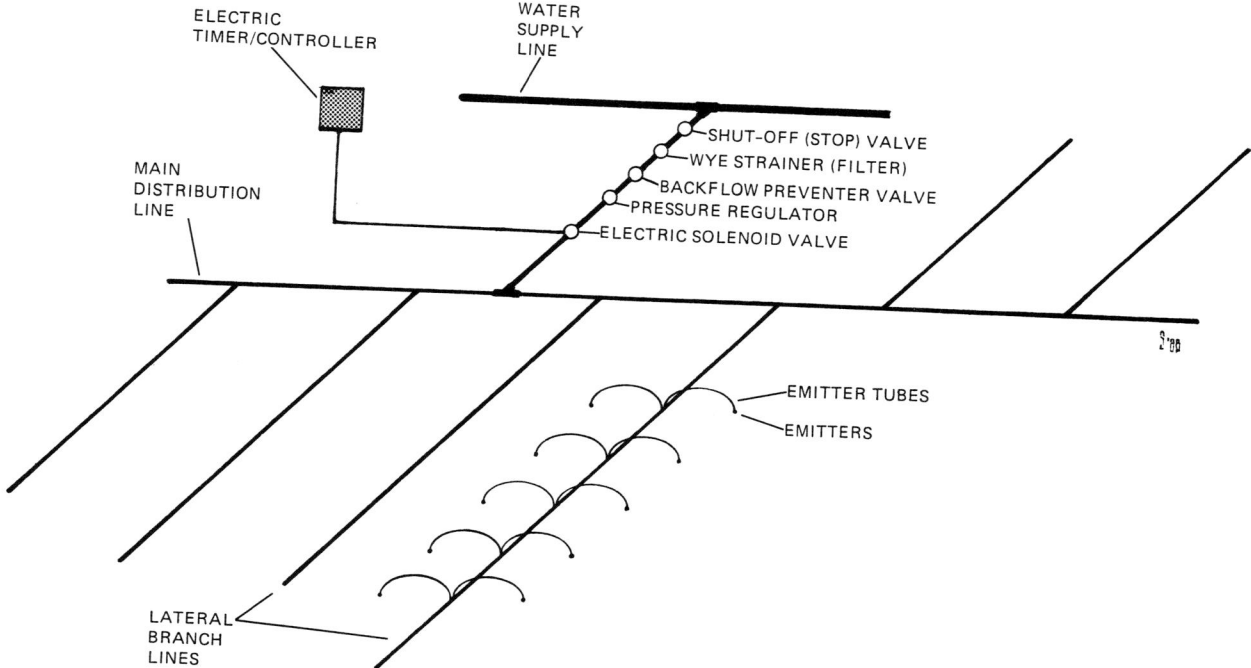

Figure 8.9 Typical drip system layout.

pressure, usually in the range of 5 to 55 psi. Certain brands are not recommended for input pressures above 40 psi and must be used with pressure-dropping regulators. Some emitters connect directly to the water supply tubing, while others are connected by means of a small-bore emitter tube. The latter arrangement seems to be the more prevalent, as it is the most flexible from a layout standpoint. Figure 8.10 illustrates various types of drip system emitters.

The type of emitter used will determine whether the installation is called a drip system or a trickle system. The difference is slight and only a matter of definition. Most emitters are of the drip type and are commonly available with flow rates in the range of 0.25 to 2.0 gallons per hour (GPH). For comparison, remember that sprinkler emitters are rated in units of gallons per minute (GPM). Soaker hoses and tubes are used in some systems. Although they are not labyrinth-type emitters, they do emit water at a slow rate to a localized area of ground. Some drip system manufacturers also offer miniature spray and misting heads to be used in conjunction with standard drip emitters. Their use, however, would no longer constitute drip irrigation technology.

Most of the tubing and piping used in drip systems are plastic, usually poly tubing (polyethylene) and some PVC pipe. These are used not only for their economy but also for ease of installation. Holes are punched into the poly tubing water supply line, and emitter tubes or pressure-compensating connectors are simply slipped into them (see Figure 8.10). These connections are crude but, because of the nature of drip system usage, generally adequate. Outdoor installations can tolerate occasional leaks. Interior installations, however, usu-

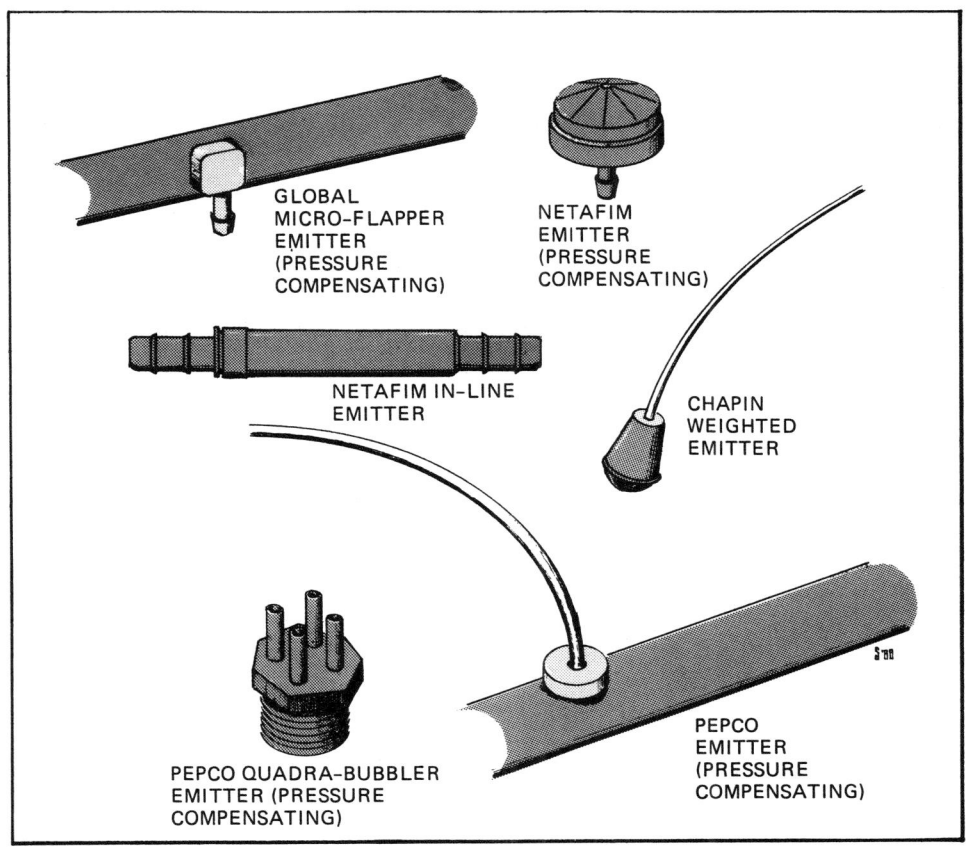

Figure 8.10 Typical drip system emitters.

ally require more secure connections. System pressures are frequently reduced to 15 to 35 psi indoors to provide a better measure of safety. Irrigation volume can be reduced indoors anyway because of lower plant transpiration rates.

The problems usually associated with drip irrigation have to do with the very small ports used to distribute and emit water becoming clogged. Dirt, debris, bugs, and mineral salts get into the system and tend to block the small openings inherent in the equipment. Emitter labyrinths have very small passages through them, and it doesn't take much to cause a blockage. Emitters are frequently placed on the surface of the ground; occasionally, they are buried. Dirt comes into contact with them, and the emitter design must be efficient enough to minimize clogging. Leaks frequently occur from rodents chewing on the poly tubing. They seem to have a liking for that kind of plastic and can cause extensive damage at times. Drip systems have also been criticized for being too sensitive to soil composition and moisture content. Because they rely on capillary action for water distribution, it doesn't work very well when the soil has accidentally dried out. Capillarity diminishes rapidly in arid soil. Frequent waterings can prevent that from happening. Poor growing media also reduce capillary efficiency, and close attention should be paid to this aspect as well.

The use of drip systems in interiorscape irrigation is a relatively recent phenomenon. The concept is being adopted slowly by the industry. There are many reasons for this but foremost are those stemming from inadequate training in the proper application, from misinformation steeped in the methods of manual plant care, and from the narrow attitude the industry has about automated interior plant care in general. There are still many uninformed souls who are convinced that automated irrigation is bad for indoor plants, in spite of the hundreds of thousands of tropical foliage plants that have been successfully watered by these modern methods for years. Much of the prejudice comes from the bad experiences that contractors have had with some systems—usually those systems involving sprinkler technology. The most common cause of malfunction is bad design and/or installation by firms inexperienced in interior work, or the fact that sprinkler technology is not the best choice for most interiorscape maintenance. This is still a problem, for the applications are relatively new and too few have the knowledge of the proper design, installation, and use of these interior systems. Some prejudice comes from the misconception in the minds of many interiorscape maintenance contractors that they will surely lose business to the automated technology. The more astute interiorscapers are embracing these new methods and are finding their real value. Much of the impetus for automated service systems is, of course, coming from the owners and managers of commercial buildings, who are looking for more efficient ways of doing things.[4]

Drip systems applied indoors are usually designed with water supply input devices similar to those used in sprinkler systems, such as shutoff valves, backflow preventers, screen filters, and pressure regulators. Branches are installed into the planter boxes and pits, with the appropriate drip, trickle, or soaker emitters used in the planting areas. Many interiorscape contractors prefer drip systems over sprinklers in building common areas because of the accident liability created by wet floors. When large interiorscapes must be serviced, the systems are designed into the structure of new buildings from the drawing-board stages, and piping can then be incorporated into the framework of the structure as the building goes up. The same applies to extensive renovation

---

[4]Judy Smith, "The Friendly Skies Add Greenery at LAX: Drip System to Reduce Maintenance Costs in Airport Interiorscape," *Interiorscape* (July/August, 1983), p. 48.

work. Piping or tubing stubs are brought up into the planter boxes during construction and will be ready for finishing connections at a later stage (see Figure 8.11). Sometimes complex layouts are necessary, and several branches or zones must be incorporated into the plan. These zones can be automatically turned on all at once by the controller, or one zone at a time can be activated for sequential watering schemes.

Because drip irrigation systems for building interiors are frequently designed by contractors and engineering firms whose training and experience is steeped in outdoor irrigation technology, there is still a tendency to provide for much more water than is necessary. Pipes that are too large, emitters that are too coarse, watering times that are too long, and other factors can lead to overwatering, where delicacy would have sufficed. It should be remembered that 15 minutes of watering with a 0.25 gallon per hour drip emitter (the smallest one available) will provide a plant with 8 ounces of water at the root zone. For most small indoor containerized plants, one watering a day with that output is more than sufficient—frequently it is too much. Irrigation cycles can be reduced to every two or three days if necessary. New problems are introduced when there is a great disparity in the sizes and water requirements of the many plants being serviced at the same time by the same system. Larger capacity emitters would be used for the bigger plants, lesser capacity emitters for the smaller ones; however, the balancing of irrigation durations and emitter sizes can be a difficult task. If the plants are direct-planted in the soil (out of their containers), then somewhat larger volumes of water may be required, as the surrounding soil absorbs a lot of moisture and cuts irrigation efficiency. It is up to the interiorscape maintenance contractor to undo some of the design excesses and to bring the watering pattern to a level more suitable for indoor horticulture. It is unfortunate that most interiorscapers don't know how to work with these automated systems yet, but that is the fact of the matter. For that reason, few of these installations are being used properly. Time, further experience, and training will overcome many of these problems.

Serious leaks frequently occur with indoor drip systems. It must be remembered that these systems were made for outdoor use where conditions are not as critical. For example, connections are coarse and many times unreliable,

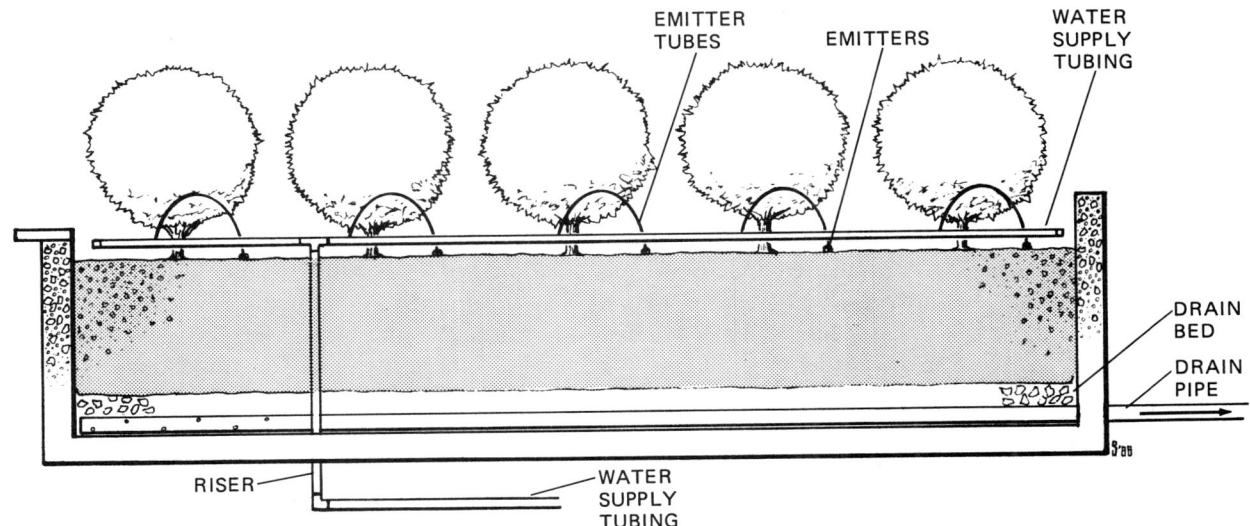

Figure 8.11 Drip system installed in a planter box.

Figure 8.12 Drip system in a shopping mall planter.

animals (particularly vermin) tend to chew on the poly tubing, maintenance technicians accidentally pull joints apart, and vandals have a habit of cutting the highly exposed emitter tubes. With openings in the water distribution tubing, the relatively long watering periods give plenty of time for unwanted leaks to make a mess of things. This is one of the major disadvantages of indoor drip systems; too much is vulnerable to water damage, and these systems must be designed, installed, and used with care. It is imperative that proper drainage be provided under the containers or planter box so that runoff can be evacuated.

Although drip irrigation is well suited to indoor planter boxes and pits in the common areas of buildings, they are not acceptable for most furnished interior areas. System pressurization is too long, the water distribution system is too coarse (esthetically unacceptable), and the integrity of the fittings is suspect. For those reasons, they are not very suitable for the interiors of houses or furnished office suites, shops, restaurant dining rooms, banks, clubhouses, or the many other indoor settings where live potted plants are used to complement interior decor.

Drip systems are well suited to the watering of containerized plants on terraces, patios, and pool decks in commercial as well as residential applications. Accidental leaks are usually not damaging outdoors. These outdoor settings generally require greater quantities of water more precisely placed than sprinklers are capable of providing. Drip technology is able to meet those needs. Irrigation cycles are longer outdoors because of the greater moisture requirements, and in many cases, the drip equipment can be simply branched off from the most convenient lawn sprinkler lines. Independent timing and control would, of course, require a separate branch for the drip system.

In summary, automated drip irrigation is seeing increasing use not only in farm and nursery applications, but in interiorscape maintenance as well. Indoors, automated drip systems are shaping up to be the preferred technology for large planters in the common (public) areas of commercial buildings. They are capable of providing a convenience and cost efficiency suitable to real estate executives, and their labor-saving and horticultural benefits are starting to become understood by the interior plantscape industry. Drip systems have limitations that will keep them relegated mainly to the open, common areas of commercial buildings and to the patios and gardens of residential units. Sensitive, furnished portions of building interiors will be increasingly serviced by the newer, more precise irrigation technologies.

Subterranean Systems

A variation of drip irrigation that can be automated and used on occasion for the watering of interiorscapes is the technology of subterranean irrigation, or *subirrigation* as it is frequently called. It has to do with the application of water underground near the root zone and sometimes below it. In theory, this is a very efficient way to water, for the moisture is applied even more directly to the roots. In practical terms, however, there are a number of drawbacks that prevent the widespread use of subterranean systems in *interior* landscaping. Those will be dealt with shortly.

Self-watering containers and buried-reservoir systems are considered forms of subirrigation, as the water is diffused into the growing medium from below. These were discussed in Chapter 7. Here we will discuss only those technologies that can be fully automated—flow devices rather than wicking devices.

Subterranean systems generally involve emitters that are tubes within tubes. The simplest versions are the early types, which were normally homemade. Garden hoses or poly tubing were punched or drilled with small holes along its length and then slipped into a heavy fabric tube or hose (see Figure 8.13). The resultant structure is connected to a sprinkler line or faucet and buried near the plants to be serviced. Water flows from the interior hose and oozes from the pores of the fabric casing into the soil. The fabric tube serves not only as a large-area emitter surface but also as a screen filter to keep the holes in the interior hose from being clogged with dirt. These systems are generally used at low pressure so as to provide low flow rates and to avoid the channeling and bursting of the soft plastic tubing.

Another version is constructed with a perforated interior hose fitted inside of a porous clay drainpipe. Water issues from the interior pipe and slowly seeps through the outer ceramic casing into the surrounding soil. Another uses a microporous, synthetic, or rubber casing to slowly exude water. Some types are simply perforated plastic pipes or hoses that are directly buried in the ground. Still another version uses in-line drip emitters installed at regularly spaced intervals in a poly tube (see Figure 8.14). Yet another type is designed with tiny holes laser-drilled into poly tubing at a reverse angle to the direction of flow (see Figure 8.15). In some models, the holes are spaced 6 inches apart and in others, 12 inches apart. Each of these configurations does essentially the same thing underground. They are similar to soaker hoses in that they emit water at

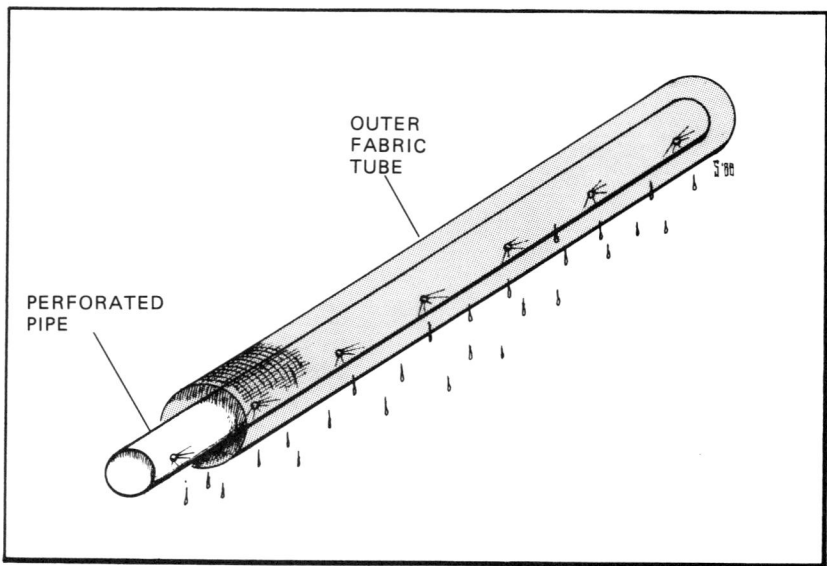

Figure 8.13 Typical ooze pipe construction.

a very slow rate along their entire length into the surrounding soil near the root zone. Water distribution is along the tubing axis.

Subterranean systems are frequently automated by connecting the water supply tubing to a means of interrupting flow. This is generally a solenoid valve operated by a sprinkler controller at timed intervals. These systems are sometimes used to irrigate interiorscapes, particularly long, narrow planter boxes that fit the natural configuration of the tubing lines. On occasion, they are circled around the root line of trees and large ornamental bushes. When subterranean irrigation is used, the foliage must be direct-rooted into the planter soil. The buried tubing keeps the plantscape's esthetics unencumbered. This has two sides to it, however, for the hidden lines also create problems when maintenance technicians have to dig out plants. The tubing gets in their way and frequently gets damaged. Another drawback is that it is difficult to monitor the condition of the soil when the tubing has been buried deeply. Most of the moisture is under the roots and the maintenance technician must check frequently and carefully to ensure that the deep soil is not saturated and thus a danger to the plants' health. Yet another disadvantage lies in the fact that the roots will

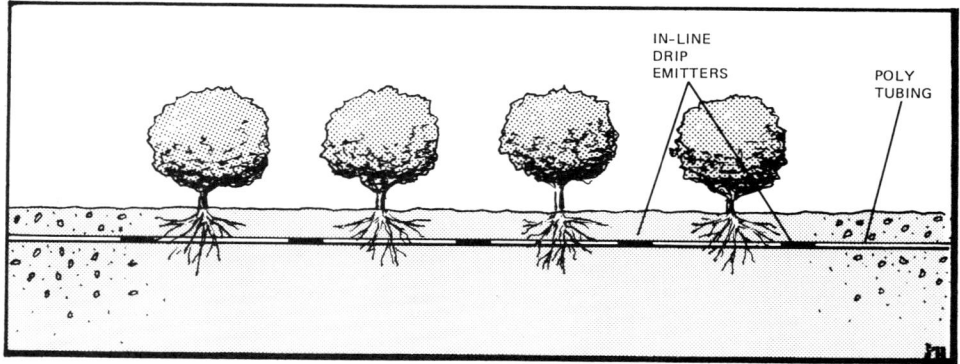

Figure 8.14 In-line drip emitter system in subterranean use (buried under plants).

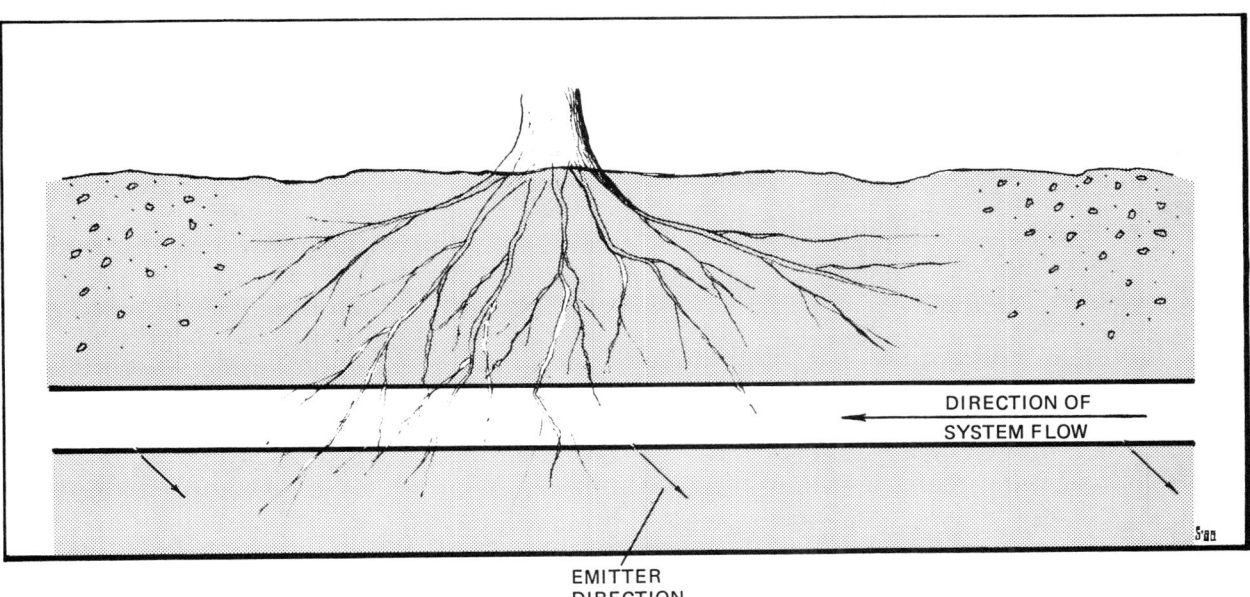

Figure 8.15 Laser-drilled soaker tubing.

quickly overgrow the submerged tubing and can easily break connections and/or choke the emitter holes. Large plants can be particularly damaging. Once this happens, it is difficult to perform maintenance on the system without also causing extensive damage to plant roots. It is essential that proper drainage be designed into the planter boxes housing subterranean irrigation.

The main disadvantage that prevents this type of subterranean irrigation from gaining wide application in interiorscape projects is that it cannot be used with smaller, containerized plants, the mainstay of the industry. Water placement is not localized enough for potted plants; therefore, watering would be into the support soil rather than the growing medium. There is little interface between them so the moisture cannot be diffused to the root zone. It should also be obvious that there cannot be any application with freestanding potted plants (those resting on the floor and furniture). That, with other obvious factors of course, eliminates uses in furnished, decorated interiors.

In summary, automated subterranean irrigation is used in limited ways indoors . . . mainly in narrow planter boxes built into common areas. It is not very versatile and, for that reason, will continue to find very limited application indoors.

Hydroponic Systems Hydroponics are quite different from other horticultural concepts in that the plants are grown without soil in a medium that is essentially inert and used primarily to support the plant. The soil substitutes are normally materials like gravel, pebbles, pearl chips, coarse sand, vermiculite, perlite, and even coal. A nutrient solution of a balanced fertilizer mix is fed the plants daily by manual or automatic means. Plant roots grow into the crevices and pores between the medium's particles and aggregates, seeking out the moisture and nutrients. The major part of the nutrient solution is allowed to drain to below the root level so as to avoid rot. This type of hydroculture is sometimes called *soil-less gardening* or *water gardening*. As an active concept, it has been around for centuries but was popularized in this century by agricultural concerns. They have found that vegetables, flowers, and fruits can be grown more quickly and with greater yields by hydroponic systems than by conventional farming. Less acreage is required; weed, insect, and disease problems are minimized; less water is used; chemical pollution is minimized; and crops can be produced year-round because most of this type of growing is done within a greenhouse. Economic and other practical problems, however, have prevented hydroponics from becoming a major agricultural tool.

Hydroponics have been used in Europe to a large extent by their interiorscaping industries to grow flowering and tropical foliage plants in indoor settings. The technology, having been developed primarily for use in sheltered areas, lent itself quite readily to decorative, rather than production, applications. The equipment used for this is much like the self-watering containers discussed earlier (see Figure 8.16). In many respects, they are similar. The Luwasa System, which is the prime European hydroculture brand, features a growing medium of expanded clay pellets that draw the nutrient solution up to the root levels by capillary action. The plant, surrounded by these clay pellets, resides in a container mounted near the surface of the planter box. The reserve liquid lies in the bottom of the container, wetting the growing medium and ready to diffuse toward the plant . . . much like the capillary diffusion of self-watering containers. The reservoirs are refilled every 3 to 7 weeks. These systems are not automated, and to our knowledge, no attempts have been made to do so in commercial interiorscape installations. Most of the automation associated with hydroponic systems has been applied in do-it-yourself home-gardening projects and in agricultural greenhouse settings. Small timer-activated pumps are fitted to inject the nutrient solution every day or so.

Hydroponics has not achieved much popularity in this country for application in interior landscapes and is not expected to because of the strong competing technology and a weak marketing effort.

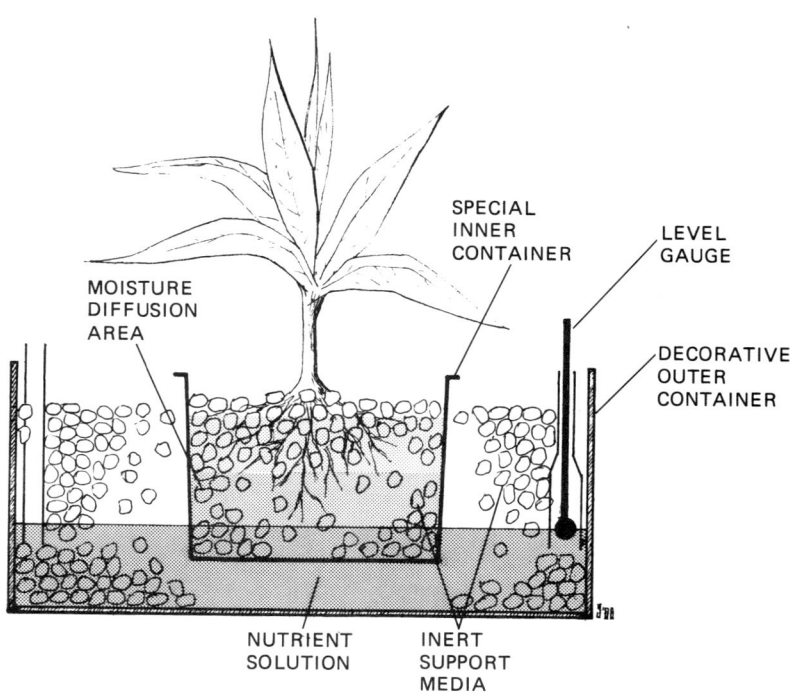

Figure 8.16 Typical freestanding hydroponic system.

Micro-Irrigation Systems

Precision Micro-Irrigation Systems use the first fully automatic technology developed specifically for the watering of containerized plants in the interiors of buildings.[5] It is more refined than drip irrigation or sprinkler irrigation, very precise in its placement of water and its control over flow rates, and designed to operate at the exceedingly low levels required by indoor plant culture. Micro-Irrigation Systems are characterized by very short irrigation cycles (seconds in duration) and a variety of flow control devices. It allows us to fully automate the watering of potted plants in the critical, furnished areas of our homes, restaurants, hotels, and offices, where it would otherwise be too risky to install drip systems and impossible to contemplate the use of sprinklers. Those technologies are simply not sophisticated enough for broad use in building interiors. Decorative plants throughout the entire building can now be serviced by an automated plant-care system. Large planters in common areas can be irrigated with the same system as small potted plants in the decorated shops and suites elsewhere in the building. Figures 8.17 to 8.32 show the many possible interior applications of precision Micro-Irrigation Systems. Figures 8.33 to 8.37 show several exterior applications of the technology.

Figure 8.17 Fast-food restaurants.

[5]Patented: U.S. Patent and Trademark Office.

Figure 8.18 Restaurant entry planters.

Figure 8.19 Lounges.

Micro-Irrigation Systems

Figure 8.20 Food service areas.

Figure 8.21 Restaurant dining rooms.

Figure 8.22 Office building atriums.

Figure 8.23 Commercial building lobbies.

Micro-Irrigation Systems 111

Figure 8.24 Office suite reception areas.

Figure 8.25 Office suites.

Figure 8.26 Shopping malls.

Figure 8.27 Mall store entries balcony planter boxes.

Micro-Irrigation Systems 113

Figure 8.28(a) Department stores.

Figure 8.28(b) Department stores.

114 Chap. 8 / Interiorscape Irrigation: Fully Automatic Techniques

Figure 8.29(a) Houseplants.

Figure 8.29(b) Houseplants.

Figure 8.29(c) Houseplants.

Figure 8.29(d) Houseplants.

Figure 8.30(a) Hotel lobbies.

Figure 8.30(b) Hotel lobbies.

Micro-Irrigation Systems

Figure 8.31 Hotel reception rooms.

Figure 8.32(a) Hallway planters.

Figure 8.32(b) Hallway planters.

Figure 8.33 Building entry planters.

Figure 8.34 Storefronts.

Micro-Irrigation Systems

Figure 8.35 Residential patios.

Figure 8.36 Apartment terraces.

Figure 8.37 Privacy lanais.

Micro-Irrigation Systems were developed around new concepts and incorporate some of the better features of the preceding methods. It promises to be the leading technology for widespread use in real estate to lighten the burden of domestic and commercial plant care. As with self-watering containers and automated drip systems (when they can be applied to interior landscapes), Micro-Irrigation Systems are capable of providing a great deal of convenience and plant-care consistency, as well as saving much manual labor, time, and money. Because it is fully automated and applicable to the entire building, rather than just in narrowly assigned areas, this new concept is able to provide convenience and cost efficiencies never before available. It will be beneficial to professional interiorscapers and to their customers. Micro-Irrigation System technology promises to solve many interior management problems for the homeowner, corporate office manager, restaurateur, retailer, and real estate executive.

When used in commercial interior landscapes, whether they be large groupings of plants in shopping malls or office building lobbies, or scattered potted plants in corporate office suites or restaurants, most, if not all, of the watering chore is removed from the responsibility of the maintenance contractor. That does not mean that automated systems can, or should, replace human plant care. There are many plant-care tasks other than irrigation that must be attended to by the contractor, but watering is the single most time-consuming chore encountered. The industry generally recognizes that between 25 and 50 percent of the on-site maintenance time is taken by manual irrigation, and irrigation is the primary reason that maintenance cycles are once a week. Without the need to water, maintenance cycles can be extended to once every two or three weeks. Using a two-week maintenance cycle as our example, labor content can be reduced by 75 percent. Two visits per month are now required, not four. That is a savings of 50 percent in labor as well as materials, travel, and fringe expenses. Further, consider the fact that roughly 50 percent of the on-site time is reduced during each of those two visits because no manual irrigation is required. That saves another 25 percent of the monthly labor cost (50 percent of the remaining 50 percent), which amounts to a total savings of 75 percent in labor costs through automated irrigation plus 50 percent in other pertinent maintenance cost elements. Plant replacement costs are also reduced considerably. Lighter physical exertion and greater concentration on the other plant-care chores increases interiorscape employee satisfaction and reduces labor turnover. New employee training is a long, expensive proposition—automation can reduce the need for much of that. Customer satisfaction is improved because there are fewer visits, and plant-care quality and consistency is at a higher level. Fertilizer costs are reduced, for precision irrigation is much more efficient. Virtually no water is wasted, as with manual irrigation, so fertilizer doesn't go down the drain with the run-off water. In total, the savings for the contractor can be considerable, and some of it must be passed on to the customer. Customer savings on interiorscape maintenance contracts of 30 to 60 percent are clearly practical. Experience has demonstrated that Micro-Irrigation Systems pay for themselves typically in a year or less.

Some of the other advantages of Micro-Irrigation Systems include the following:

- Because of the precise flow control and placement of water, drain-off (overflow) from containers is eliminated.
- Because standby irrigation water is contained in tubes within the room environment, its temperature at the time of irrigation cycles is essentially the same as that of the room and the plants.
- Fertilizer and other systemic chemicals can be dissolved in the irrigation water for automatic feeding to the plants.
- Because the soil is moist at all times, irrigation water is not wasted wetting out dried growing media. As a result, less water is required, plants are not subjected to water stress, and mineral salt buildup is reduced.
- Comprehensive systems can be designed to service large planter boxes in common areas at the same time as freestanding potted plants in furnished areas.
- Special containers are not required.

Uses of Micro-Irrigation Systems

The versatility of this concept permits its use in servicing virtually every containerized plant in a building interior . . . no matter how remotely or delicately placed. Its broadest applications are possible in new and renovated construction where water distribution networks can be designed into the building structure and conveniently installed while partitions are open. Retrofit installations can generally be made almost as well, but of course cannot be done as easily or as inexpensively.

The following list is a summary of the potential uses for precision Micro-Irrigation Systems:

Residential uses

Private Homes and Apartments
- Furnished interior living areas
- Greenhouse windows
- Sunrooms
- Atriums
- Outdoors on patios, terraces, balconies, lanais, pool decks

Condominiums (Public Areas)
- Lobbies
- Hallways
- Recreation centers
- Outdoors on patios, terraces, balconies, pool decks

Commercial uses

Corporate Offices
- Furnished office suites
- Outdoor office balconies

Sales Offices
- Furnished offices
- Display rooms
- Conference areas
- Built into displays

Office Buildings
- Lobbies
- Atriums
- Administrative offices
- Leasing offices

Restaurants
- Furnished dining areas
- Built-in planter boxes
- Hanging planters
- Waiting areas
- Entry planters (indoors and out)
- Lounges

Banks
- Furnished office suites
- Furnished open plan areas
- Built-in planter boxes
- Hanging planters

Hotels
- Lobbies
- Atriums
- Furnished offices
- Terrace planters
- Pool decks
- Guest rooms and suites
- Ballrooms

- Hallways
- Restaurants/lounges
- Sunrooms
- Fitness rooms
- Shops

Retail Stores
- Display areas
- Showrooms
- Entries and exits
- Executive offices

Shopping Malls
- Promenade areas
- Shop entrances
- Food service areas
- Executive offices

Hospitals/Medical Centers
- Solariums
- Waiting rooms
- Administrative offices
- Doctors' offices
- Lobbies
- Hallways

Country Clubs
- Clubhouses
- Restaurants
- Pro shops
- Exercise and game rooms
- Hallways
- Sun and pool decks

Libraries, Colleges, and Other Institutions
- Hallways
- Reading rooms
- Study halls
- Dormatories
- Outdoor walkways

Marine
- Luxury yachts
- Yacht clubs/marinas

In summary, Micro-Irrigation Systems represent the newest concept in fully automated interior plant care and a new form of building automation. The technology is very different from sprinkler and drip systems and is specifically designed for interior applications, with its very low flow requirements as well as its critical esthetic and reliability specifications. It is the most suitable of all technologies for broad application in building interiors and is expected to ultimately become the standard of the industry. In the following chapters, we will discuss the technology of Micro-Irrigation Systems in detail, explaining the differences between it and the preceding methods, how it works, and how it is designed, installed, and used in buildings. We will try to be as thorough as possible within the constraints of this publication.

9 THE CONCEPTS OF MICRO-IRRIGATION SYSTEMS

Overview Micro-Irrigation™ Systems are different from preceding technologies in a number of ways, all of which have to do with refining the methods by which water is delivered to plants, the quantities of water involved, and the techniques of installation. The technology is designed specifically for the irrigation of potted plants in the furnished interior areas of buildings . . . the most critical applications for any irrigation system. However, because the technology is highly refined, these systems can also be used in other areas of the building where sprinkler and drip systems might normally be installed, as well as outside the building in patio and balcony applications. This permits containerized plants in the common and furnished interior portions of a building, as well as some outdoor areas, to be serviced by the same automatic irrigation system.

The technology used in precision, Micro-Irrigation Systems was originally developed by Aqua/Trends in response to the need for better ways of caring for interior foliage plants in the South Florida vacationland. Many homeowners and condominium owners are away from their dwellings a good deal of the year, and their potted houseplants suffer from either neglect or well-meaning but ineffective plant-sitters. From these beginnings, more sophisticated systems were developed to service the heavy-duty requirements of commercial interiorscape installations. Many versions had to be designed to meet the needs of a broad variety of interior applications. What they all have in common, however, is the use of what we call *Pulse-Flow*™* technology, which is the reduction of watering cycles to mere seconds, rather than minutes or hours as in other systems. That, coupled with the use of frequent, periodic watering cycles, adjustable mini-valves at each plant, and other flow control devices, makes for a system of the highest refinement and flexibility that is capable of watering small dry-loving plants on the same system and at the same time as larger plants that require much more water. Micro-Irrigation Systems can be used on different building levels, so that multifloor installations are easy to accomplish. Of equal importance indoors, Micro-Irrigation Systems can be installed in such a way as to be almost completely hidden, so as not to upset the esthetics of the interior decor. These are precision systems that provide a degree of versatility and convenience never before available to the homeowner, interiorscaper, property manager, or facilities manager.

Pulse-Flow is a trademark of Aqua/Trends.

The Pulse-Flow Concept

In order to accommodate the much smaller volumes of water required by ornamental plants indoors, particularly those in containers, the concepts of short-interval watering were developed. These we call *Pulse-Flow technology* . . . the flow of short pulses of water, seconds in duration, used repetitively at fairly short intervals. The resulting irrigation pattern provides a high degree of accuracy and flexibility.

The first thing that had to be resolved was the widespread notion that indoor plants must be allowed to dry out for an extended period before the next watering is applied—and then that watering be done with a "heavy hand." The idea has become so dogmatic that most consumers, plant shops, and interior plantscapers still believe that this is the only valid regimen for interior plants and that no other irrigation scheme can keep plants thriving. Following are some excerpts from various publications on the subject. As expected, most consider only manual irrigation techniques, and some even warn against use of automated systems.

> When you do water your plants, do so thoroughly, so that the water reaches all parts of the pot. But do not give so much water every time that some always drains from the bottom of the pot, or valuable nutrients will be washed away.[1]
>
> Happily, none of the vagueness of the "how often" question beclouds the "how much" question. Whenever a plant needs watering, it must be watered thoroughly . . . until water drains out of the holes at the bottom of the pot.[2]
>
> Feel the soil to a depth of 1" below the surface; if it's dry to the touch, add tepid water to the soil surface. Continue until you see water seeping from the drainage hole. Allow the plant to drain (either into a sink or drainage saucer) for at least 10 minutes. . . . When the top inch of soil again becomes dry to the touch (a few days to a few weeks, depending on the pot size and type), repeat this procedure.[3]
>
> The first and most important rule when watering any plant is: water it thoroughly. . . . Thorough watering means to saturate the soil . . . from top to bottom.[4]
>
> It is always recommended that plant watering be done manually, not by automatic watering systems. . . . Basically, the movement of water in a soil is vertical and not lateral. Because of this, it has always been a rule of thumb to water well when watering. . . . Do not shallow-water the plants. Frequent light sprinklings are usually injurious, in that the surface of the soil remains moist, while the strata below remain dry. It is better to lengthen the period between waterings and then water the plant well, than it is to water frequently only the top soil surface.[5]
>
> The growing medium should be watered when it needs it, not according to a predetermined schedule. . . . Plants require a constant supply of water to maintain their normal processes. . . . Apply a sufficient quantity of wa-

---

[1]Michael Wright, ed., *The Complete Indoor Gardener*, (New York: Random House, 1979), p. 220.

[2]Maggie Oster, ed., *The Green Pages*, (New York: Ballantine Books, 1977), p. 216.

[3]*Sunset House Plants—How to Choose, Grow, Display*, (Menlo Park, CA: Lane Publishing Co., 1983), p. 19.

[4]Joan Lee Faust, *New York Times Book of Houseplants*, (New York: Quadrangle/The New York Times Book Company, 1973), p. 12.

[5]Richard L. Gaines, *Interior Plantscaping*, (New York: Architectural Record Books, 1977), p. 10.

ter to thoroughly wet the growing medium from top to bottom with some draining from the growing container. Small quantities of water applied at frequent intervals will not disperse evenly throughout the entire mass of growing medium.[6]

These passages contain several inconsistencies and inaccuracies. They are typical, however, of the information given to homeowners as well as the training given interior landscape professionals. It has led to much confusion about irrigation practices, and NOW THAT AUTOMATED TECHNIQUES ARE AVAILABLE, THESE OLD BROMIDES NO LONGER APPLY TO ALL SITUATIONS. THEY ARE ORIENTED ONLY TOWARD MANUAL IRRIGATION, AND MUST BE ACCEPTED IN THAT CONTEXT.

The first real glimmer of light comes when we consider that the tropical foliage plants used indoors to a large extent are native to habitats where growing conditions are always moist; for example, where the rain forest floors remain damp at all times. At least one author has recognized this reality:

> The amount of water a given plant needs is determined by the conditions it found in its natural habitat. A plant which in nature grows by a waterfall, where it is continually bathed by the spray, or in a tropical jungle where rain is a daily occurance, will need constant moisture and without it may dry up and die. But even with a moisture-loving plant, the word is moist, not soggy.[7]

Most plants flourish under moist soil conditions, and it is not until we get them into our homes or offices that the dry-wet-dry syndrome is started. Even the tropical foliage nurseries that produce these plants water the containerized foliage at least once a day through drip, mist, or sprinkler irrigation. The potting soil is never allowed to dry out; it is kept evenly moist, at a level with which the plant is comfortable. "Evenly moist" are the key words. They denote keeping the soil as close to the proper moisture level as possible at all times and not permitting broad fluctuations from wet to dry. Horticultural recommendations call for most plants to be kept evenly moist (see the "Mini-Encyclopedia of Houseplants" in Appendix A). If Mother Nature can do it and if the large wholesale nurseries can do it, we know that we can too. It's just a matter of having the right means available to accomplish it accurately and conveniently. That explains the technique of allowing potted plants to become dry before the next watering. The tendency with manual irrigation techniques is for potted plants to be dangerously overwatered, displacing life-giving oxygen from the soil mass. Long ago, gardeners found that in letting the plants dry out for a time, oxygen was allowed to reenter the soil. This also minimized the chances of overwatering. That technique, however, promotes broad swings in moisture levels that can have adverse effects on the plant and the type of care it's given. Excessive dryness can create water stress in the plant, a form of shock. Some of the cells actually die off, stunting growth. It is too easy for careless or inexperienced maintenance technicians to overlook a plant or two during the course of their maintenance visits, promoting moisture starvation in portions of the root system. The resultant shock affects the plant's health and, if severe, could be terminal.

Partial drying also creates problems when the soil is irrigated. Remember

---

[6]George H. Manaker, *Interior Plantscapes*, (Englewood Cliffs, NJ: Prentice-Hall, Inc., 1987), p. 179.

[7]"101 Ways to Love, Grow and Care for House Plants," *Woman's Day* Creative Series, *Woman's Day* (March, 1981), p. 80.

that enough water must be introduced to rewet the soil, mulch, etc. This is not as easy as it sounds. Surface tensions must be overcome before the moisture can penetrate amongst the grains and fibers of the organic matter in the growing mix. The use of small concentrations of surface-active agents (*surfactants*) in the irrigation water helps to break down these tensions and promote wetting. Household detergents are a common form of surface-active agent, and they do in a wash what other surfactants accomplish in planter soil. They promote wetting as part of their job. Fine sand and peat moss fibers are particularly difficult to rewet when dry. By spreading irrigation cycles out over one- or two-week periods, the surface soil can become particularly dry. During the next irrigation, much of the water will simply filter through the soil or run off the surface without much benefit. Some of this drains to the saucer or, worse, onto the carpet or tile floor. In order to keep planter soil in an evenly moist condition with a proper balance of moisture and aeration, small-dose applications of water would have to be made at frequent intervals merely to replace the small quantities of moisture lost between waterings. To ask someone to do this manually is to inflict an inconvenience and attention to detail that few would tolerate. It is simply not practical for most homeowners or professional interiorscapers to follow such a demanding schedule. So, because automated techniques were not available to accomplish this, they found that the next best way was to allow the soil to dry out some between waterings. Oxygen can then permeate the root area before being displaced again at the next dousing. Some plant experts advocate that small plants be placed in large pots of water until they are fully saturated and large plants be inundated with water until the excess drains out of the bottom holes (see Figure 9.1). Most specimen plants are large and sometimes ponderous, and it is impractical to move them often, so this heavy watering must be done in place. Can you imagine careless or inexperienced attendants trying to do that in a furnished living room, restaurant, or office suite?

To summarize, the methods recommended for manual irrigation have been devised as much for reasons of convenience as for plant health. THEY APPLY ONLY TO MANUAL IRRIGATION AND ARE NOT TO BE CONSTRUED AS BEING THE ONLY WAY TO IRRIGATE INTERIOR DECORATIVE PLANTS. We've made the point before, but it is important and worth repeating.

During the course of a day, moisture is lost through transpiration and by evaporation from the surface of the planter soil (see Figure 9.2). The loss is slow; therefore, its replacement should be slow. Studies show that only fractions of an ounce of water are daily lost to the soil of most small containerized plants. Even larger specimen plants do not require much more water indoors. The technology of Micro-Irrigation Systems is based on the principle of providing small doses of water at frequent intervals, so as to replace only the moisture just lost and to evenly maintain the level of soil moisture best suited for that particular plant. Under this scheme, there is no need to douse-and-purge, douse-and-purge, for oxygen is always in sufficient quantity around the roots. The soil is never permitted to become dry, nor is it permitted to become saturated. It maintains a good balance of moisture and oxygen at all times. Because soil particles are always somewhat moist, the problem of constantly breaking surface tensions to rewet them does not exist. Because large quantities of water are never used, run-off problems become a thing of the past.

The application of small doses of water permits slow diffusion into the soil before the next watering. Pulse-Flow techniques permit operating cycles typically only 10 to 15 **seconds** in duration, generally feeding only a fraction of an ounce or so of water to each plant. For very small plants, that is enough; but for larger plants, the irrigation cycles must be repeated once or twice more each

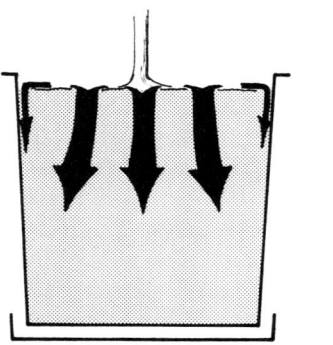
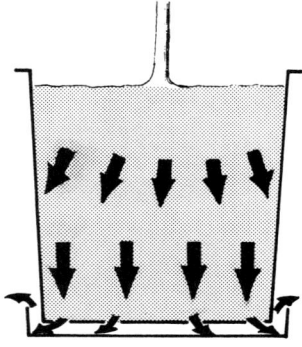

LARGE DOSES OF WATER APPLIED AT INFREQUENT INTERVALS DRAINS THROUGH RAPIDLY AND INEFFICIENTLY—SOMETIMES CAUSING DAMAGING OVERFLOW.

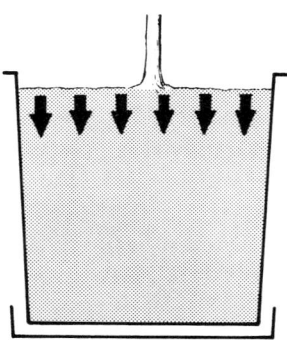
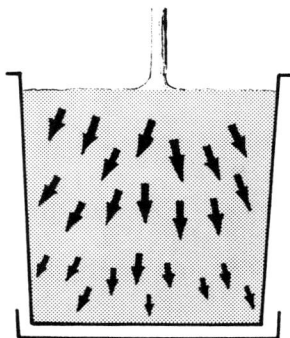

SMALL DOSES OF WATER APPLIED AT FREQUENT INTERVALS DIFFUSES SLOWLY THROUGH THE SOIL MASS BEFORE THE NEXT DOSE IS APPLIED.

Figure 9.1 Results of plants being fed large amounts and small doses of water.

Figure 9.2 Moisture loss through transpiration and evaporation.

day. By turning down the adjustable emitters on the smaller plants, even potted cacti and other dry-loving varieties, can tolerate multiple irrigation cycles. In a given installation, irrigation cycles can be repeated as many times as necessary to give all of the plants being serviced their prescribed amounts. However, there is seldom a need for more than two or three cycles per day. By automating the operation of this delicate watering technique, we are able to provide an accuracy, consistency, and convenience only dreamed of by manual irrigation practitioners . . . not to mention the best irrigation regimen for plant health. Because of the methods involved, automation is a necessity with Micro-Irrigation Systems.

Aqua/Trends' Micro-Irrigation Systems were designed to be highly flexible in their ability to control the small amounts of water fed during the irrigation cycle. In addition to the variability of flow adjustment at each plant and the use of multiple cycles daily, the length of the irrigation cycle can also be easily varied to provide more or less water during each operating period. The duration of the basic irrigation cycle for interior plants is 10 or 15 seconds. Provisions have been made at the control center to easily double the duration of flow to 20 or 30 seconds whenever necessary. These multiple controls give the system a high degree of flexibility and enable it to cope with the moisture needs of most combinations of plants on a common installation.

In Pulse-Flow technology, watering cycles are measured in terms of **ounces of water per 10 seconds** to reflect the small quantities issued during each cycle and the very short cycles used. This is contrasted with **gallons per minute** as the unit used to measure emission in sprinkler systems and **gallons per hour** with drip systems.

ONE OF THE MOST IMPORTANT THINGS TO REMEMBER ABOUT MICRO-IRRIGATION SYSTEMS IS THAT THEY ARE DESIGNED TO CONSTANTLY MAINTAIN A GIVEN, OPTIMUM LEVEL OF MOISTURE IN A PLANTER, NOT TO WET OUT DRIED PLANTER SOIL.

Moisture Diffusion Patterns

The emitter most commonly used in Micro-Irrigation Systems is a small adjustable valve that issues a narrow stream of water onto the surface of the soil above the root zone. That stream actually impinges onto an area no larger than 1/4 inch in diameter. From that small section, the moisture must diffuse throughout the planter (see Figure 9.3).

The mechanisms are similar to those found in drip irrigation systems. Capillary action is relied on to diffuse moisture from a point of concentration throughout a root area. The efficiency of this diffusion is therefore highly dependent on the composition of the planter media, which is soil in most cases. A well-balanced planter mix will promote good moisture retention as well as efficient capillary transfer. Capillary action is aided by very small spaces, or pores, between soil particles or fibers. Larger pores aid in soil aeration and drainage, but smaller ones promote capillarity. This is one of the reasons for the good moisture diffusion in clay—the pores are very small. Conversely, the drainage characteristics of clay are bad because of the tiny pores. Fibrous materials in the soil tend to aid capillarity because they have microscopic pores between the individual fibers and even smaller pores in the amorphous regions between their long-chain molecules. In fiber technology, these pores are frequently called *interstices*. They are relatively long and very narrow and run in the direction of the fiber axis. Water seeps into these pores, swelling the structure. It then travels along the length of the fibers in a highly efficient, natural-wicking action, following the elongated pores as if they were railroad tracks. That is the reason wicks are made of narrow woven, twisted, and braided fi-

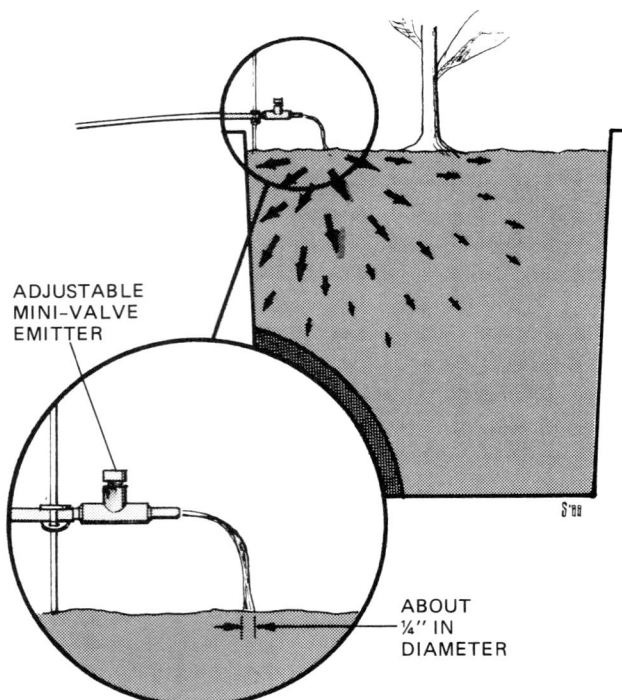

Figure 9.3 Moisture diffusion pattern in a Micro-Irrigation System. (Courtesy of Aqua/Trends.)

brous materials. They exhibit a high degree of capillarity along their axis and readily transfer water from one end to the other.

Natural fibrous materials used as soil modifiers in horticulture, such as sphagnum peat moss, composted wood chips, strawy manure, ground fir, and redwood and cypress bark, exhibit these properties and enhance the wicking properties of the soil, making it easier to diffuse moisture throughout the entire area of a planter. The use of a properly formulated growing medium in interior landscape work cannot be overemphasized. It is particularly important with drip irrigation systems and Micro-Irrigation Systems, which rely so heavily on the capillary diffusion of moisture. We might mention that these technologies work acceptably even in poor soil but not quite as efficiently . . . nothing works as well in poor soil.

Moisture diffusion in a properly formulated soil mix is lateral as well as vertical. There are many in the horticultural business who maintain that irrigation water is simply pulled down by gravity and does not have a chance to fan out laterally. While this is essentially true of heavy doses of irrigation on coarse sandy soils, it does not describe the diffusion patterns in well-balanced mixes.

Moisture Diffusion Studies

In order to define and record what was happening in planters serviced by Micro-Irrigation Systems, a detailed study was made of the diffusion patterns. It spanned a period of about 9 months and recorded literally thousands of moisture readings. For the main part of the study, a fairly large planter containing a three-stalk corn plant (*Dracaena fragrans*) was chosen as the subject because we knew it provided a challenge to a system relying on natural capillary diffusion. This plant had been serviced by an automated Micro-Irrigation System for more than 5 years. The container was of a decorative plastic, 16 inches in diameter and 12 inches deep. The irrigation emitter was branched off into two legs so that the water impinged onto two small sections of the planter for better distribution. Figure 9.4 illustrates the corn plant used, and Figures 9.5 and 9.6 illustrates the emitter arrangement used in this study.

Readings were taken at seven different positions and five levels down 5 inches into the soil mass of the container (see Figure 9.8). Measurements were made with an Instamatic Moisture Meter, modified for greater accuracy as described in Chapter 7. In order to follow the pattern of moisture diffusion, a set of readings (at each of the different positions and levels) was taken (1) just before a watering, (2) just after a watering, (3) 4 hours after a watering, (4) 8 hours after a watering, and (5) 12 hours after a watering (just before the next one). A small pump- and reservoir-sourced, Micro-Irrigation System was used to provide the watering. These operate at low pressure. Two irrigation cycles, equally spaced and 20 seconds long each, were used per day. The adjustable valve at the plant was initially set so that during the 20 seconds of operation, 2 ounces of water were fed to the plant, 1 ounce from each side of the emitter tee. The study was started in the winter of 1987 (South Florida winters are relatively mild) and concluded in the fall of the year. During the period of more active

Figure 9.4 Corn plant used in the moisture diffusion studies. (Courtesy of Aqua/Trends.)

Figure 9.5 Emitter arrangement used in the moisture diffusion studies. (Courtesy of Aqua/Trends.)

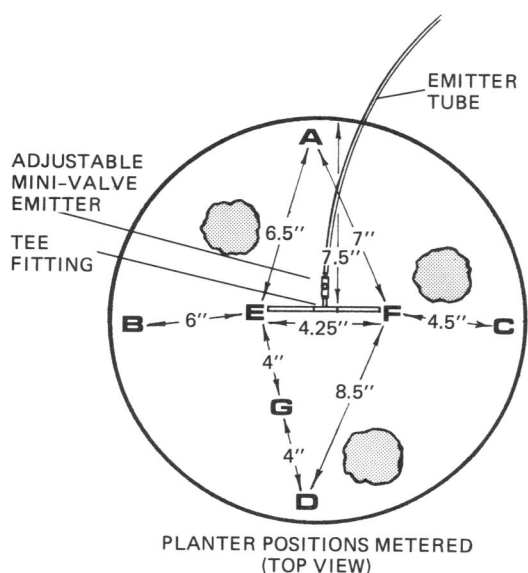

PLANTER POSITIONS METERED
(TOP VIEW)

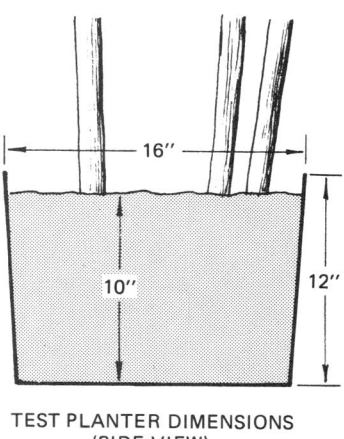

TEST PLANTER DIMENSIONS
(SIDE VIEW)

Figure 9.6 Moisture diffusion study. (Courtesy of Aqua/Trends.)

growth, it was necessary to increase the dosage to 2.25 ounces of water per cycle. The recorded readings were carefully analyzed and various conclusions were reached. Following are the details of the study. If this is too technical for the reader, simply skip to the summary section for the bottom-line results.

During the course of the studies, it was noticed that in the 1-hour period just after a watering, an irregular pattern of moisture was evident on the soil surface. It emanated from, and was centered around, Positions E and F where the water impinged on a patch of soil 1/4 inch in diameter. The moisture pattern appeared to gravitate toward the nearest plant stalks (see Figure 9.7). It clearly demonstrates a strong lateral movement of water in the planter medium. After 2 hours, the pattern had disappeared, having been absorbed by subsurface soil as well as partially evaporated into the atmosphere.

During the early portion of the study, it was also recognized that greater amounts of surface moisture were being evaporated than was desirable. Overall moisture levels were declining noticeably because of it. A 3/4-inch layer of sphagnum peat moss was added to the surface as a mulch, and the readings were carefully scrutinized for moisture level changes over the next several months.

Although the thousands of individual readings resulting from this study will not be reported to save the reader from complete boredom, Figure 9.8 illustrates the typical variations found from one level to the next as well as between positions. Although Figure 9.8 is a simplistic summary of the sets of readings taken at two points in time, it nevertheless illustrates the types of readings typically found at various times of the diffusion cycle. It will be noticed that only the small areas of soil just below the watering points become very moist, and that happens immediately after a watering. The heavy pockets of moisture dissipate to other areas and are absorbed by the plant during the 12-hour period following the watering. Surface levels show low-to-modest amounts of moisture. The moisture gradually increases toward the lower levels. If one consults the moisture meter calibration table in Chapter 7 (Table 7.2), it will be seen that

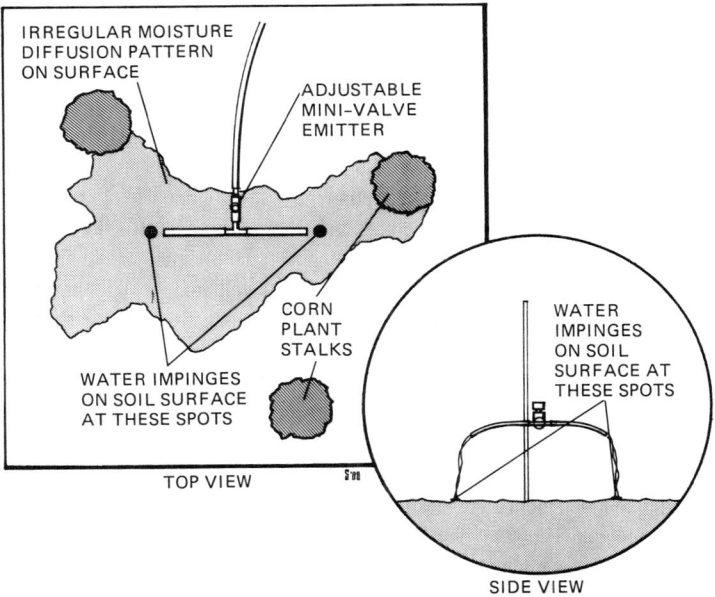

Figure 9.7 Typical diffusion pattern on soil surface after a watering cycle. (Courtesy of Aqua/Trends.)

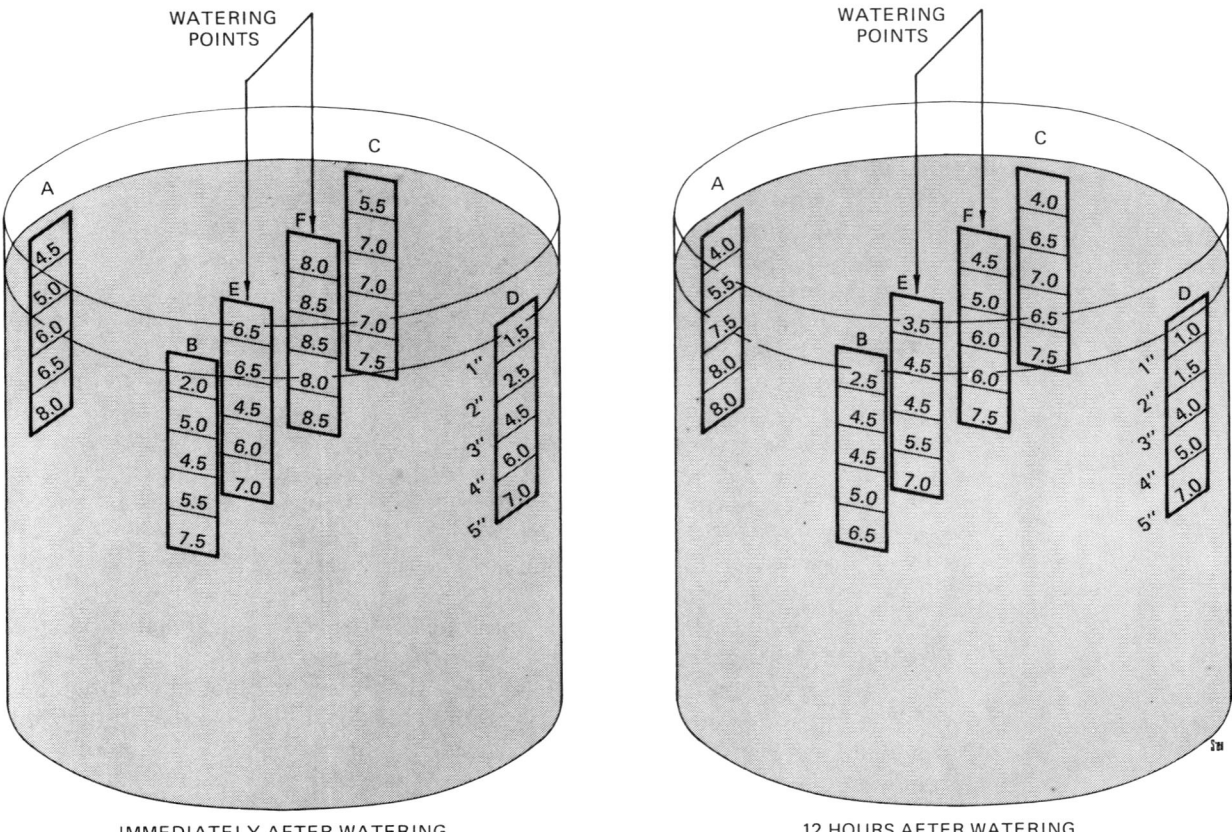

Figure 9.8 Moisture diffusion cycle—typical measurements at various positions and depths. (Courtesy of Aqua/Trends.)

small differences between readings are inconsequential in terms of percentage of moisture content in the soil. The moisture at any location in the planter (except the watering points) stays fairly constant throughout the entire diffusion cycle. This supports the contention that the plant absorbs enough moisture from the last watering to keep the soil mass at a relatively constant dampness level.

Because the 3 inches or so just below the surface are in a lightly damp state, they contain large amounts of oxygen available for plant use. There is never any inundation of these crucial layers with water, as there would be with manual irrigation. Furthermore, compaction of the top soil layers does not occur from this type of overhead watering, as it does with manual and sprinkler irrigation. With properly adjusted systems, the lower levels never become waterlogged, and there is no excess to drain from the bottom of the container. The full moisture diffusion cycle shows a very slight increase in dampness at any level just after watering, then declining again slightly over the next 12 hours until the next gentle watering replenishes the mass again (see Figure 9.9). One must remember that very small doses of water are applied at each cycle. It was also seen that more moisture is depleted during the day than at night.

In taking the thousands of readings for this study, it was noticed that meter reaction was strongly influenced by the soil composition in a given subsurface location and by the proximity to a root mass. The homogeneity of the planter mix used in the study was found to be less than ideal, and pockets of soil were encountered that diffused and retained moisture better than others. Position D

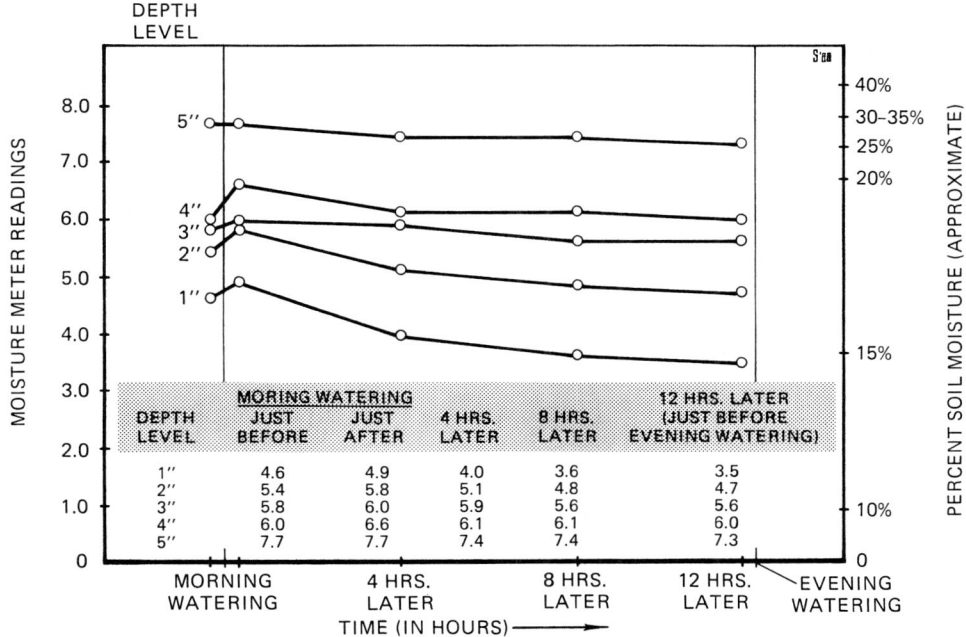

Figure 9.9 Moisture diffusion cycle matrix showing moisture at various depths throughout the diffusion cycle; taken from integrated meter readings (average of all positions at each level). (Courtesy of Aqua/Trends.)

exhibited particularly poor diffusion properties (see Figure 9.8). The subsurface areas around the roots were found to be better moisture absorbers than the outlying sectors. This was particularly true of finely divided root systems with many root hairs. Moisture tends to diffuse into those areas more readily because its fibrous nature promotes better wicking activity. Pockets of root/soil mass may exhibit higher moisture readings than other sections devoid of roots, regardless of their depth.

Supplementary studies were done with the following:

1. A dwarfed corn plant (*Dracaena fragrans*) in a clay pot 13 1/2 inches in diameter by 4 inches deep, containing a well-balanced soil. This plant had been serviced by a Micro-Irrigation System for more than 3 years, receiving 0.7 ounces of water at each of two daily irrigation cycles (see Figure 9.10).

2. A golden evergreen plant (*Aglaonema commutatum*) in a plastic container 10 inches in diameter by 9 inches deep, filled with a very sandy soil. Installed in a location with a fairly high light level, this plant had been serviced by a Micro-Irrigation System for more than 5 years, receiving 1.0 ounces of water at each of two daily irrigation cycles (see Figure 9.11).

3. A ming aralia plant (*Polyscias fruticosa*) in a plastic container 12 inches in diameter by 12 inches deep, filled with a well-balanced soil. This plant had been serviced by a Micro-Irrigation System for more than two years, receiving 2.8 ounces of water at each of two daily irrigation cycles (see Figure 9.12).

4. A pothos plant (*Epipremnum aureum*) in a plastic container 10 inches in diameter by 9 inches deep, filled with a well-balanced soil. This

Figure 9.10 Dwarfed corn plant used in studies. (Courtesy of Aqua/Trends.)

Figure 9.11 Golden evergreen used in studies. (Courtesy of Aqua/Trends.)

Figure 9.12 Ming aralia used in studies. (Courtesy of Aqua/Trends.)

  plant had been serviced by a Micro-Irrigation System for more than five years, receiving 0.8 ounces of water at each of two daily irrigation cycles (see Figure 9.13).
5. A dwarfed ponytail palm (*Beaucarnea recurvata*) in a clay pot 19 inches in diameter by 5 inches deep, filled with a well-balanced soil. This plant had been serviced by a Micro-Irrigation System for more than seven years, receiving 0.7 ounces of water at each of two daily irrigation cycles (see Figure 9.14).

Micro-Irrigation Systems require less water for the needs of a given plant than manual irrigation because of the gentle application at frequent intervals. The technique used prevents the soil from drying yet allows the thorough diffusion of moisture before the next watering cycle and eliminates wasteful runoff. Because less water is used, fertilizer requirements are reduced and mineral salt buildup in the soil is slowed considerably. It is largely a function of the amount of water (and waterborne fertilizer) used over a period of time. An interesting sidelight came to our attention during the studies, due in part to the

Figure 9.13 Pothos used in studies. (Courtesy of Aqua/Trends.)

Figure 9.14 Dwarfed ponytail palm used in studies. (Courtesy of Aqua/Trends.)

aforementioned factors. The dwarfed corn plant showed white accumulations of mineral salts on its soil surface 6 or 7 inches from the emitter. Their presence there clearly demonstrated the lateral movement of moisture in the soil (see Figure 9.15).

In summary, although further experimentation is necessary to fully define these mechanisms, the initial series of moisture diffusion studies has demonstrated conclusively that Micro-Irrigation Systems are capable of maintaining moisture levels in planter soils in a remarkably accurate manner. The gentle watering cycles permit the slow diffusion of moisture throughout the root ball and keep the plant system at evenly moist levels at all times . . . something that manual irrigation can accomplish only at a great inconvenience and expense.

Mulches and Moisture Retention

During one period of the diffusion studies, average moisture levels in the test planter slowly declined. With the hot weather coming on, a more rapid decline was expected. When using Micro-Irrigation Systems, the normal course of action that would be taken under such circumstances would be to adjust the emitter valve for a slightly greater flow or to switch the system into its longer operating cycle; for example, switching from 10 seconds of operation to 20 seconds, its "high gear". In this case, however, the 20-second watering period was already being used, and it was decided to use a top dressing of sphagnum peat moss to reduce surface evaporation, which was suspected to be a major cause of the moisture loss. A 3/4-inch layer of peat moss was spread across the planter surface, except for the small areas immediately below emitter outlets . . . where irrigation water impinged. Without changing the emitter adjustment, moisture level readings were continued for several months. The downward trend did, in fact, reverse, confirming the use of mulch to be quite effective in reducing moisture loss from surface evaporation. The effect, of course, is not as dramatic as it would be outdoors where direct sun would cause very rapid surface drying. Indoors, the effects are more subtle, but they do alter the irrigation schedule and cause the use of more water (and effort where manual irrigation is being used). It took about two months under the Micro-Irrigation System regimen of 2.25 ounces of water every 12 hours, for the subsurface moisture to return to its earlier levels. This study was carried out in South Florida during a period when that area was experiencing its hottest summer on record. Although the interior that housed the test plants was air-conditioned, it was nevertheless warmer than normal, thus naturally increasing plant transpiration. The curve of soil moisture indices illustrated in Figure 9.16 graphically portrays the turnaround. The accompanying Table 9.1 supports the data and provides a summary of the moisture diffusion studies.

In this chapter, we have discussed the concepts behind precision Micro-Irrigation Systems and how this new technology is different from those methods that preceded it. Chapter 10 will describe the operating equipment and technical systems that were developed to implement these new concepts of plant care—the elements that make it all happen.

Figure 9.15 Salt buildup in dwarfed corn plant. (Courtesy of Aqua/Trends.)

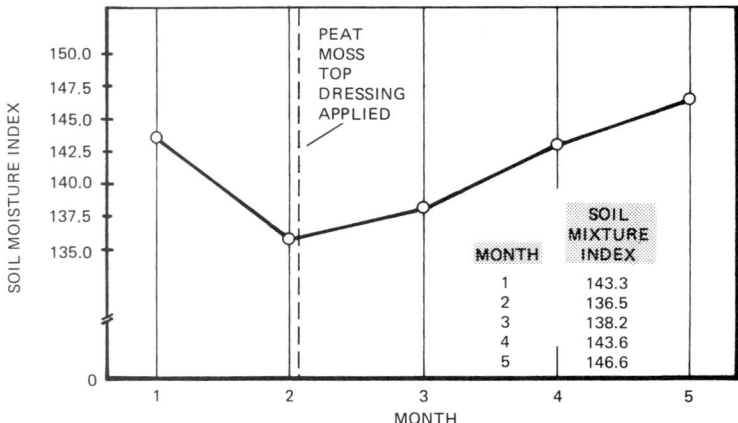

Figure 9.16 Soil moisture indices over a 5-month period before and after the addition of a peat moss top dressing. (Courtesy of Aqua/Trends.)

Mulches and Moisture Retention

TABLE 9.1a
Integrated Moisture Meter Readings*
One Month Before Mulch Was Added

	Condition**				
Depth	1	2	3	4	5
1"	3.4	4.6	4.1	3.9	3.5
2"	4.9	5.9	5.4	4.8	4.6
3"	6.1	6.6	5.6	5.6	5.5
4"	6.5	7.0	6.7	6.4	5.9
5"	6.9	7.6	7.4	7.4	7.0

Soil Moisture Index = 143.3***

TABLE 9.1b
Integrated Moisture Meter Readings*
Just Before Mulch Was Added

	Condition**				
Depth	1	2	3	4	5
1"	3.4	4.6	4.1	3.3	2.8
2"	4.4	5.6	5.0	4.9	4.1
3"	5.2	5.8	5.7	5.3	4.9
4"	5.9	6.5	6.4	6.4	5.5
5"	7.2	7.9	7.4	7.3	6.9

Soil Moisture Index = 136.5***

TABLE 9.1c
Integrated Moisture Meter Readings*
One Month After Mulch Addition

	Condition**				
Depth	1	2	3	4	5
1"	3.1	4.6	3.7	3.4	3.1
2"	4.9	5.9	5.0	4.4	4.6
3"	5.3	6.6	5.9	5.4	5.4
4"	5.7	6.8	6.3	6.1	5.8
5"	6.9	7.7	7.4	7.1	7.1

Soil Moisture Index = 138.2***

TABLE 9.1d
Integrated Moisture Meter Readings*
Two Months After Mulch Addition

	Condition**				
Depth	1	2	3	4	5
1"	4.6	4.9	4.0	3.6	3.5
2"	5.4	5.8	5.1	4.8	4.7
3"	5.8	6.0	5.9	5.6	5.6
4"	6.0	6.6	6.1	6.1	6.0
5"	7.7	7.7	7.4	7.4	7.3

Soil Moisture Index = 143.6***

TABLE 9.1e
Integrated Moisture Meter Readings*
Three Months After Mulch Addition

	Condition**				
Depth	1	2	3	4	5
1"	4.1	4.8	4.9	4.4	3.7
2"	4.1	6.1	5.3	4.9	5.0
3"	5.5	6.3	5.9	5.6	5.6
4"	6.4	6.9	6.5	6.1	6.0
5"	7.6	7.9	7.8	7.6	7.6

Soil Moisture Index = 146.6***

Source: Aqua/Trends

*Readings from all seven locations at a given depth level were totaled and averaged.

**Conditions: 1 = Readings just before morning watering.
2 = Readings just after morning watering.
3 = Readings 4 hours after morning watering.
4 = Readings 8 hours after morning watering.
5 = Readings 12 hours after morning watering (just before evening watering).

***Soil moisture index is the sum total of all integrated readings on a given day and indicates the soil moisture retention level at that time.

10 THE EQUIPMENT OF MICRO-IRRIGATION SYSTEMS

Overview The equipment supplied by Aqua/Trends for precision Micro-Irrigation™ Systems is quite varied, as many different system configurations are necessary to meet a broad variety of installation needs. Some elements are similar to those used in drip and sprinkler systems, but most are different. The need for specialized control systems and water distribution networks was fostered by the unique and demanding requirements of indoor applications, particularly in the context of their broadest uses. As we have learned in previous chapters, tropical plants used in interior decor require relatively small amounts of water to sustain them. That means smaller tubing and pipe sizes can be used in the water distribution network. In order to achieve the highest degree of control over the flow and placement of water, as well as to minimize the risk of serious leaks, the Pulse-Flow concept was developed, very short watering cycles at fairly frequent intervals. That, in turn, necessitated the development of specialized controllers capable of providing low flow rates and extremely short operating cycles. The rigorous esthetic requirements of having to install these systems into highly sensitive, furnished areas of a building interior, like living and dining rooms, executive office suites, restaurant dining rooms, and all the other conceivable nooks and crannies of a decorated building, made system leak resistance of extreme importance. Outdoors, a leak, major or minor, generally has little serious effect. Indoors, of course, such accidents can have very costly consequences. However, the very short periods of pressurization used in Micro-Irrigation System technology, during which flow occurs, make it highly improbable that large amounts of water could leak from a system of this type—nothing close to the magnitude that might be encountered with drip or sprinkler systems. Nevertheless, tubing and fittings had to be chosen carefully to assure complete system integrity, so that nothing could happen. Some of the specifications must meet building codes, for many of these systems are integrated into the framework and partitions of a building; and local codes, along with normal inspections and testing, take jurisdiction over such installations. These and other considerations made this new technology unique in many ways. This chapter will delve into the details of Micro-Irrigation Systems equipment and provide the reader with an understanding of the elements that make the system work.

The Basics   As with all other automatic irrigation systems, there must be (1) a source of water, (2) a means of delivering the water to the planter locations, (3) a device somewhere between the two to cause flow to start and stop, and (4) a timing and control device to cause the activating device to start or stop, open or close. These elements are usually widely separated and must therefore be remotely controlled.

The source is generally city water, when it can be accessed, or pump-drawn water from a well or reservoir of some kind. Irrigation systems for building interiors tap into the city water at cold water piping running through the structure (cold water mains). Pump systems used in interior irrigation draw water from an exterior well or from a reservoir installed somewhere in the building.

The means of delivering water to the planter locations is always a tubing or piping network leading from the water source to the various plant placements. This network can be simple or complex, depending on the nature of the installation. It can be short or extremely long. Indoors, this could mean a 10-foot stretch of tubing to water a couple of potted plants in an office workstation or a run of several hundred feet to service an entire building floor; there are many diverse applications between these two extremes as well.

Electrically operated solenoid valves are placed at appropriate locations in the tubing line to interrupt the flow of water-main-sourced systems. This is true indoors and out. Water will flow through the tubing network only when the solenoid valve is open. Electrically operated pumps are used in the other types of systems to create a flow of water. Water will flow through the system only when the pump is operating.

Electromechanical or electronic time-switch/controllers are used to control the action of the solenoid valves and pumps. They time the period between operations and the length of time the system is activated and watering plants. They are called controllers because they provide and remove power to the activating devices at the proper times.

These explanations are a simplistic view of the technology involved. Other devices are used as well, but the aforementioned elements are essential for automatic systems. Other elements will be mentioned as we go through the detailed discussions in the balance of the chapter, and from here on we will be concentrating solely on precision Micro-Irrigation Systems.

Categories of Micro-Irrigation Systems   High-pressure and low pressure systems are the major division of precision Micro-Irrigation Systems technology. They are different in the ways they deliver water to the tubing network; other than that, they are essentially the same. Each of these major divisions is, in turn, subdivided into a number of other categories according to the way it is used, that is, its application. There are various levels of duty, for example. Combined, they are capable of servicing a broad variety of installation configurations. The equipment used for each also has its own peculiarities. Although this precision plant-care technology was designed mainly for critical interior applications, it finds noncritical indoor and outdoor uses as well, and the equipment used for each application had to be designed with its particular needs in mind. An illustration of the various configurations of Micro-Irrigation Systems can be found in Figure 10.1.

High-Pressure Systems   High-pressure systems, in the context of this technology, are sourced by the building's cold water lines, which, in turn, are usually sourced by the local water department. Occasionally, the building may have its own well and pump

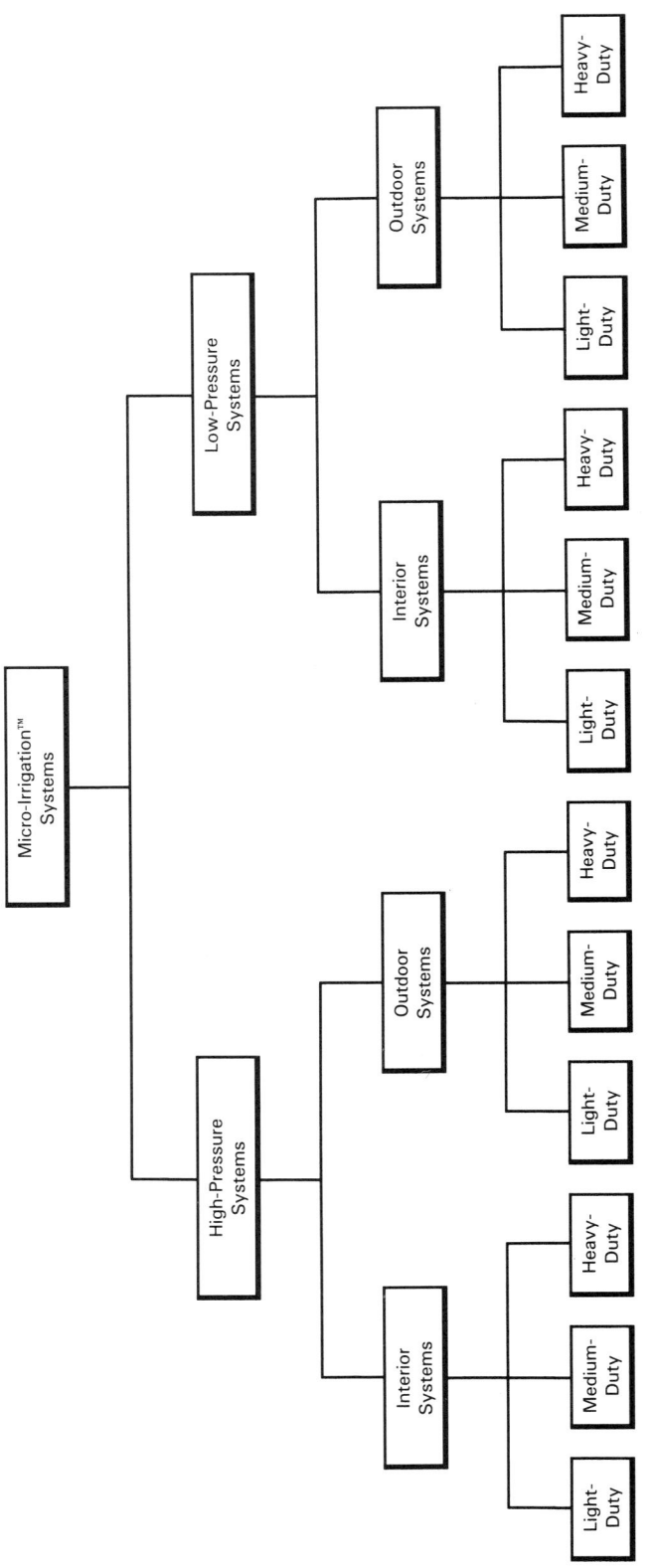

Figure 10.1 Categories of precision Micro-Irrigation Systems.

system to furnish its needs. Input pressures to buildings are most commonly in the range of 55 to 70 psi. Many local building codes require a reducing-pressure regulator in the input line if pressures exceed 80 psi, since plumbing fittings and appliances in building interiors are subject to excessive wear or damage by high pressures.[1] Pressures are not fixed but fluctuate throughout the day, more so at certain times than at others. For example, most city water systems are under the heaviest use in the morning when people are preparing for work and in the evening when they are showering or running washing machines or dishwashers. Summer weekends create a heavy demand on local water systems, as lawn sprinkler systems are operated more frequently and cars are washed. Sometimes local industry creates a heavy, fluctuating demand that also affects the pressure in residential and commercial buildings. Building and zoning codes are meant to control system aberrations such as that, but the reality is that local controls don't always work as planned.

The general configuration of a high-pressure version of Micro-Irrigation System is shown in Figure 10.2. The system is connected to the cold water line by means of a hose bibb (faucet), as shown in the illustration, or by means of a soldered fitting installed at a juncture of the cold water pipe. The irrigation system becomes another branch of the plumbing, but under independent control. Immediately after the connection, a shutoff valve of some sort is installed to shut down the irrigation branch in emergencies or for repair or for maintenance of the system. Sometimes, the connection and shutoff valve are one and the same, such as with the hose bibb shown in the illustration, or they are an in-line globe valve. Many plumbing codes call for an anti-siphon device installed between an auxiliary water branch (such as an irrigation system) and the city water supply to prevent the accidental pollution of the local water system. There is always a filter of some kind to remove dirt and other contaminants from the water supply. This is generally installed just after the connection so as to reduce wear on all system parts and prevent the small orifices from clogging. There is always some contaminant present in the water supply, so chances should not be taken. In smaller systems, a screen filter is integral with the input

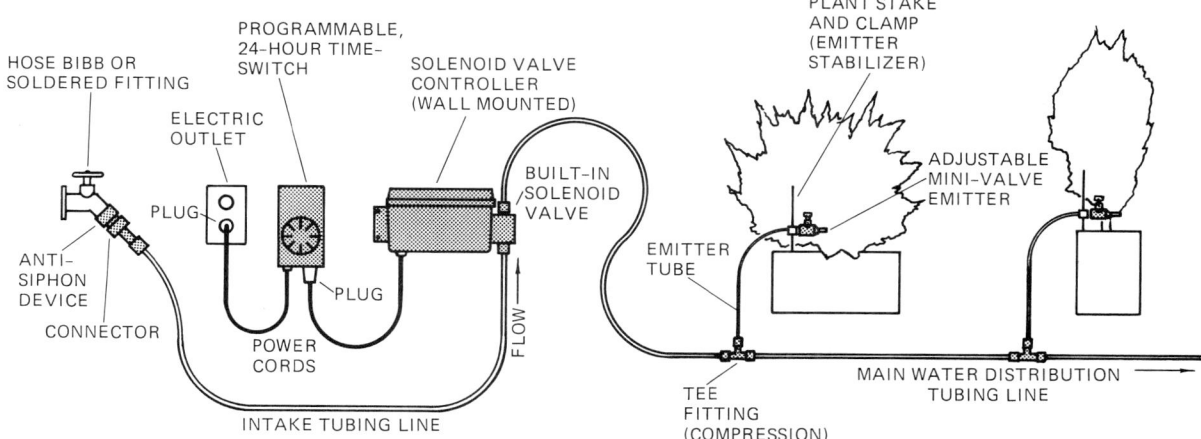

Figure 10.2 Micro-Irrigation Systems—a typical high-pressure configuration. (Courtesy of Aqua/Trends.)

[1]Robert M. Hettema, *Mechanical and Electrical Building Construction* (Englewood Cliffs, NJ: Prentice-Hall, Inc., 1984), p. 74.

fitting of the solenoid valve so no additional filtration is necessary. In light- and medium-duty systems where just a few plants have to be serviced, it may be necessary or desirable to reduce the system pressure to lower levels. In that case, a pressure reducer or pressure regulator is introduced just after the filter unit or in some instances, after the solenoid valve. The type of pressure regulator will, in large measure, determine its positioning. Integral screen filters can be found in some of these devices as well. Next in line comes an electrically operated solenoid valve that controls the flow of water through the system at timed intervals and durations. A solenoid valve is an open-and-shut type of device with no other flow regulation involved. It opens to permit water through for very short durations and shuts down for long intervals to prevent water from flowing through the system. The timing and control of the solenoid valve is accomplished through time switches and solenoid controllers. In Aqua/Trends' Micro-Irrigation Systems, most models have solenoid valves mounted in their controller for a compact, easy-to-install unit. The solenoid valve controller is plugged into a 24-hour timer (time switch) that operates it according to the time schedule programmed for the system. The timer determines only when the system comes on, not when it shuts off. That is determined by the solenoid valve controller, which permits the valve to open for only very short periods of time. Downstream from the solenoid valve the water distribution network is connected, which is a network of tubing, fittings, and flow-control devices that direct water to the individual plants. At intervals throughout the network are connections to small-diameter tubing, which carries water to the individual potted plants. At the end of these tubes are fitted miniature adjustable-valve emitters that can be turned down for very low flow rates or that can be opened up for more volume. The emitter assembly is held in place by a clamp and stake that are anchored into the potting soil.

The system is considered high pressure because it operates essentially under building pressures of around 55 psi. While that level of pressure can be looked at as being only modest when compared with truly high-pressure fluidic systems operating at a few hundred pounds of pressure, when compared to the pump-operated versions of Micro-Irrigation Systems, these pressures are higher. For those reasons also, tubing and pipe fittings must be more secure, and are chosen accordingly.

High-pressure Micro-Irrigation Systems are used in a broad variety of installations, covering just a few to hundreds of plants both indoors and out. Because of the higher pressures involved, they are capable of moving greater volumes of water into the far reaches of a building. The low-pressure systems have some limitations in that regard. To implement a high-pressure installation, easy access must be available to the building's water line. As this is not always the case, there are situations where low-pressure versions must be substituted.

Low-Pressure Systems

The low-pressure versions of Micro-Irrigation Systems use a reservoir as the water source. Pumps are used to draw the water and deliver it throughout the tubing network at a very low pressure of 3 to 15 psi. Some models reach a pressure as high as 30 psi, but those are exceptions. The water distribution network used in low-pressure systems is basically the same as the network used for high-pressure configurations except that tubing connections are not as critical. Aqua/Trends' systems use units called *pump/reservoir modules* (see Figure 10.3). These are integrated units, having a small pump built into the water container for convenience and economy.

The pump is powered by a controller designed for that purpose. Wires interconnect the two. The pump controller is, in turn, powered by a 24-hour

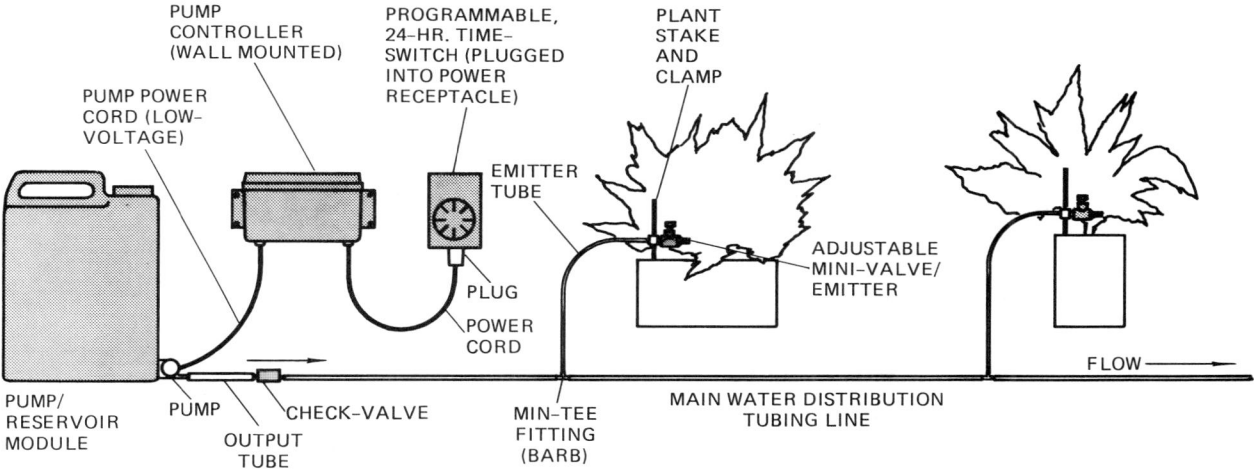

Figure 10.3 Micro-Irrigation Systems—a typical low-pressure configuration. (Courtesy of Aqua/Trends.)

time switch that is programmed to switch the pump on at predetermined intervals (usually one to three times per day). Here again, the 24-hour timer determines only when the system comes on. It is the controller that determines how long the system is permitted to operate. That duration is normally 10 to 30 seconds. Some of Aqua/Trends' models have a 24-hour timer integrated into the pump or solenoid valve controller.

Low-pressure versions of this technology are used whenever convenient connections to a water line are not available for high-pressure systems, as well as in certain critical interior areas where the safest type of installation must be made. The greatly reduced pressures of these versions lend themselves to such situations.

Because some attention is necessary in refilling the reservoirs, these low-pressure systems cannot be classified as fully automatic. There are techniques available, however, for automatic reservoir refilling, and they extend the capabilities of the installations.

Sequence of Operation

The sequence of operation is basically the same for both high- and low-pressure systems. It is demonstrated graphically in Figure 10.4. The 24-hour time switch runs continuously. At a preset time of day (typically one, two, or three times per day), it switches the output power on and applies it to the pump or solenoid valve controller. The controller then applies power to the pump or solenoid valve for only 10 to 30 seconds, causing water to flow through the distribution tubing network for that short period of time. The water reaches the plant locations and passes through emitter tubes, then the emitters, and into the planter soil. The emitters are preadjusted for the proper amount of flow for each plant (each has its own emitter). When the system stops, pressure is released from the tubing and stays essentially at zero until the next activation many hours later. This provides an important measure of safety for the interior environment. Total pressurization of these systems is no more than about 40 seconds per day. Although occasional leaks do not need to be feared as they do with other technologies, low-pressure Micro-Irrigation Systems provide the safest situations. Flow rates are lower, irrigation time is much shorter, and minimal water pressure puts less strain on the connecting joints. High-pressure ver-

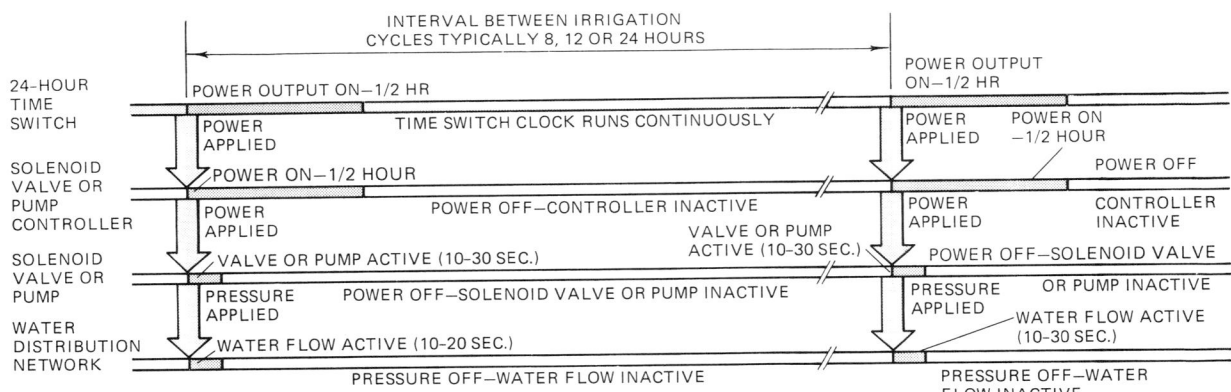

Figure 10.4 Typical sequence of operation using a 24-hour electro-mechanical time switch to activate the sequence.

Control Centers The group of equipment around the source of water is considered the control center. In the case of high-pressure systems, it involves the actual connection to a cold water line, shutoff valve, filter, pressure regulator, timer, and solenoid valve controller. In the case of low-pressure systems, it involves the pump/reservoir module, timer, and pump controller. The control center is the heart of the system, providing water and the means for its overall control. Downstream, there are other control devices that perform more subtle chores.

Timers The timers used in most current versions of Micro-Irrigation Systems are simple, electromechanical 24-hour time switches. There are light- and heavy-duty versions of these; the use dictates which are the practical choices. The timers are capable of turning the system on as many as 12 times per day. In most installations, however, only 2 or 3 times per day are necessary.

There are many variations of these timers used in Aqua/Trends' product line to accommodate different installation needs. Some models simply plug into a wall outlet, while others are wired into an electrical junction box. Some are housed in weatherproof cases to be used outdoors or in wet interior locations, such as in planter boxes (see Figure 10–5). Some models are simple, inexpensive utilitarian units, while others provide a more decorative and professional appearance to the installation. Some models are meant for medium- to heavy-duty use, while others are light-duty versions.

In most cases, pump or solenoid valve controllers plug directly into the timer, and this provides an easy, versatile means of interconnection that can be done by anyone. Neater, more professional installations can be made by using a timer that has all electrical connections made in a hidden junction box. Input, as well as output power wiring is kept out of view. These, of course, require the services of an electrician.

Another means of timing is offered by Aqua/Trends for its systems. It is a programmable timer/controller based on carrier-frequency, remote-control technology. It transmits control signals through the building's electrical wiring to remote locations where various Micro-Irrigation Systems can be activated once or twice daily. That schedule is enough for most light- to medium-duty

Figure 10.5 Weatherproof timer used with Micro-Irrigation Systems. (Courtesy of Aqua/Trends.)

installations. The important feature of this configuration is that any number of satellite Micro-Irrigation Systems scattered throughout the building can be controlled remotely as a group from a central location. The concept will be discussed again in the following chapters on design, installation, and interfacing with other technologies.

In some of the newer models, timers have been designed into the digital control circuits. That makes for a compact control center. The configuration has other advantages as well.

There are, however, many good reasons for choosing separate timers and controllers. One reason has to do with the fact that an independent timer can be used to control supplementary lighting for planters, and other accessories at the same time it is operating automated irrigation equipment. That subject will be dealt with in more detail later in the book.

Solenoid Valve Controllers The purpose of the solenoid valve controller is to provide a timed application of power to the activating element of the system . . . which is the solenoid valve in high-pressure versions. In keeping with the concepts of Micro-Irrigation System technology, the controller activates the system electronically for very short periods, generally 10 to 20 seconds indoors and occasionally up to 30 or 40 seconds outdoors (patio/terrace installations). One of the advantages of low-duty cycles (very short operating periods and very long dormant periods) is that minimal operating times provide unusual longevity for the electronics and electromechanical devices involved (pumps and solenoid valves).

Aqua/Trends has developed a broad variety of controllers to meet a wide

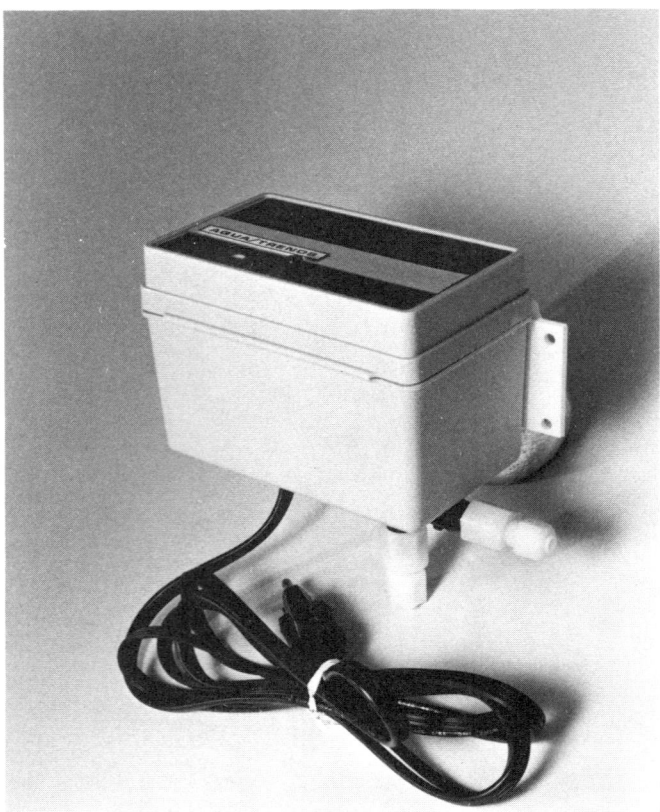

Figure 10.6 Light-duty solenoid valve controller. (Courtesy of Aqua/Trends.)

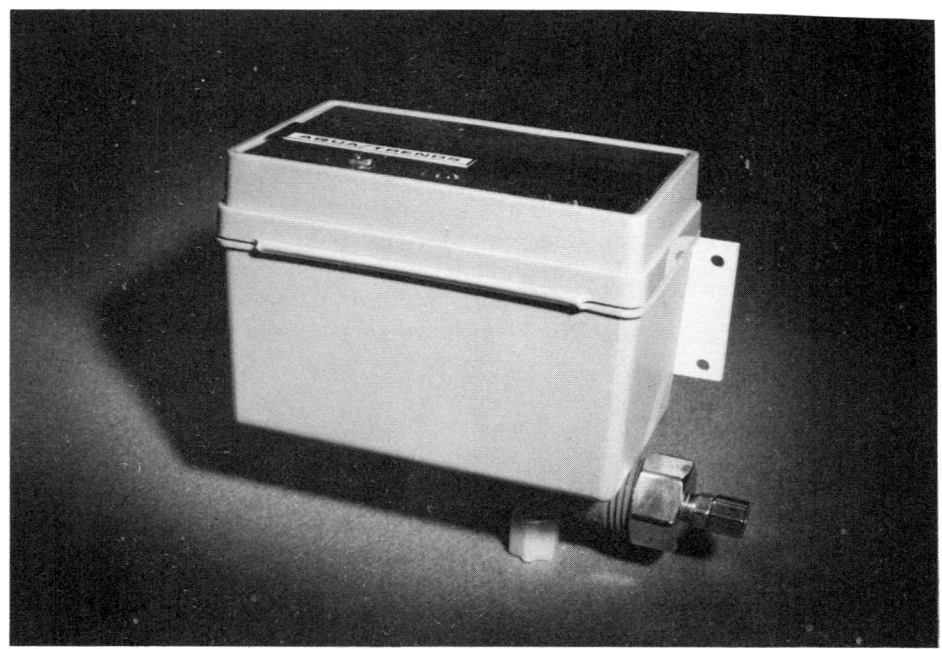

Figure 10.7 Medium-duty solenoid valve controller. (Courtesy of Aqua/Trends.)

Solenoid Valve Controllers

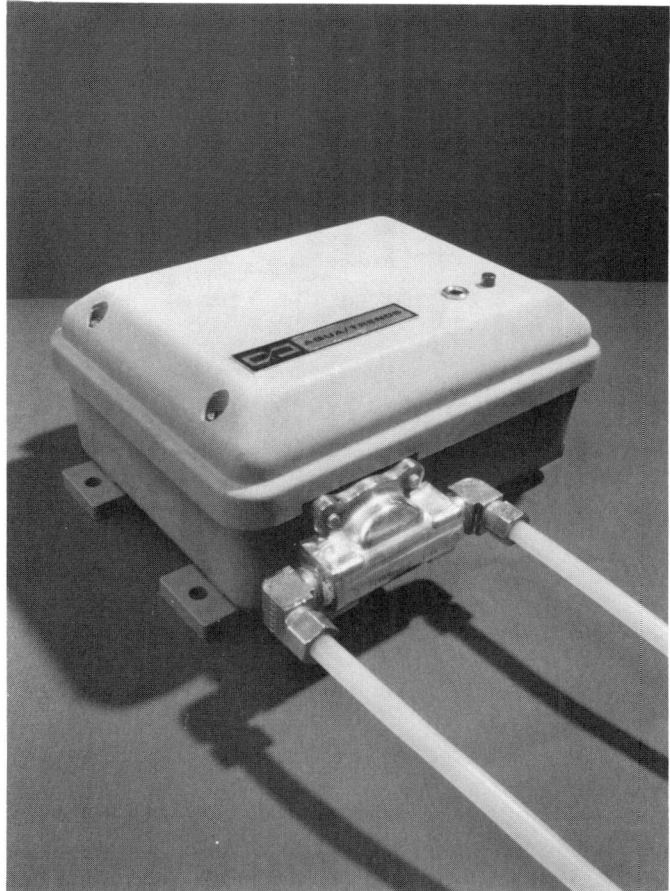

Figure 10.8 Heavy-duty solenoid valve controller. (Courtesy of Aqua/Trends.)

range of installation needs. In virtually every model, the solenoid valve is mounted integrally into the controller case, making a compact, easy-to-install unit. Most models are designed for indoor use where conditions are dry, but a few are housed in all-weather cases for outdoor and very damp indoor settings. The size and ruggedness of the unit determine the type of duty to which it can be put. Light-duty solenoid valve controllers are capable of servicing only small interiorscapes with relatively few plants, generally up to 12 or 15. Medium-duty units have heavier electronics and valves to feed water through larger tubing networks and are capable of servicing a greater number of plants, up to about 50. Heavy-duty solenoid valve controllers are even sturdier. They are able to feed irrigation water into large-diameter tubing and piping configured into more extensive and complex arrangements. As many as 200 plants can be serviced by some of these larger units. For even more extensive and complex interiorscapes, multiple systems can be installed to care for a single zone. Multizone models are also available.

With a few exceptions, solenoid valve controllers are simply plugged into electromechanical or electronic time switches to provide their input power of 115 volts AC (alternating current) at programmed, recurring intervals. A couple of models are wired into receptacle boxes for hard-wired connections to time switches. Still other models have timers integrated into the controller's elec-

Figure 10.9 All-weather solenoid valve controller for outdoor use. (Courtesy of Aqua/Trends.)

tronic circuits. All controllers are wall mounted and can be considered a type of specialty appliance, which, in fact, they are in the eyes of many local building codes. The variations provide a broad degree of versatility in meeting installation needs.

Most models have switches to conveniently change irrigation cycle timing. For example, with the simple throw of a switch, the duration of each operating cycle can be changed from 10 to 20 seconds, or from 15 to 30 seconds. Also built into these units is a remote-control port (jack) into which a long manual-control cord or a radio-controlled activator can be plugged. This allows the operator to turn the system on manually for very short periods while adjusting emitters at remote plant locations. The utility of such an arrangement will become more apparent later when we discuss system installation and start-up. Most solenoid valve controllers in the Aqua/Trends line are also designed with multiple safety features to prevent accidents.

Pump Controllers   In the same way that the solenoid valve controllers activate electrical valves, pump controllers power small pumps and time their operation for very short periods . . . generally 10 to 20 seconds in interior applications and 15 to 30 seconds in outdoor systems. The input power cords of these devices are simply plugged into electromechanical or electronic time switches (timers) to provide 115 volts AC (alternating current), while output power cords connect to the terminals of pump/reservoir modules. Output of most units is a safe 12 volts DC (direct current).

Pump Controllers

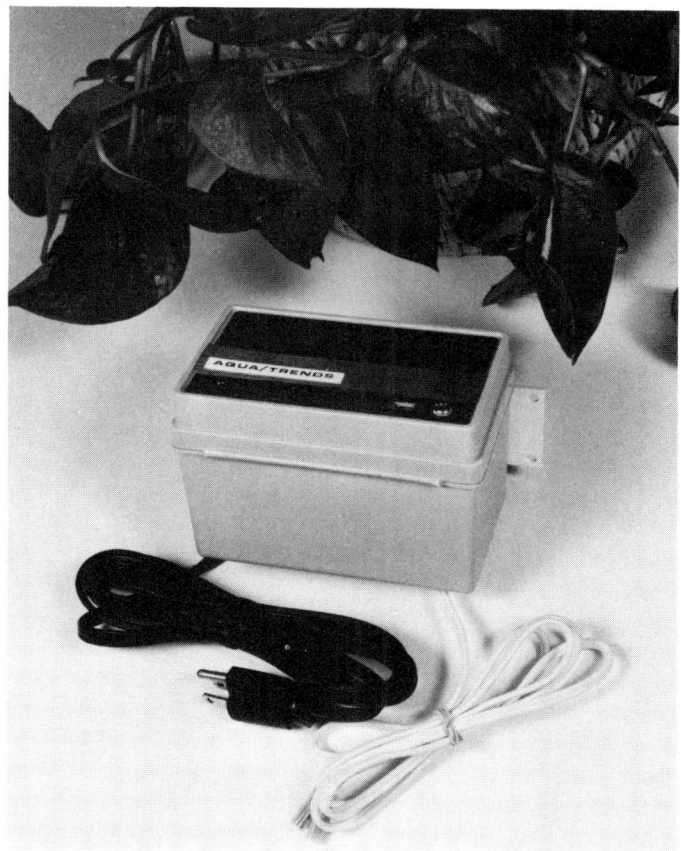

Figure 10.10 Light-duty pump controller. (Courtesy of Aqua/Trends.)

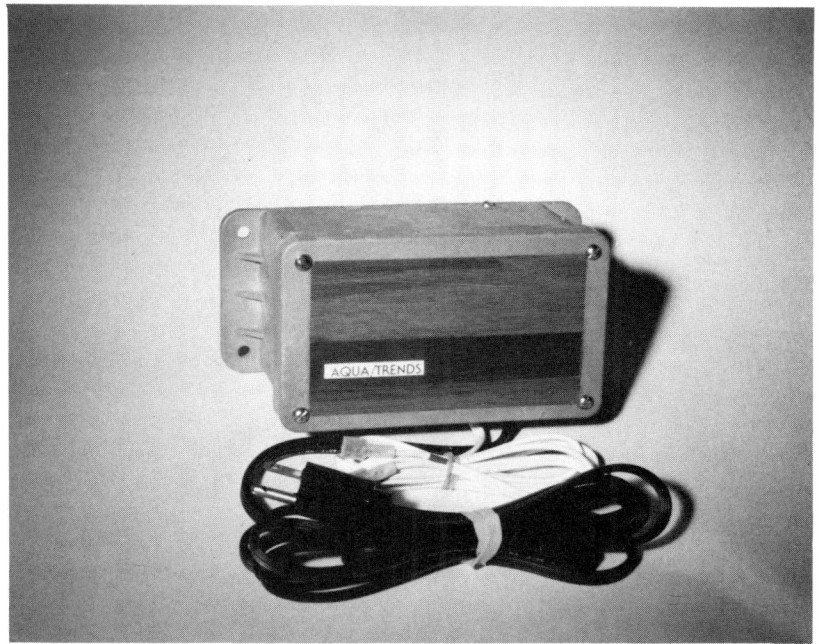

Figure 10.11 All-weather pump controller for outdoor use. (Courtesy of Aqua/Trends.)

On rare occasions, pumps are integrated into the controller for special purposes. Other models provide timers integrated into the controller itself for a compact, space-saving unit. The size and ruggedness of a pump controller has to do with its duty requirements. Larger pumps servicing extensive and complex plantscapes require a heavy-duty controller, while smaller pumps can get by with light-duty controllers. Most models are for indoor use, but a few are housed in all-weather cases for outdoor use or for use indoors in damp locations. Pump controllers are simply wall mounted, and these, too, are considered a type of specialty appliance.

All models provided by Aqua/Trends have a switch for the convenient control of the watering cycle duration, as do the solenoid valve controllers. Also provided is a remote-control port for convenient manual operation of the unit from remote locations while setting up.

Pump/Reservoir Modules   These are plastic containers of 2-1/2 gallon to 32 gallon capacity with self-contained pumps; they are used in conjunction with electronic pump controllers. They function as water sources. Actually, the size of the reservoir can be much larger for special applications. Within some of these containers submersible pumps of various duty ratings are mounted. In other models, pumps are mounted on the outside wall of the reservoir. In all cases, the pump connects

Figure 10.12 Light-duty pump/reservoir module. (Courtesy of Aqua/Trends.)

Figure 10.13 Outdoor patio pump/reservoir module. (Courtesy of Aqua/Trends.)

to a controller appropriate to its size and duty rating by means of a low-voltage electric power cord. Most of these pumps are operated by the application of 12 volts DC for short periods of time, in keeping with the Pulse-Flow concept of Micro-Irrigation Systems technology.

Small 2-1/2 gallon pump/reservoirs are inexpensive units capable of caring for up to about 10 to 12 plants. Also available are 5, 10, 20, and 32 gallon pump/reservoir modules that can accommodate the needs of larger or drier (outdoor) installations. Patio installations, for example, require the use of 20 or 32 gallon reservoirs in order to dispense the relatively larger volumes of water for reasonably long periods of time without refilling. At least six weeks of continuous operation is sought in designing any low-pressure installation.

The pumps used in these low-pressure systems are small in size (some miniature) and provide water flow at very low pressures, generally below 12 psi. A few units are exceptions and reach 30 psi for special needs, particularly where plants are at different levels (for example, upstairs/downstairs or in hanging planters).

Tubes and Pipes Tubing and pipes make up the distribution networks that carry water from the control center (source) location throughout the building area that is to be serviced. The subject can become quite complex and bewildering, for there is more than one system in determining product sizing. The first system is rooted in the nomenclature of the plumbing industry, where pipes are sized by describing their inside diameter. The second system comes from the instrumentation and control segments of industry, where small-diameter tubing of various descriptions is commonly used. That tubing is sized by describing the outside diameter. The third system comes from various industries using hoses as part of a multitude of mechanical devices. The hoses are sized by describing their inside diameter. Installations of Micro-Irrigation Systems sometimes use a combination of relatively small-diameter tubing, hoses, and piping. Most systems, however, require only tubing. It would be wise to fully understand the nomenclature involved, for it will also be an important factor when we consider fittings, flow-control devices, and accessories later in this chapter. It is something that anyone dealing with Micro-Irrigation Systems at the design and/or installation levels must work with intimately.

The flexible tubing, hoses, and much of the piping used in the installation of Micro-Irrigation Systems are plastic, with some metal pipe used on occasion. The reasons for the use of plastics are mainly (1) the economy of product, (2) the ease of use, and (3) the growing scarcity of copper pipes and tubing. Copper pipes and tubing are slower to install, thus increasing installation labor costs. They are generally cut with a hacksaw or tubing-cutter, deburred, fluxed, and soldered ("sweat") into fitting sockets. The process is time-consuming and not cost-efficient. Flexible plastic tubing, on the other hand, can be cut quickly with a utility knife and connected by means of easy-to-use compression and slip-on barb fittings. It can be bent into fairly tight curves, saving some of the time and cost of installing angle fittings. Plastic pipe is cut with a hacksaw or tubing-cutter, deburred, cleaned, and solvent-welded into fitting sockets. Hose products are used in small quantity and will not receive much attention in the following discussion. It is recommended that flow velocities in plastic tubing be kept below 5 feet per second.

Various types of plastics are used in the manufacture of flexible tubing. The main ones adopted for Micro-Irrigation System installations are as follows:

- polyethylene, otherwise called PE or poly tubing.
- polyvinyl chloride, called PVC.
- polybutylene, called PB.

Various types of plastic piping are also used:

- polyvinyl chloride (PVC).
- polybutylene (PB).
- chlorinated polyvinyl chloride (CPVC).

Copper is the only type of metallic pipe or tubing used in Micro-Irrigation System installations, although that could change if the shortage of copper becomes more acute in this country. It is used mainly where building codes restrict the use of plastics.

Flexible tubing is manufactured in sizes as small as 1/8" OD (outside diameter), 3/16" OD, and 1/4" OD. These small sizes are commonly called *capillary tubes*. Larger tubes are sized in increments of 3/8" OD, 1/2" OD, 5/8" OD ... up to 1" or more. Table 10.1 shows a breakdown of the flexible tube types and sizes most commonly used in this technology.

TABLE 10.1
Flexible Tubing Used in Micro-Irrigation Systems

Material	Nominal Tubing Size
Polyethylene (PE)	$\frac{1}{4}''$ OD
	$\frac{3}{8}''$ OD
	$\frac{1}{2}''$ OD
Polyvinyl Chloride (PVC)	$\frac{3}{16}''$ OD
	$\frac{1}{4}''$ OD
	$\frac{9}{32}''$ OD
	$\frac{1}{2}''$ OD
Polybutylene (PB)*	$\frac{1}{8}''$ ID
	$\frac{1}{4}''$ ID
	$\frac{1}{2}''$ ID

*PB tubing is generally sized by describing inside diameter because the manufacturers cater mostly to plumbing trades accustomed to this nomenclature.

The pipe sizes most commonly used in Micro-Irrigation Systems are 1/2" ID and 3/4" ID (inside diameter). Table 10.2 details pipe sizes and types.

The relationship between inside and outside tubing and piping diameters is important because it determines the wall thickness, which, in turn, may determine the maximum water pressure to which the product can be subjected. The inside diameter also determines the flow rates and pressure losses that water encounters from internal friction. The plastic composition also bears on these factors. Outside and inside tubing diameters also determine the types of fittings that can be used, particularly when dealing with compression types. This subject will be discussed in detail in the next section. Table 10.3 shows the relationship between inside and outside diameters among the commonly used products.

There are industry standards for pipe sizes, but such is not the case with small-diameter flexible tubing. It is subject to variations in size from manufacturer to manufacturer, as well as from any given source. This can cause serious problems when it comes to properly connecting the tubing into a network by means of tubing fittings. Disconnections can occur, causing leaks. It is strongly recommended that all tubing and fittings be obtained from the same reputable source of supply. These will have been pretested for compatibility and pur-

TABLE 10.2
Pipes Used in Micro-Irrigation Systems

Material	Nominal Pipe Size*
Polyvinyl Chloride (PVC)—Sch40	$\frac{1}{2}''$ ID
	$\frac{3}{4}''$ ID
Polybutylene (PB)	$\frac{1}{2}''$ ID
	$\frac{3}{4}''$ ID
Chlorinated Polyvinyl Chloride (CPVC)	$\frac{1}{2}''$ ID
	$\frac{3}{4}''$ ID
Copper—Type L (medium)	$\frac{1}{2}''$ ID
	$\frac{3}{4}''$ ID

*Pipes are also sized by inside diameter because most are used by the plumbing and landscaping irrigation trades accustomed to this nomenclature.

TABLE 10.3

Tubing and Piping Diameters and Other Specifications
(Select Types Common to Micro-Irrigation Systems)

Tubing or Pipe Description	Nominal Size	Actual Outside Diameter	Actual Inside Diameter	Actual Wall Thickness	Rated Friction Loss/100 Ft. at 2 GPM	Color
Polyvinyl Chloride (PVC)	3/16" OD	0.21"	0.165"	0.0225"	N/A	clear
	9/32" OD	0.29"	0.160"	0.065"	N/A	black
Sch40	1/2" ID	0.840"	0.620"	0.109"	1.3 psi	white
Polyethylene (PE)	1/4" OD	0.25"	0.16"	0.045"	N/A	black/natural
	3/8" OD	0.375"	0.25"	0.0625"	N/A	black/natural
Polybutylene (PB)	1/8" ID	0.25"	0.126"	0.062"	N/A	grey/blue
	1/4" ID	0.375"	0.25"	0.0625"	N/A	grey/blue
	3/8" ID	0.500"	0.366"	0.067"	20.5 psi	grey/blue
	1/2" ID	0.625"	0.491"	0.067"	4.9 psi	grey/blue
	3/4" ID	0.875"	0.705"	0.085"	0.9 psi	grey/blue
Copper—Type L	1/2" ID	0.625"	0.545"	0.080"	3.8 psi	copper
	3/4" ID	0.875"	0.785"	0.090"	0.8 psi	copper

Sources: Data sheets—Vanguard Plastics, Inc.; Qest (U.S. Brass); Shell Chemical Co.; and LCP Chemicals and Plastics. Bench Testing—Aqua/Trends.

chased only from reliable manufacturers. Much grief can be avoided by adhering to this advice.

Tubing colors are also important in this technology. Small-diameter clear PVC tubing is used in room interiors where it might be visible. Clear tubing easily blends in with the surrounding colors and makes the installation less obtrusive. One of the problems with clear tubing, however, is that algae tend to build up in the tubes when they are near windows or other bright, sunlit areas. In that case, black, grey, or an opaque tubing of some other color should be used to shield the water from algae-promoting rays. This problem is relatively minor but can cause infrequent disruptions in service when the algae clog small orifices and tubing lines. Algae usually manifest themselves in low-pressure systems that carry dissolved fertilizers in the irrigation water.

Tubing that is hidden under carpeting or furniture or in partitions can be of any color. Tubing that is used outdoors, where it is subject to weathering (particularly exposure to sunlight), is easily degraded unless it has been protected by pigmentation or chemical modifiers that absorb or screen out ultraviolet rays. Tubing products suitable for outdoor use are generally black or grey, as the dark pigmentation helps protect the plastic.

Laser Soaker Line®[2] is a special type of plastic tubing made to dispense water along its length in the manner of a soaker hose. Tiny holes are drilled every 6 or 12 inches by means of a controlled laser beam. Its size is slightly over 1/4" OD. Laser Soaker Line finds specialized applications in Micro-Irrigation Systems that will be dealt with later.

Fittings The subject of tubing and pipe fittings is one that could easily be relegated to a chapter of its own. It is extensive, complicated, and potentially quite confusing. In this section, we will try to clarify the issue as best we can in this short space. Take a deep breath, and wade into what follows with courage.

Fittings used to connect tubes, hoses, and pipes in Micro-Irrigation Sys-

---

[2]Laser Soaker Line is a registered trademark of Pepco Extruded Products, Inc.

tems cover a broad variety of types, materials, and sizes, each designed for a specific application. The main categories are (1) barb types, (2) compression types, (3) threaded types, (4) soldered types, (5) solvent-welded types, and (6) combination types. Barb and compression fittings are the ones that find the most use in this technology.

Barb fittings are the easiest to use but are considered the least secure. The tubing is simply slipped onto a leg of a fitting, with friction generally relied on to hold the connection securely. Friction is usually not enough, so the legs of most barb fittings are manufactured with ribs and/or notches to grab the softer plastic tubing as securely as possible (see Figure 10.14). These fittings are made of plastic or brass. Various materials are used in the manufacture of plastic barbs. Nylon, ABS, acetal, and high-density polyethylene are the most common compositions. A broad range of fittings can connect tubing from the smallest to the largest sizes. Clamps are used to secure tubing to barbed fittings whenever higher water pressures are to be encountered. Some crimp over the connection with a special tool, while others exert spring tension or are tightened down with screw adjustments. It is very important that the right type of clamp or crimp ring be used. Most modern building codes permit polybutylene tubing to be used in plumbing systems with barb-fitting connections. The stipulation, however, is that properly applied crimp rings secure the connections. In applications where lower water pressures are encountered, barbed connections can be secured with adhesives, selected for their holding power as well as their compatibility with the materials being bonded (fitting/tubing interface).

Compression fittings are made to exert a clamping action on the tubing after it is slipped into the fitting socket. The clamping effect is accomplished by tightening down a nut, either by hand or wrench depending on the type of fitting. The nut, in turn, crimps a gripping collar onto the softer tubing—just enough to hold it securely. Properly designed plastic compression fittings that

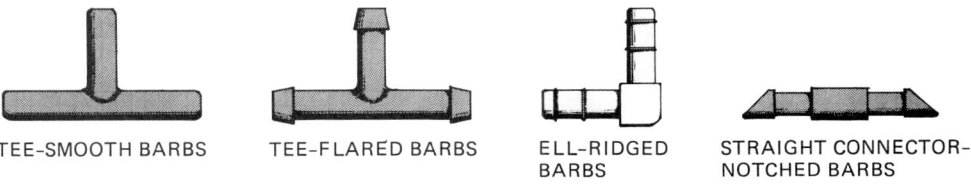

TEE-SMOOTH BARBS   TEE-FLARED BARBS   ELL-RIDGED BARBS   STRAIGHT CONNECTOR-NOTCHED BARBS

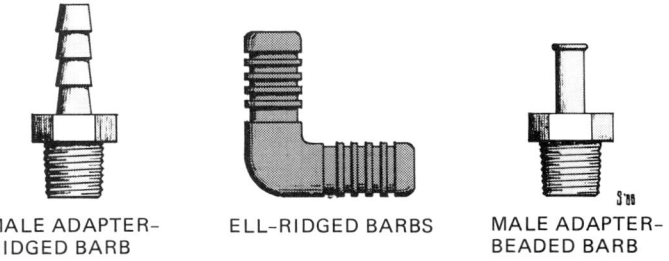

MALE ADAPTER-RIDGED BARB   ELL-RIDGED BARBS   MALE ADAPTER-BEADED BARB

Figure 10.14 Typical barb fittings.

require only a hand tightening to provide secure connections are available (see Figure 10.15). Various means of clamping down and gripping the tubing are designed into the products, with each manufacturer having its own method. Some means work better than others with a given type of tubing, so the proper fitting must be selected. Compression fittings are made of plastic or brass. The types of plastics used are similar to the materials injection molded into barb fittings. With most compression fittings, soft, flexible tubing must first be supported by a brass insert to prevent wall collapse when the nut is tightened. Many sizes of compression fittings are available to accommodate a tubing range of about 1/8" OD to 3/4" OD. Most compression fittings are sized according to the outside diameter of the tubing lines they are to connect. Some manufacturers that deal with industries used to pipe sizes (inner diameter specifications) will naturally market their products sized with that nomenclature. Such is the case with plumbing or irrigation-oriented product lines. When crossing fields of technology, dual-sizing methods will be encountered. That is frequently the case in products used for Micro-Irrigation Systems, where parts have been borrowed from a number of other technologies. Aqua/Trends has attempted to simplify the situation by cataloging parts under a common nomenclature. Compression fittings can connect dissimilar materials. For example, plastic fittings can be used to interconnect various plastic tubes, as well as aluminum or copper. Brass fittings can also interconnect copper as well as plastic tubing.

The threaded fittings used in Micro-Irrigation System installations are generally hybrids, combining one or more legs (or branches) threaded with a standard pipe or hose thread and the other leg (or legs) with a compression, slip, or barb connector. Such fittings are commonly used to interconnect a pipe system with a tubing system; for example, connecting flexible tubing to a PVC sprinkler pipe. Threaded fittings are also used to adapt between fitting types. For example, a tee fitting might have a center branch molded with a pipe thread, onto which is screwed another fitting having a barb connector for small-bore tubes. Pipe threads are of two basic types, male and female. Male fittings have their threads on the outer circumference, while female fittings have their threads on the inner circumference (see Figure 10.16). Threaded fittings used

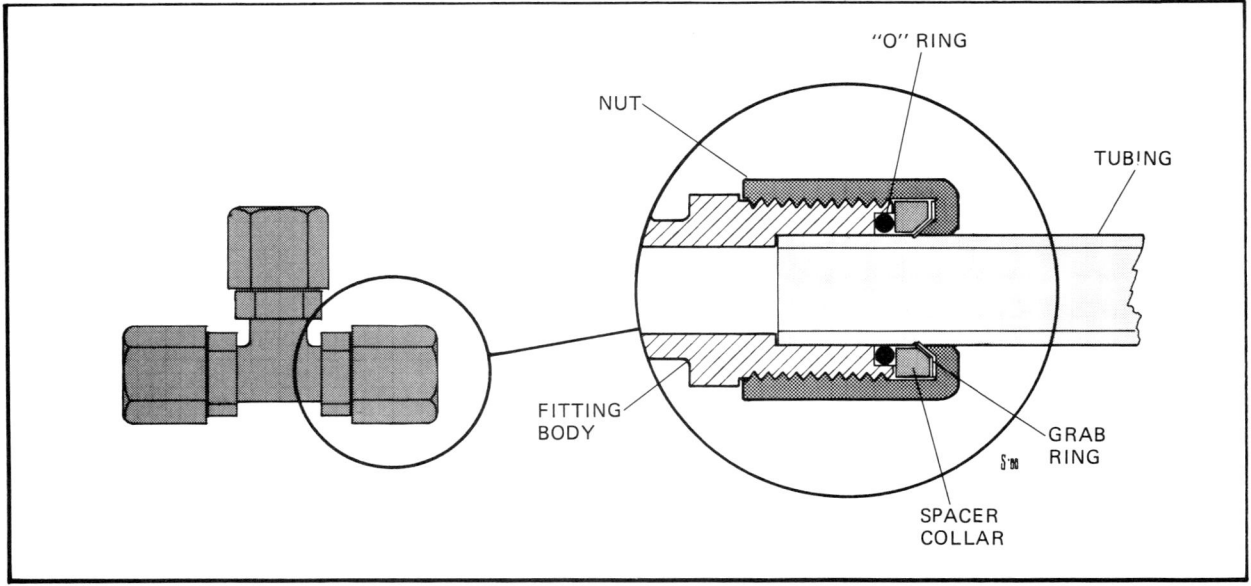

Figure 10.15 Typical style compression fitting used with plastic tubing.

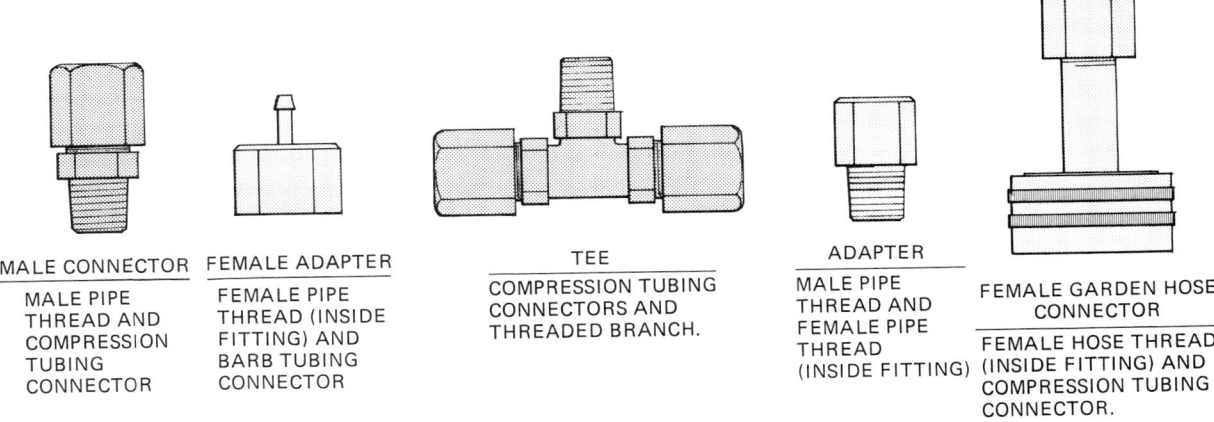

Figure 10.16 Typical threaded fittings.

in this technology are of plastic or brass. The better quality plastic fittings are molded of the harder materials, such as nylon, ABS, or acetal. There are many types of threaded fittings to fit a wide variety of requirements.

Soldered fittings (or "sweat" fittings) are used only infrequently in this technology. Their main application is in connecting an automatic plant-care system to a building's water line. A tee fitting is soldered into the water line, and a branch is extended from it leading to the irrigation system input fittings and controls. Soldered fittings are made of brass.

Solvent-welded fittings (or "slip" fittings) are used mainly in sprinkler system piping. The pipe is simply slipped into fitting sockets after appropriate solvent has been applied to the interface. It actually dissolves some of the fitting and pipe-surface plastic and, when dried, rehardens the joint. The bond is permanent, and when done properly, is watertight. These fittings and pipes are made of inexpensive PVC. In Micro-Irrigation Systems, they are used primarily outdoors in patio systems and indoors as manifolds in planter boxes and the like (generally in noncritical areas). CPVC pipes are also solvent welded.

There are many varieties and configurations of fittings, each having its own nomenclature. Anyone involved in the design or installation of Micro-Irrigation Systems should be well conversant with the nomenclature. The illustrations shown in Appendix B depict the various types used in installations of this technology . . . it is quite extensive. Of course, only a portion is applied in each installation.

Fitting Specifications   Fittings are specified by describing their type and the sizes of tubing or pipe that they interconnect, as well as the way each end makes the connection. Take tee fittings, for example. They are unique in that three connections are made to them, and a conventional description had to be worked out by the industry. The straight-through part of a tee fitting is called the *run*. The leg off to the side is called the *branch* (see Figure 10.17). Such a fitting is specified by listing the sizes of the two ends of the run first, then the size of the branch; for example, 3/8" OD tube × 3/8" OD tube × 1/4" OD tube for a compression tee fitting. The 1/4" part connects to a smaller side branch line. The 3/8" parts are connected to the main tubing, providing flow straight through the fitting (the run). Here is another example: 1/4" OD barb × 1/4" OD barb × 1/4" MPT. In this case, the run consists of barb fittings that slip into 1/4" tubing, while the branch is a

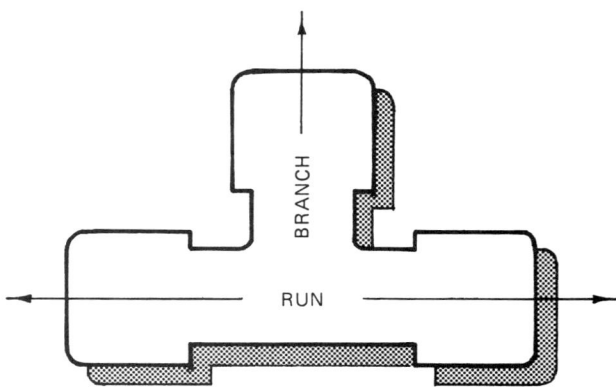

Figure 10.17 Tee fittings nomenclature.

1/4" male pipe thread that would screw into an adapter fitting of some sort, providing connection to another part. The same size and type of connectors all around a tee fitting ... 1/4" compression, for example ... would simply be specified as a 1/4" compression tee or 1/4" union tee-compression type. Similar expressions would be used for other types of fittings. Elbows with 1/4" compression fittings on both ends would be called a 1/4" ell compression fitting and would be suitable for interconnecting 1/4" OD tubing. All compression fittings are specified in terms of the outside diameter of the tubing to which they connect. Barb fittings are sized according to outside diameters by some sources and according to inside diameters by others. The larger fittings generally specify the latter, as that terminology is common in the markets they usually service. Careful attention must always be paid to the tubing and pipe diameters (outside and inside) so as to be able to choose the proper fitting.

The sizing of pipe threads is fairly straightforward. Male pipe thread sizes are designated XX" MPT or XX" MNPT. They both stand for "Male, National Pipe Tapered Thread," a convention standard. An example of the spec for a male pipe thread would read 1/2" MPT. That might be on the branch of a tee fitting or one end of an elbow fitting or male adapter, as shown in Appendix B. Female pipe thread specifications are similar. They are designated as XX" FPT or XX" FNPT. Alternate designations for pipe threads are sometimes encountered as Mipt or Fipt (male iron pipe thread or female iron pipe thread). Keep in mind that threaded pipes (generally iron) are not used in this type of work—only fittings and connectors that use pipe threads are used.

Flow Control Devices

As implied by the subject category, these devices control the flow in one manner or another in the water distribution networks of Micro-Irrigation Systems. Among them are shutoff valves, check-valves, vacuum breakers, flow reducers, and pressure regulators. They will be quickly and simply described in this section, along with their functional task.

Shutoff (or "stop") valves are used in high-pressure systems at the irrigation branch input or connection to the building's cold water line. Its purpose is to disconnect the irrigation branch from the rest of the plumbing in case of trouble. They are generally globe-type valves, which refer to the configuration of the moving part of the valve (the part that cuts off the flow). Shutoff valves are placed in-line with various means of connection—some with threaded fittings, some with barb fittings, and some with compression or soldered fittings.

Check-valves are small devices that are installed in-line at various critical locations in the water distribution system, permitting flow in one direction only. Although there are several types of check-valves in common use, the only ones used in Micro-Irrigation Systems are spring-loaded types. They employ a spring to close off the flow through the device when the forward-motivating pressure has been released (the pump stops or solenoid valve closes). Forward pressure opens the check-valve for flow, but when that has ceased, the spring closes it against backward (or reverse) flow, as well as any additional forward flow. Water can thus flow in one direction and only when sufficient pressure is applied. (see Figure 10.18).

In installations where tubing runs at various levels, without check-valve control, water drains by gravity to the lowest levels when pressure is released. Plants at the bottom of the installation will continue to get drain-off even after system shutdown; consequently, overwatering results unless a check-valve in the tubing lines prevents it. Check-valves also prevent unwanted forward flow when the system is dormant. Figure 10.19 is a fairly complex example of how check-valves can closely control the flow of water, making it do what is desired when it is desired, and keeping it from taking unwanted directions at unwelcomed times.

Imagine the figure without check-valves at Positions F, I, and K. During the brief operating cycle in which water flows to the plants, it is unimpeded, flowing only through tubing. In reaching up to the ceiling of the second floor, a distance of perhaps 16 or 18 feet, the water flows against gravity. As the irrigation cycle is completed, the water pressure drops to nearly zero within the tubing network. That permits the forces of gravity to take over. Water filling the portions of the tubing at Position E is most subject to movement as that is the highest point in the system. From Position E, the water will drain forward, down into the hanging planter at Position D. At the same time, some of the water from Position E and all of it filling the riser tube (two floors of it) will

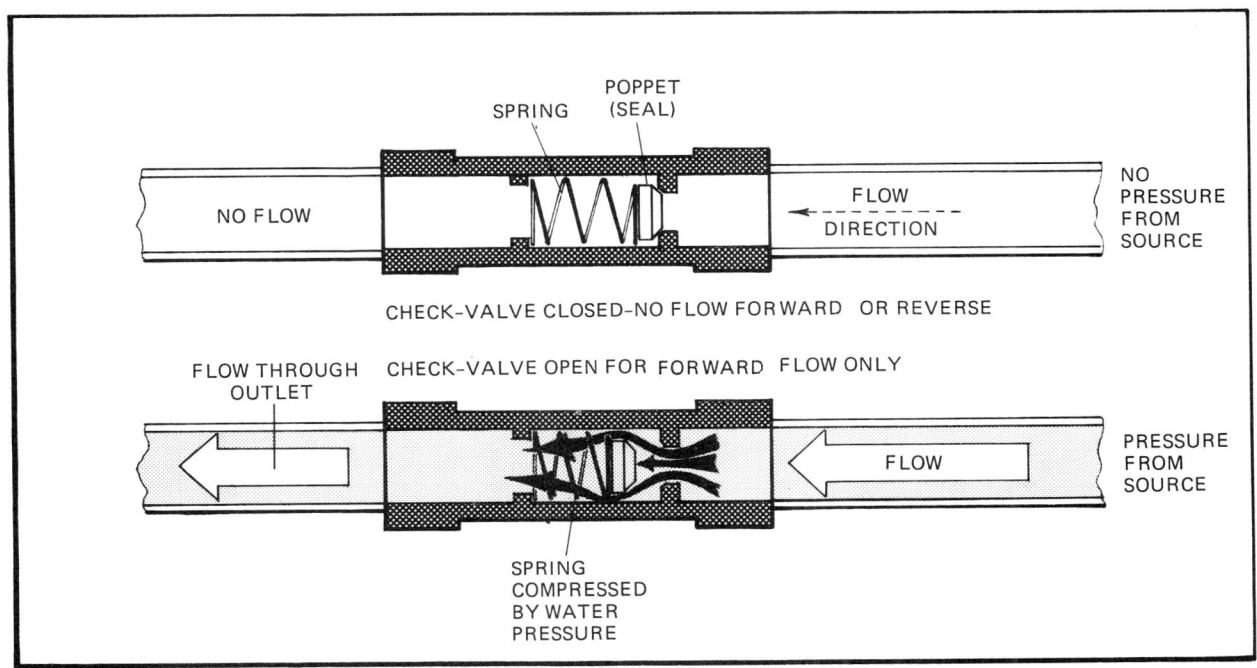

Figure 10.18 Operation of spring-loaded check-valves.

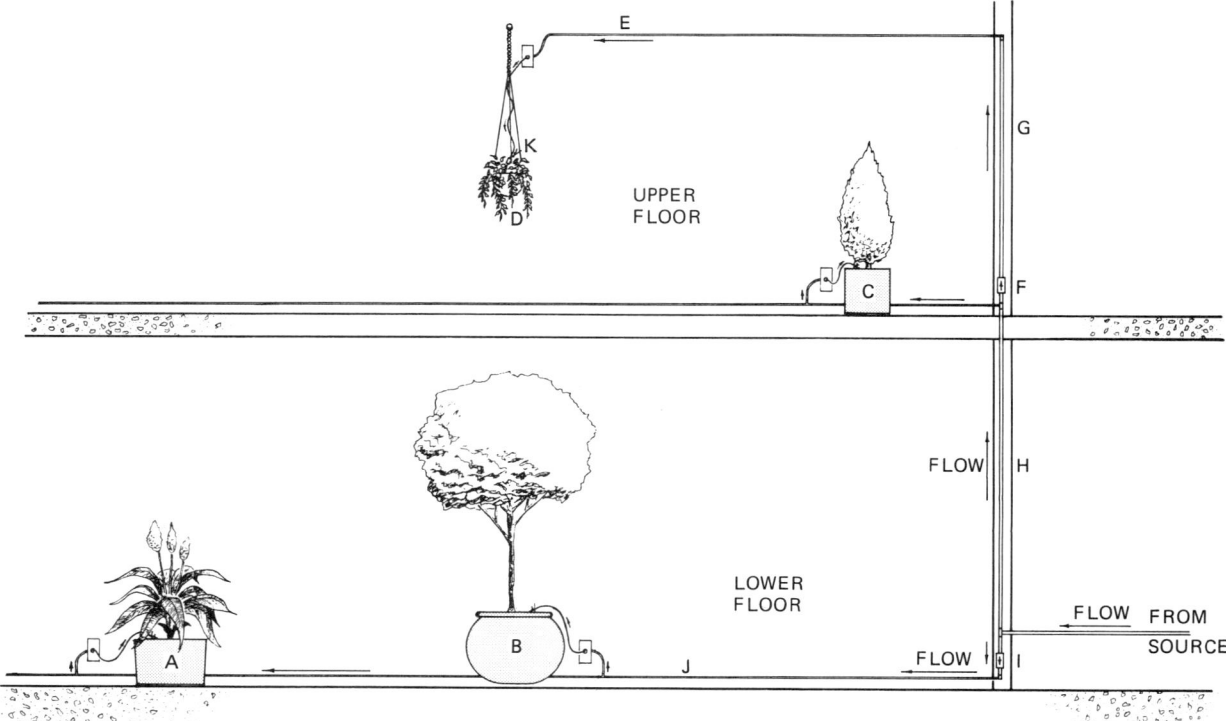

Figure 10.19 Typical application of check-valves in a Micro-Irrigation System installation.

drain backward down the tube into Planter C and then farther down into Planters A and B. Remember that, unless unchecked, water tends to flow to the lowest point in the system. The result is flooded planters and possibly flooring as well. Another unwelcome result is that the entire tubing system will have been purged of water. That means it must be refilled during the next operating cycle before anything can reach the planters. To solve these problems, we find ways to foil Mother Nature. Check-valves, when properly chosen, placed, and oriented, can hold water against the forces of gravity. In this example, we would install spring-loaded check-valves at strategic positions in the tubing network. One is placed at Position I, oriented with its one-way flow path pointing down. This will permit flow to the plants being serviced on the lower floor yet prevent backdrain into those plants after the pressure has been released. Another check-valve is installed at Position F, which is just above the branch servicing grounded planters on the upper floor. Its orientation has the one-way flow path pointing upward, so water can flow through it to Position E and beyond. Being above it, that check-valve can effectively prevent backdrain into Planter C. A miniature check-valve is placed in the small tubing line leading down the macrame hanger to Planter D. Its one-way orientation is down so that water can flow to the plant, yet the spring tension will be enough to hold afterflow from the planter when pressurization has ceased.

These functions also serve to keep water trapped in the tubing lines between watering cycles. That concept is very important in Micro-Irrigation Systems, as empty tubing must be refilled during the next operating cycle before it can irrigate anything. Air must be purged from the tubes as they refill. This takes time, and operating times are very short in this technology; so it is possible for the system to spend all of its operating cycle simply refilling tubes unless we design check-valves into the network at critical locations to prevent this from happening.

Other gravity-flow problems are frequently encountered that are simpler to solve, but they, too, would utilize check-valve flow controls. Figure 10.20 illustrates a common problem. Decorative plants are frequently placed on different levels in the same room. As water tends to seek its own level by means of the natural laws of physics, any water in the highest tubing would tend to drain down to the lowest point, including into the lower planter. That again causes overwatering of some plants at the lower levels and the probable underwatering of others at the upper levels. In Figure 10.20, by installing a miniature check-valve in the small tubing line at Position C, water would be permitted to flow to Planter B yet would block the gravitational backflow from Planter B to Planter A.

The use of check-valves will be discussed further in Chapter 11, which covers systems design.

Another type of flow control device used in Micro-Irrigation System installations is a vacuum-breaker or backflow preventer. It is a type of check-valve

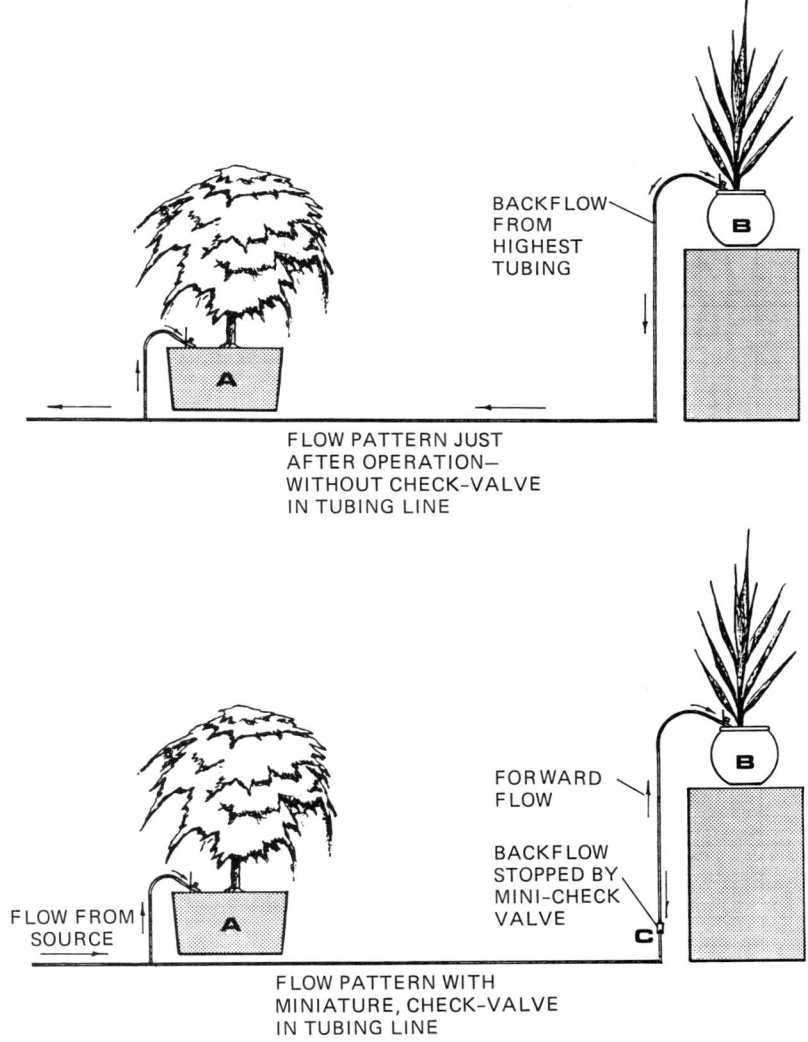

Figure 10.20 Typical use of miniature check-valves with plants on different levels.

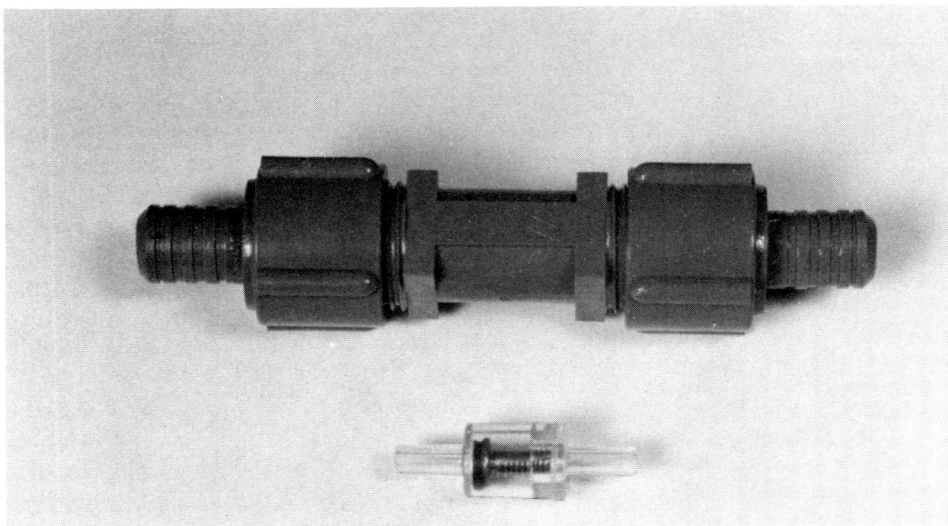

Figure 10.21 Two commonly used check-valves.

used in high-pressure systems to prevent water in the irrigation tubing from being sucked back into the building's main water supply, with the possibility of resultant contamination. This might be particularly important if fertilizer injection was being used or if dirt was allowed to enter the tubing lines through emitters. This protection is most often mandated by local building codes, which call for such devices to be placed in the system at the connection between the main water supply and the irrigation branch. We have all seen outdoor hose connections fitted with little cylinders on the hose bibb outlet. Those are backflow preventers. The hose connectors are just one type. Others connect in-line with other pipe fittings. Some versions are meant to be subjected to constant high pressure; others are for use at intermittent pressure. These factors determine their heft or construction, as well as their position in the water distribution network.

Flow reducers are small rubber or synthetic rubber discs that fit into other flow control devices, such as solenoid valves, to lower the flow rate of the system. Because interior plants need relatively small amounts of water for their sustenance, the supply lines are frequently oversized for the need at hand and permit too much water to flow. Supplementary controls, such as flow reducers, bring flow rates down to more manageable levels. Some types are designed to maintain a constant flow rate through broad fluctuations of input pressure. These pressure-compensating types are desirable in Micro-Irrigation System installations.

Pressure regulators are other devices used to bring pressure and flow rates down to manageable levels. Keeping in mind that with all other factors being equal, the higher the system water pressure, the greater the flow rate. That can be seen in Figure 10.22. As water pressure increases, so does the flow.

Although the figure demonstrates this relationship in 1/2" ID tubing, it is essentially the same with other tubing sizes and types, but at different levels. The input water pressure from building supply lines is in the range of 55 to 70 psi for most city water systems. That supplies the motivating force for quite a bit of water to flow through the system. Even by reducing the time of flow to very short periods, as we do in Micro-Irrigation System technology, the volume is frequently not restricted enough. Flow reducers can be used in some cases,

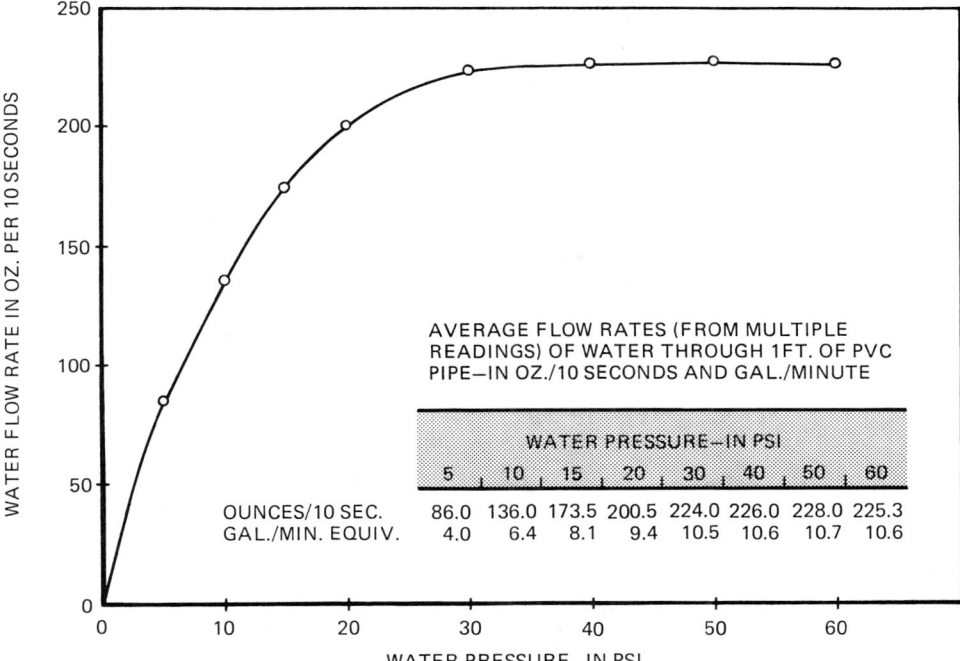

Figure 10.22 Pressure/flow relationship in ½" PVC pipe (1 ft. long) (empirically derived data—courtesy Aqua/Trends).

and pressure regulators in others. There are basically two types of pressure regulators: adjustable and fixed. The adjustable type permits the regulation of input pressures to provide a range of output pressures. The more common types permit an input pressure as high as 300 psi, and output pressures can be regulated between approximately 25 and 75 psi. The fixed regulators are preset at the factory for output pressures of 6, 10, 13, 15, 20, 25, 30, 35, and 45 psi.

Emitters Emitters are the "business end" of Micro-Irrigation Systems. They issue the flow of water to plants. Some are adjustable and some are fixed in flow rates. Adjustable types permit the fine-tuning of the volume of water going to individual potted plants. The Micro-Irrigation Systems by Aqua/Trends use a miniature adjustable valve as the emitter in most installations (see Figure 10.23). It permits very fine adjustments to be made at the plant location so that during each operating cycle, the plant can receive as little as a few drops of water to as much as several ounces.

The actual range of flow rates with any emitter will depend on the water pressure and flow rate of the system at the emitter location. That is not the same as input (or source) water pressure. Many things happen to the water during its journey from the irrigation branch input to the emitter locations, particularly in extensive installations. Every foot of tubing the water goes through creates a drag on it due to friction along the tubing walls and intermolecular friction within the fluid itself. Every bend and fitting connection does the same. A gradual reduction in the flow rate results—the farther down the line the water goes, the lower the rate. Each active outlet along the way reduces it even more. The data in Table 10.4 illustrates the range of flow rates at various pressure levels from Aqua/Trends' adjustable mini-valve/emitters. It gives a clear indication of the precision watering capability of this technology. As little as two drops of water can be dispensed during a 10-second operating cycle.

Figure 10.23 Adjustable mini-valve/emitter and stabilizer. (Courtesy of Aqua/Trends.)

That sensitivity makes the system useful for maintaining dry-loving plants, such as small cacti, or any other type of plant with low moisture needs for that matter. By adjusting the same emitter to its fully open condition, as much as 10.5 ounces can be made to flow during a 10-second irrigation cycle (at 65 psi), making it useful as well, for large plants in need of more generous doses of water. Most decorative containerized plants require an adjustment on the low end of the range.

Another type of emitter used in Micro-Irrigation Systems is the Laser Soaker Line, previously described in the section on subterranean systems in Chapter 8 and in the section on tubing in this chapter. Its use in this technology is mainly in planter boxes where the plants are grown directly in the ground and are fairly close together. It is also used to form a loop around the base of

TABLE 10.4

Emitter Profile—Flow Rates at Various Pressure Levels from Aqua/Trends' Adjustable Mini-Valve/Emitters

	Flow Rates		
Water Pressure	Emitter Almost Closed	Emitter About Half Open	Emitter Fully Open
3.3 psi	2 drops/10 sec.	N/A	4.0 oz./10 sec.
10 psi	2 drops/10 sec.	1.5 oz./10 sec.	4.0 oz./10 sec.
20 psi	2 drops/10 sec.	2.0 oz./10 sec.	5.5 oz./10 sec.
30 psi	2 drops/10 sec.	2.5 oz./10 sec.	7.0 oz./10 sec.
40 psi	2 drops/10 sec.	3.0 oz./10 sec.	7.7 oz./10 sec.
50 psi	2 drops/10 sec.	3.5 oz./10 sec.	8.6 oz./10 sec.
55 psi	2 drops/10 sec.	3.5 oz./10 sec.	9.2 oz./10 sec.
65 psi	2 drops/10 sec.	N/A	10.5 oz./10 sec.

Source: Aqua/Trends.

Accessories

Figure 10.24 Laser Soaker Line—as used around bases of containerized trees and shrubs.

trees and other large plants to distribute water more evenly in their containers (see Figure 10.24). In Micro-Irrigation Systems, the Laser Soaker Line is generally used in conjunction with an adjustable in-line valve to vary flow rates. With this regulation, it is capable of a wide flow range (see Table 10.5).

A third type of emitter used with Micro-Irrigation Systems is called a *fountain plate*. It is installed in the wall next to a planter and fed through a water distribution line installed within the wall partition. The stream of water traverses the space between the wall and the plant container (in the manner of a small fountain) without the use of a tubing conduit. It has many applications in heavily trafficked public areas both indoors and out (see Figure 10.25), particularly where vandals are likely to destroy tubing lines.

Accessories   Water distribution manifolds are an important accessory in irrigating planter boxes as well as outdoors in patio installations with this technology. They pro-

TABLE 10.5
Emitter Profile—Flow Rates from a Laser Soaker Line with Variable Pressures and Six-Inch Hole Spacing

Water Pressure	Flow Rate per Foot	
	Oz./20 Sec.	Oz./10 Sec.
5 psi	0.833 oz./ft.	0.417 oz./ft.
10 psi	1.25 oz./ft.	0.625 oz./ft.
20 psi	1.75 oz./ft.	0.875 oz./ft.
30 psi	2.08 oz./ft.	1.042 oz./ft.
40 psi	2.33 oz./ft.	1.167 oz./ft.
50 psi	2.58 oz./ft.	1.292 oz./ft.
55 psi	2.58 oz./ft.	1.292 oz./ft.

Source: Aqua/Trends.

Figure 10.25 Fountain plate emitter.

vide a multiple connection for emitter tubes, those short small-diameter tubes reaching the plant and fitted with adjustable mini-valve/emitters (see Figures 10.26 and 10.27). Manifolds are available factory prepared to save time and labor, but they can also be assembled on-site.

Special irrigation receptacles have been developed to provide convenient access to water distribution lines running within wall partitions in those integrated Micro-Irrigation Systems installed as part of the building infrastructure. They are mounted in wall partitions in much the same way that TV antenna plates or phone jacks are installed. Within the wall are connections to the water distribution tubing lines threaded through it. Into these special irrigation receptacles are plugged small emitter-tube assemblies that are directed to the potted plants on the floor and furniture nearby (see Figure 10.28).

At least one irrigation receptacle is installed on each wall for convenient access to all plants in a room. If necessary, receptacles can be installed in ceiling panels, movable partitions, office workstations, residential kitchen countertops, and vanity tops—in a wide variety of locations. Inactive receptacles are simply plugged until needed (see Figure 10.29). It is not necessary for these

Accessories

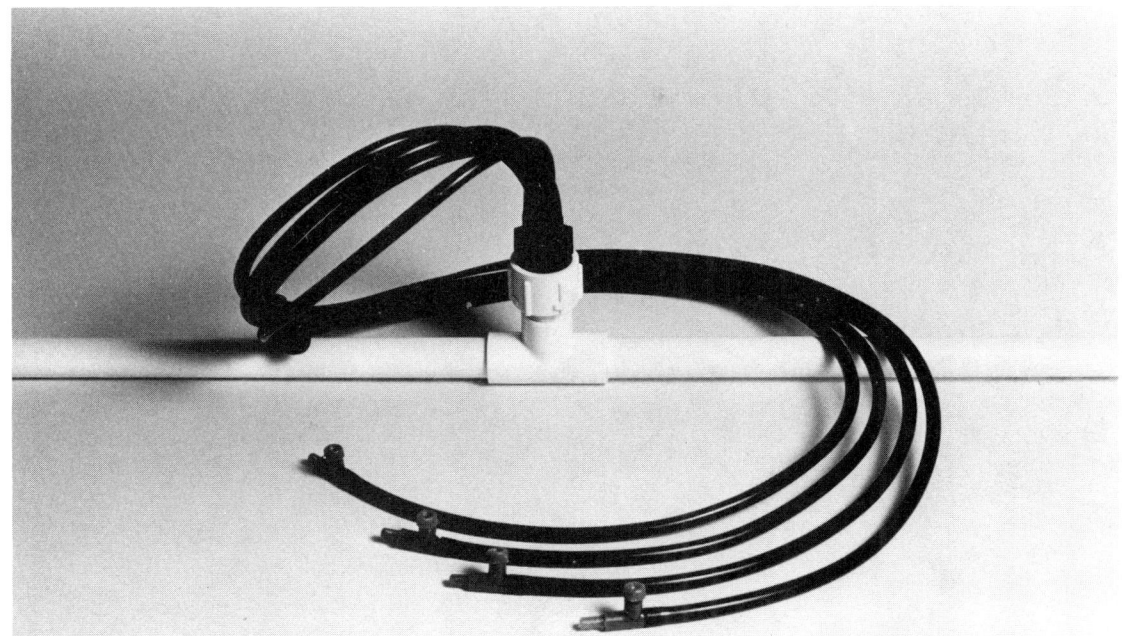

Figure 10.26 Irrigation manifold showing a manifold head, emitter tubes, and adjustable mini-valve/emitters. (Courtesy of Aqua/Trends.)

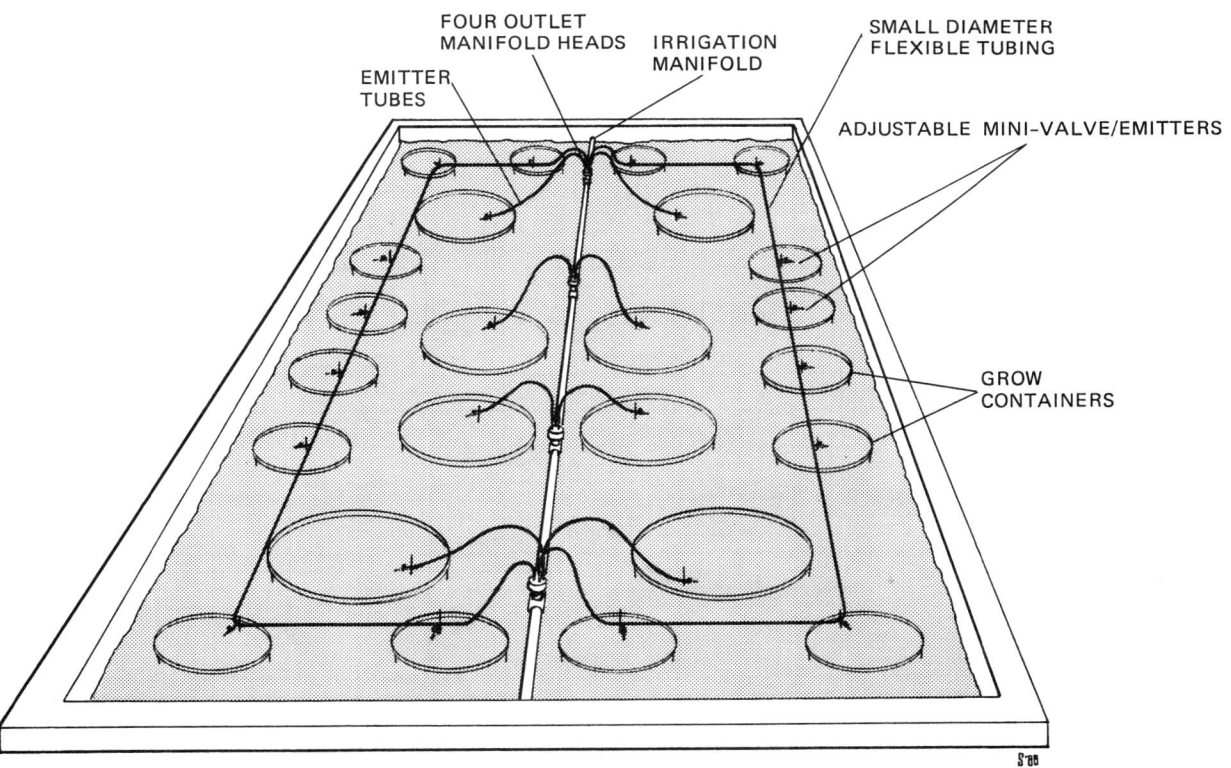

Figure 10.27 Typical installation of an irrigation manifold in a built-in planter bed.

Figure 10.28 Special irrigation receptacle with emitter-tube assembly.

Figure 10.29 Irrigation receptacle deactivated with a plug. (Courtesy of Aqua/Trends.)

Figure 10.30 Emitter stabilizer assembly. (Courtesy of Aqua/Trends.)

special irrigation receptacles to be placed wherever potted foliage plants will be located. Emitter tubes can be installed to carry water as far as 100 feet from the receptacle, hidden, where possible, behind furniture or under carpeting and around the perimeter of rooms. Each small receptacle can service as many as 12 to 15 plants.

Emitter stabilizers are used to hold emitters in place in the plant container. They consist of a stake that is poked into the soil and a small tubing clamp that slips onto the stake (see Figure 10.30).

Packaged Systems   Aqua/Trends offers a variety of prepackaged Micro-Irrigation Systems that contain all of the equipment, fittings, and accessories needed to meet a wide variety of installation requirements. By choosing the proper prepackaged system, all (or almost all) of the components needed for a job are included. In many cases, of course, an installation must be customized by purchasing equipment and parts individually, but packaged kits make it easier to purchase system components, since everything is preselected for compatibility.

11 AUTOMATIC IRRIGATION AS APPLIED TO INTERIORSCAPES

Overview It would be a mistake to say that no longer is there a reason for any interior plant to suffer through improper care. That would be an oversimplification of the facts, but with the advent of this new technology, we have come very close to that ideal. From a practical point of view, with proper advanced planning and design, a building can now be constructed with enough technology incorporated in its innards to automatically provide major care for virtually every decorative plant likely to be used within it, either at the time of construction or throughout its lifetime. That does not mean that human plant care is no longer required (we sincerely hope that fact was made crystal clear in earlier chapters), but the labor requirement can be drastically reduced. Also, from a practical standpoint, the technology is now available for installation in *existing* buildings, systems that would provide automated care to most decorative plants gracing their inner confines. In this latter case, the limitations are not those of technical capability but of practical and cost considerations.

Until recently, interiorscapers, interior designers, architects, and real estate owners and managers, as well as homeowners, thought of technical aids only in terms of systems for landscape plant outdoors and in open, public areas of building interiors. Only agricultural and landscaping irrigation systems have been available for use indoors, and the adaptations have lacked control as well as the esthetic sophistication to meet the special needs of furnished residential and commercial areas. Precision Micro-Irrigation Systems™ now provide the basic technology and esthetic attributes around which comprehensive automated plant-care installations can be designed. Combined with other technologies, when necessary, the ideal of complete interior coverage can be realized. This chapter will discuss how the most common and needy interior spaces can benefit from this state-of-the-art building and facility management technology.

Commercial Buildings and Facilities It is within the commercial sectors of our society that automated building management systems have the greatest economic impact. That is particularly true of automated interior irrigation systems. Outdoor sprinklers have permitted the design of extensive and beautiful landscaping around commercial and institutional buildings, minimizing concern about the cost and bother of manual irrigation for the turf and foliage plantings. Most extensive landscaping would, in fact, not have been possible to maintain cost-effectively today were it not for automatic sprinkler systems and would therefore have been scratched from many project plans. The beauty and esthetic harmony they now engender would have been missing from our lives. In a similar way, much of the beauty and commercial benefit of indoor plantings has been passed over for lack of a viable maintenance technology. Sprinkler and drip systems have been applied to planters in open common areas, and that has provided many projects with the impetus needed to make these interior landscapes practical. Most other areas of the building have been neglected, however, because of the absence of a practical means to cope with their special demands. That led to the use of self-watering containers in hallways, lobbies, office suites, and other furnished building sections. The limitations of these devices were soon recognized, and the need for fully automatic systems persisted. Precision systems are now available to fill that void, and practical, complete coverage of building interiors can be effected.

Commercial buildings and facilities present many challenges to the designer of interior irrigation systems; there are so many permutations involving different needs for office building lobbies, hotel suites, corporate office suites, restaurants, shops, etc. The one thing they all have in common is the need for a hidden, highly controlled system for dispensing water to a wide variety of scattered plants, most of which are in relatively small containers. In some cases, one technology can't do it all, and combinations must be installed. This is particularly true of retrofit (add-on) projects where the cost of installing some of the hidden tubing for a central system may be considered prohibitive. In these cases, nonautomatic technologies (such as self-watering containers) must then be utilized to care for a portion of the installation.

Because of its advanced attributes, Micro-Irrigation Systems can provide the major functional capability to most commercial projects. In many situations, nothing else is needed for complete building coverage. In other cases, independent systems can be installed in a needy portion of a building to service a restaurant, shop, bank, fitness center, etc. Building developers that want to take full advantage of automatic interior irrigation will have an integrated version of Micro-Irrigation Systems designed into the project, which will provide coverage for all public areas as well as all furnished tenant areas. The illustration in Figure 11.1 shows one way Micro-Irrigation Systems can be integrated into the structure of a building, with branches reaching every corner, if needed.

Aqua/Trends calls these integrated versions, Mirage III Systems™. They feature special *irrigation receptacles* installed on the surface of the walls, ceilings, and cabinets of rooms, much like TV antenna plates or phone jacks, to provide convenient access to the irrigation water distribution lines threaded through wall and ceiling spaces. Into these receptacles are plugged emitter-tube assemblies, small-diameter tubing that is directed to the potted plants on the floor and furniture of the room. An adjustable mini-valve/emitter is provided at each plant for fine control. Figure 11.2 illustrates how an integrated Micro-Irrigation System might be laid out on each floor of a commercial (or residential building). Extensions from this basic layout can generally be made to reach portions of another floor or exterior areas, such as decks, patios, balconies, etc. Branches can also be designed to care for planters in open public areas of the building. Micro-Irrigation Systems are therefore multifunctional and able to accommodate furnished tenant areas as well as open common areas.

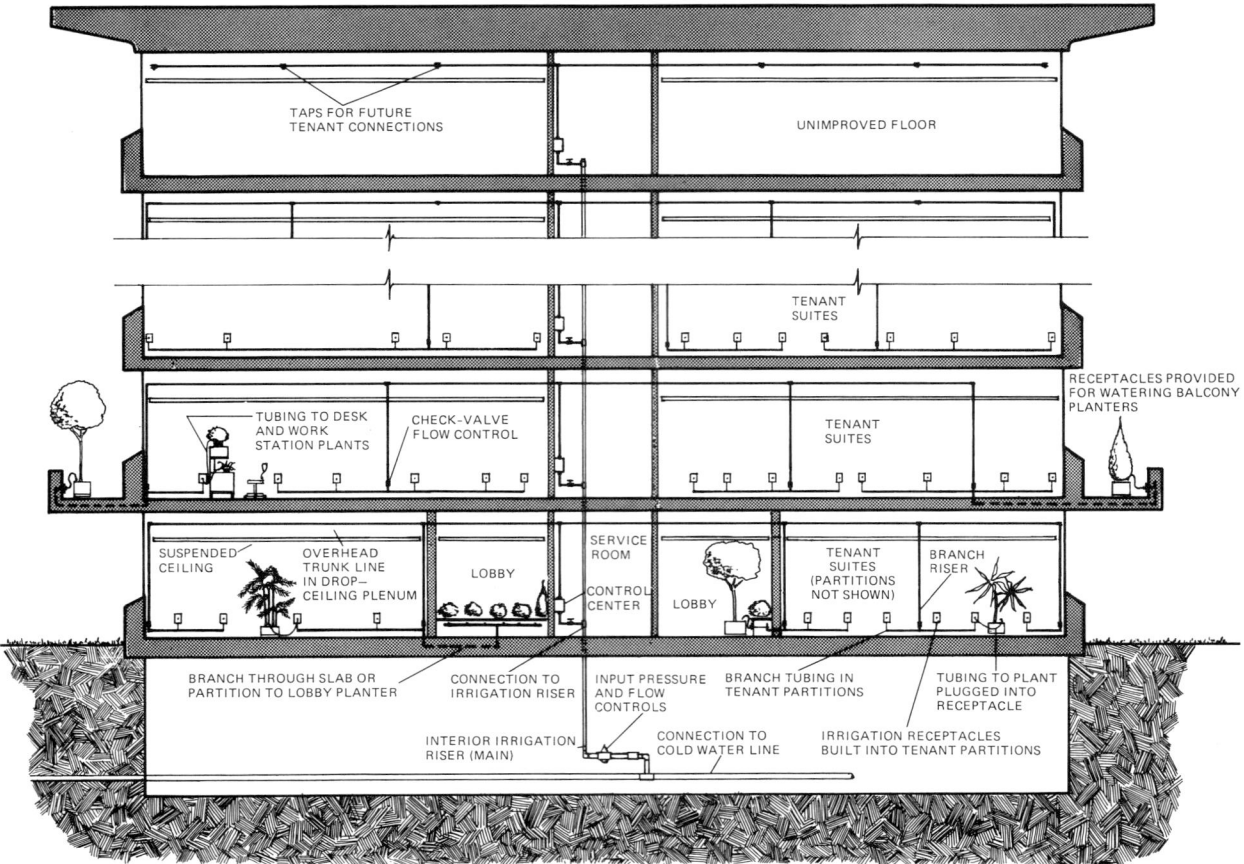

Figure 11.1 Multistory office building showing typical Micro-Irrigation System tubing distribution. (Courtesy of Aqua/Trends.)

Office Buildings   Office buildings harbor very diverse plant-care requirements, and there are many categories of this commercial real estate, so the ramifications are extensive. All office buildings have an entryway and foyer or lobby or reception area. Not all have plants in these public areas. Many are stark to the point of looking uninviting, without any human appeal. Some of the newer building projects, however, are being constructed with built-in planters in the lobby areas to provide a natural, softer look to the building's greeting place, attempting to provide beauty and a welcome, relaxed feeling to the building's tenants or visitors as they come in the main entrances. Some architects go further and design extensively verdant public areas that carry traffic through the building or provide resting or dining areas. Many of these are in the form of atriums and promenades. Planter boxes and pits are used extensively in these situations, with free-standing containerized plants rounding out the interior landscape. The more enlightened architects, developers, interior landscapers and property/facilities managers have begun to specify some form of plant-care automation for these installations to provide a cost-effective management tool, as well as to help assure the preservation of the interiorscape's beauty. Reliable plant maintenance services are not always at hand. Because sprinkler, subterranean, and various configurations of drip systems were the only technologies available, they were the ones specified. From a practical viewpoint, these versions of auto-interigation can be quite acceptable for many common-area planters, if they are **de-**

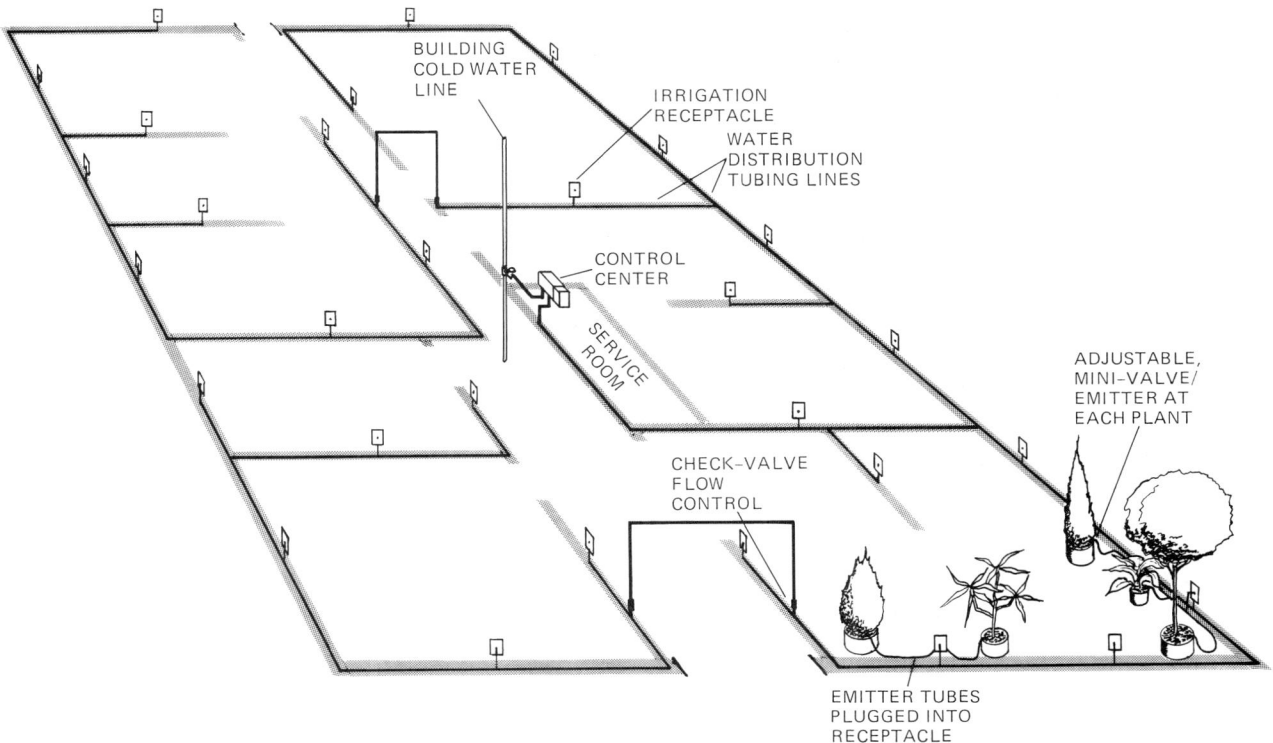

Figure 11.2 Typical layout of a built-in (integral) Micro-Irrigation System on one floor of a structure. (Courtesy of Aqua/Trends.)

signed, installed, and used properly. These installations are more or less standard irrigation systems with slight variations to accommodate the interior setting. The supply is either city water or a proprietary well. The simpler systems are merely branches of the outdoor sprinkler network, generally set up as a separate timed zone to keep the irrigation cycle as short as practical with the timer available.

More complex systems in large buildings have their own interior systems. Controls are installed outdoors or in a core area of the building, with piping running underground from there to the planter locations. At the input, there is a shut-off valve, a backflow preventer, a filter, a pressure regulator, a solenoid valve, and a time clock/controller. In the more complex system designs, the latter is a multizone type that is able to control a number of irrigation zones sequentially. Because of the large volumes of water possible through these systems, properly drained planter boxes are a necessary preventative measure. In spite of their shortcomings, these automated agriculturally based systems are capable of considerable labor savings and will quickly pay for themselves.[1] Their use is still the exception rather than the rule in commercial building projects, but the application rate is increasing as more architects and interior landscapers become more comfortable with their design, use, and benefits. It is incredible to consider the number of foliage plants and trees being installed in large commercial buildings without the benefit of modern irrigation. These installations will suffer with high maintenance costs throughout their useful lives,

---

[1]Stephan Scrivens, *Interior Planting in Large Buildings*, (New York: Halsted Press/John Wiley and Sons, 1980), p. 79.

not to mention occasional plantscape deterioration due to inadequate manual care. It must be remembered that with technical aids available, manual irrigation is the least reliable of all methods.[2]

As we've discussed previously, the other areas of the building have been neglected because refined technologies for service to furnished areas have not been available. With the advent of Micro-Irrigation Systems, that has changed, and building owners or managers now have the option of providing automated plant-care services to their tenants as well—a cost-saving amenity. In times of slack commercial real estate markets, this gives the more progressive developer a leg up on the competition. In large office buildings, most of the decorative plants are upstairs in the furnished suites and not in the first-floor lobby, and that is where most of the technical assistance is needed for their maintenance. Micro-Irrigation Systems can now provide that in small office facilities or modern office towers scraping the clouds.

The most cost-effective way of incorporating Micro-Irrigation Systems into buildings of this type is to integrate them into the framework during construction. The source is usually the city water supply with a special irrigation branch (or branches) tapped off of the main water line. Multistory buildings are accommodated by installing a separate, independent system on each floor of the building. The details of this arrangement will be discussed more fully in the chapters on design and installation, but suffice it to say that it provides a more flexible system than trying to incorporate one huge operating system. The controls are installed in the core area of each floor and the tubing is distributed from there through the partitions of the individual office suites, restaurants, shops, etc. Control can frequently be integrated with the building's computerized energy management system, so that time switches are not necessary to energize the individual Micro-Irrigation System branches.

The construction of office buildings is mostly done on speculation, with only minor parts of the building leased prior to completion. During these periods of overconstruction and excess of real estate inventory (the condition we currently face), banks and other construction lenders require larger portions of the building to be pre-leased before they will grant a construction loan; but under the best of circumstances, seldom does a new building face more than 40- or 50-percent occupancy when it is completed. For these reasons, some developers are reluctant to expend the capital for automation or tenant amenities of any kind that might have to wait years for initial use. Partial installations of automated plant-care systems are made in these instances. Low-cost *skeleton* installations are made in the building areas that will be vacant. These consist of control centers (timers, solenoid valve controllers, and input fittings) and a roughed-in piping network, which consists of main lines installed overhead in what will become the drop-ceiling plenum (see Figure 11.3). These pipes are fitted with plugged outlets for later connection to tubing networks that would service office suites when they are finally constructed in the vacant space. Other vacant areas of the building are handled in much the same way. For example, those that would eventually house banks, shops, restaurants, or any other commercial establishment likely to use live plants in their decor would benefit by having an automated irrigation system available for their use.

Pre-leased areas would, of course, be designed with automated Micro-Irrigation Systems integrated into their partitions, with special irrigation receptacles mounted on the walls, floors, ceilings, and/or built-in cabinetry of the rooms and ready to plug into with emitter tubing that would run to the various

---

[2]Stephan Scrivens, *Interior Planting in Large Buildings*, (New York: Halsted Press/John Wiley and Sons, 1980), p. 47.

Office Buildings

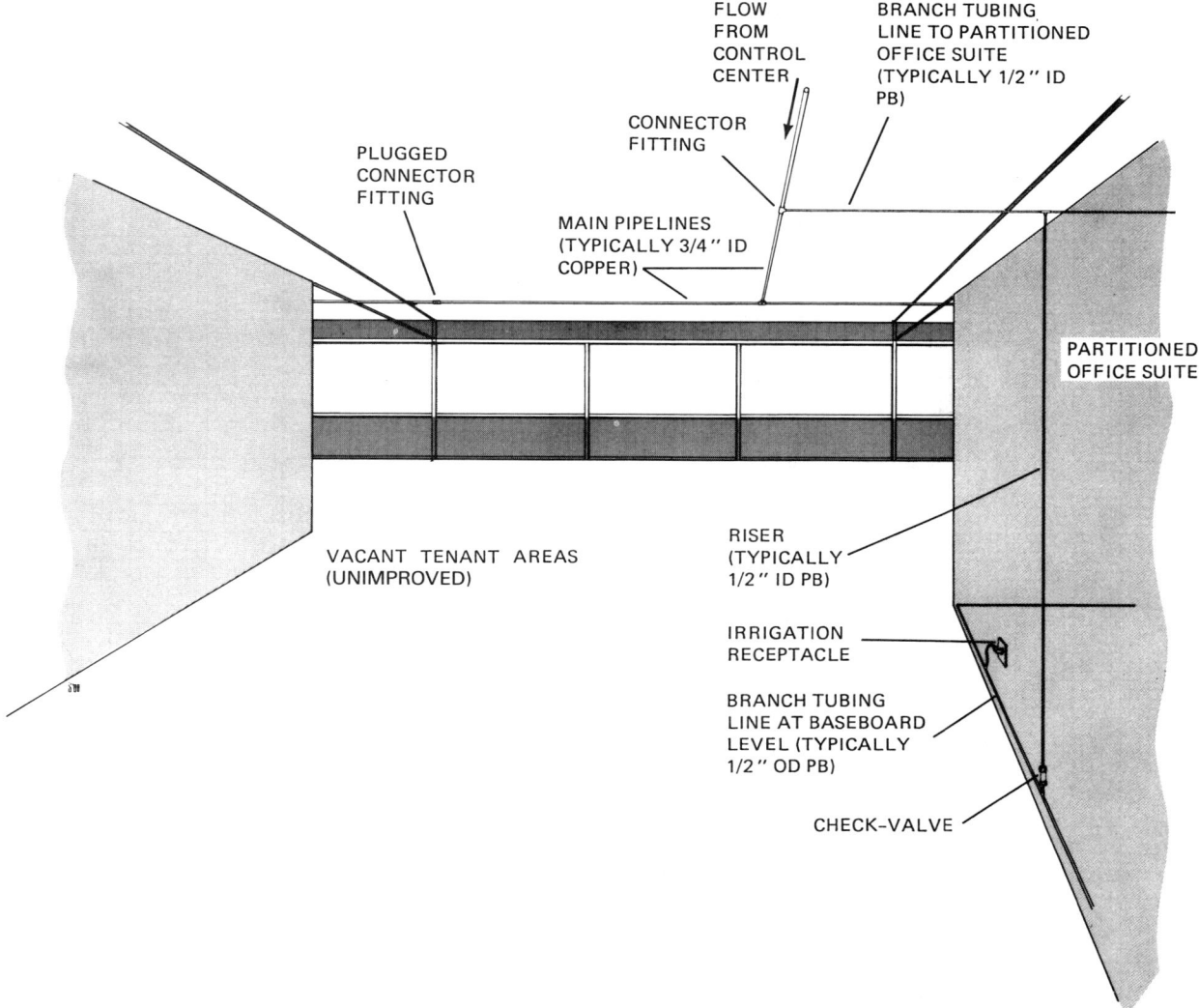

Figure 11.3 Typical skeleton installation in vacant areas of a commercial building.

container plants nearby. Keeping in mind that the decorative plants do not have to be placed next to or even near the irrigation receptacles, the tenant space can be designed and constructed without knowledge of final decorator schemes and still obtain full benefit from the system. Plants can be placed anywhere and moved at any time.

Building developers may pre-install Micro-Irrigation Systems into their projects and offer the automated plant-care services to tenants as an option, with additional monthly charges added to the lease arrangement. The tenant, in turn, benefits from being able to save generally between 30 to 60 percent on the cost of contract plant-care services. Need for contractor visits can be extended from every week or 10 days to every two, or even three, weeks. That means quite a labor savings to the contractor, and a good bit of this should be passed on to the customer—the building tenant. Unfortunately, many interior-scapers are still hostile to all manner of modern irrigation systems and will offer excuses for not being able to extend their maintenance cycles, in spite of advanced expertise to the contrary. Enlightened, cooperative interiorscapers can always be found who will work with automated systems. They enhance their professional image by being able to work with state-of-the-art plant-care systems. Some developers have provided supplementary plant-care services for their tenants by hiring interior landscape people to work in-house. The building services can then include plant leasing and maintenance, along with the other amenities. Building management companies and building maintenance contractors can do the same for their clients. Automated Micro-Irrigation Systems make these complementary real estate services practical.

Another plan for including automated Micro-Irrigation Systems in projects involves the building owner charging the first tenant for the cost of installation in that office suite. It can be a one-time charge. The tenant can recoup that investment, usually during the first year or so, through monthly savings on plant-care contracts. The overall savings during the course of a five-year lease can be significant, and the savings over the next five-year lease can be even more significant, as the cost had already been amortized during the first five years.

Tenants realize other benefits as well. For an office manager working on a fixed budget, more plants can be installed and maintained for the same amount of money, thus improving the decor, augmenting the corporate image, and providing an environment more suitable for employee efficiency, health, and satisfaction. Conversely, when a fixed number of plants is suitable, the cost of leasing and care is also reduced. The manager and company executives will also find:

- Fewer interruptions by plantscape technicians.
- More consistent plant-care, as less skill is required; there is also less foliage deterioration if the contractor "misses a beat" due to employee illness, vacations, or turnover.
- Less office mess from the inevitable spillage that manual irrigation engenders.
- An enhanced corporate image due to the high-tech facility management techniques.

Not all tenants would opt for the new amenity. Where pre-installed, however, it would be in place to offer the service to new tenants when space turnover occurs. Most office space turns over many times during its useful lifetime, and each tenant has a different set of needs. The automated plant irrigation systems would be useful to some, but of only marginal interest to others. The fact that it was in place provides the building with versatility and a market appeal that less-sophisticated structures lack. The inclusion of technology increases the value of the building, as is the case with any other investment in capital equipment. By leasing the equipment or paying for the initial installation up front, tenants are buying systems for the owner and, in doing so, adding to the value and marketability of the building. Tenants augment net worth for project investors.

An important benefit of integral Micro-Irrigation Systems has to do with the fact that *comprehensive* interior irrigation systems can be designed around them to care not only for the tenant suites but also for the common areas: those portions of the building normally relegated to sprinkler and drip irrigation (that is, when automation is used at all). The new technologies can do these other chores better than the old. Because most of the system is upstairs in the main part of the building, the common areas are then secondary irrigation zones. They can be branched off from the main system for their role in automated building maintenance. Lobbies, atrium balconies, eateries and relaxation areas, elevator landings, hallways, promenades, and passageways are just some of the common areas that are targets for interior plantscaping and automated care. Coverage of the building's administrative and sales offices are other important applications for the developer.

Partial building coverage is also sometimes considered. There are projects designed with highlight areas that are to be the center of attention. These areas are usually well decorated with plants and frequently fitted with accent lighting. Smaller zone systems may be considered for these applications, giving limited-area coverage. Occasionally, only one or two tenants in a small office building may require automated plant care. The developer may decide to install integrated systems only in those isolated suites to accommodate those tenants. Sometimes, the responsibility for installation is left up to the tenant—subject, of course, to approval by building management.

Automated Micro-Irrigation Systems can also be retrofit into office buildings, but the costs of installation are slightly higher because of the greater difficulty in threading tubing networks from one area to another. Retrofits are technically feasible and, in most cases, practical. Since each installation is different, they must be evaluated on a case-by-case basis.

There are many outdoor uses for Micro-Irrigation Systems in office building projects. Many are designed with balconies off the upper floors or patios extending from the main floor. Frequently, these are landscaped with potted trees and bushes. It is difficult to water foliage with conventional irrigation systems under these conditions, but the new technology is well suited to this because of its precise placement of water and its low flow rates. Another application outdoors is the servicing of hanging planters sometimes used overhead under covered walkways between buildings and canopied entryways.

Figure 11.4 Retrofit installation in the atrium of a small office building.

Corporate Offices Corporate office managers handle the use (or disuse) of plants in a multitude of ways. Plants are becoming more appreciated for their ability to enhance the decor and provide a more natural and relaxing workplace. More recently, industrial psychologists and NASA researchers have found employee efficiency and health benefits as well. Most office managers hire an interiorscaper to do the office landscaping and take care of the plants too. This is generally under a lease/maintenance contract, but sometimes the office owns the plants and contracts for maintenance only. In many cases, it is the interior designer who specifies the use of plants to complement the newly established decor.

Containerized foliage plants are used in a multitude of ways in entryways, reception areas, hallways, executive offices, sales offices and display rooms, conference rooms and boardrooms, open work areas, lunchrooms, relaxation areas, fitness rooms, and rest rooms. The plants hang from ceilings and beams. They are placed on filing cabinets, desks, workstations, bookcases, partitions, wall units, end and coffee tables, kitchenette countertops, and on the floor amongst the furniture to give color and break up the starkness of the architectural and interior design elements. They also add a bit of brightness and the calming effects of nature's gifts.

Such diverse settings for potted plants create many challenges for the installation of technology to care for them. As already mentioned, the vast majority of potted plants are still watered manually. The labor expended in this endeavor around the world is enormous. Micro-Irrigation Systems are able to reduce related effort and costs to minimal levels, but the practicality in any given situation will depend on the ease of installation. Sprinkler and drip systems are not practical in these settings and thus cannot be considered. Micro-Irrigation Systems can make up the major portion of a plant-care installation, supplemented with self-watering plant containers wherever needed.

In buildings that have already been fitted with an integral automated plant-care system, there is usually a connection nearby to tap into and route tubing to the office suite for irrigation service. These connection points are generally overhead in the space above the suspended ceiling. Most modern offices are constructed this way. These large overhead spaces are used for routing air-conditioning and heating ducts, wiring, plumbing, and a multitude of other things. That overhead plenum also provides the convenient opportunity for us to retrofit Micro-Irrigation Systems. Tubing and fittings are installed into that space and dropped down into office partitions to be further spread out to other locations (see Figure 11.5). In such cases, the control center on the floor occupied by the office suite will regulate flow to it and other neighboring suites.

Emitter-tube assemblies are plugged into the special irrigation receptacles and routed to wherever the potted greenery happens to be. This works well in enclosed offices but becomes more of a problem in open-plan work spaces. These applications can be handled in other ways (see Chapter 12). There are times when it is not practical to route tubing across flooring; for example, across tiled floors. Consequently, self-watering containers become the technology of choice to take care of isolated plants. While Micro-Irrigation Systems are technically capable of servicing planters positioned like this, the tubing installation is best accomplished before the concrete floor slab has been poured. Irrigation receptacles can be installed in the floor at planter locations. To do it later as a retrofit would be much more difficult and expensive. Systems of this type can be added more easily when floors are renovated.

There are also occasions when passing tubing under carpeting to workstations in open-plan areas is not practical. The passage of electrical or data processing wiring and other services to these island work areas is complicated

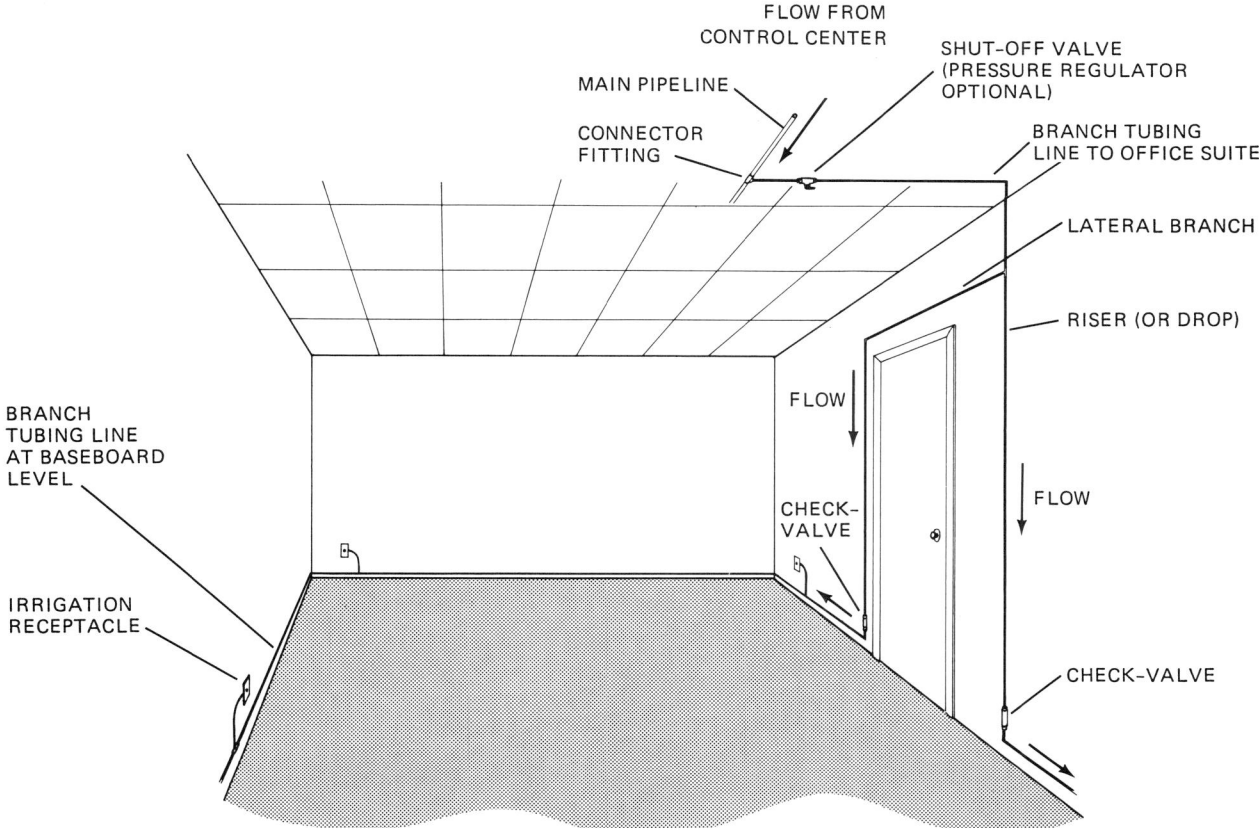

Figure 11.5 Typical Micro-Irrigation System installation in an office suite.

by the difficulty in passing wiring (or tubing) under the carpeted passageways without creating an unsightly bump or ridge on the floor. This is particularly true where commercial carpeting is used. It is cemented to the concrete floor surface. One solution is to install small, inexpensive low-pressure Micro-Irrigation Systems into the structure of the workstation furniture to service the plants on it (and around it), as shown in Figure 11.6. Each of these units is independent of the main building system and would be capable of servicing all plants on workstations up to four modules in size. Tubing is threaded through the furniture partitions from one location to another. The equipment can be installed prior to furniture delivery by the dealer or by the office furniture manufacturer. Irrigation receptacles can be mounted in desk, cabinet, and partition surfaces for a neat, convenient installation. Techniques are available to provide safe, leak-free systems. In this way, all workstation plants can be watered automatically without overflow and with only infrequent attention.

In buildings that are fitted with integral Micro-Irrigation Systems, tubing can frequently be plugged into wall-mounted irrigation receptacles at the perimeter of the open workroom and passed under carpeting to island workstations where they are connected to another small hidden network of tubing that distributes the water. In some of the newer office buildings, floors have been raised several inches above the concrete floor slab, which creates a floor plenum used for service wiring and, of course, tubing runs. This arrangement produces a

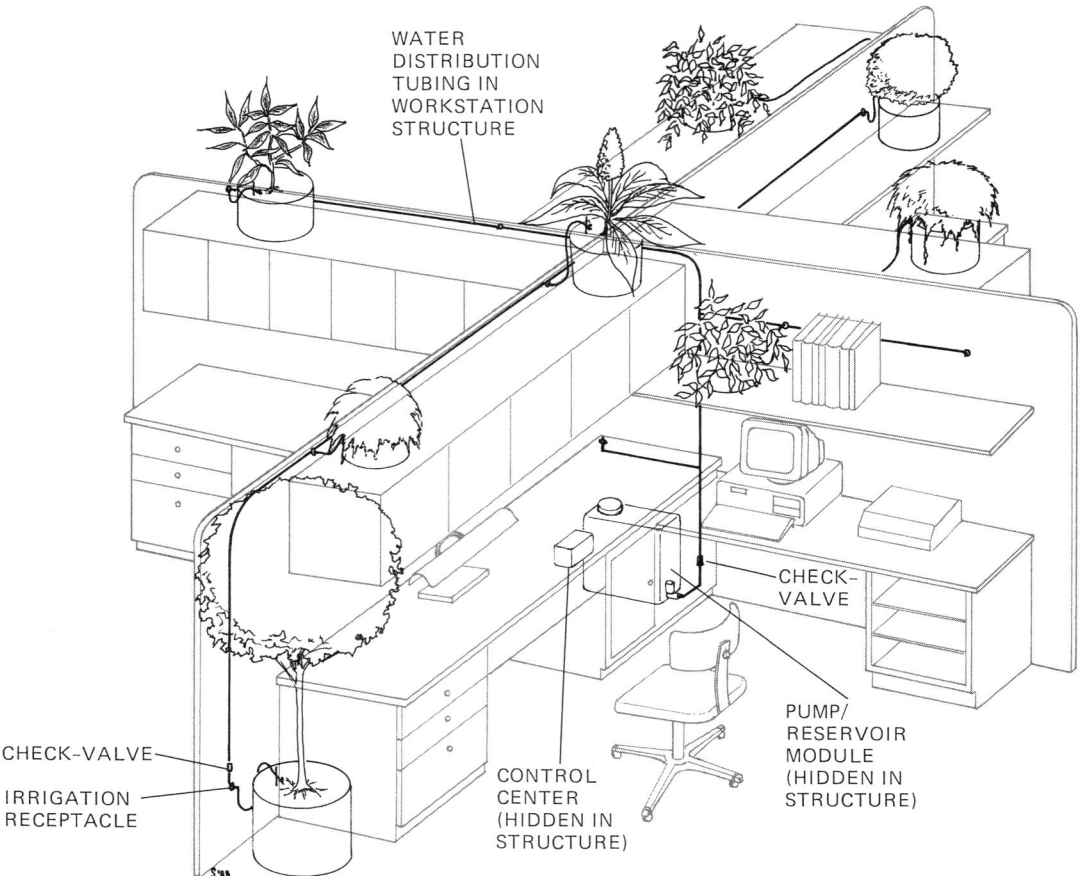

Figure 11.6 Typical layout for multiple workstations serviced by a small Micro-Irrigation System.

high degree of flexibility as well, because laterally run tubing is only semi-permanent and can be removed and rerouted when the floor plan is changed.

Another practical technique in offices is the use of carrier-frequency remote-control systems to time the interval between operating cycles. They are simple, inexpensive energy management systems that inject electronic control signals onto the office's electrical wiring. Receiving modules are plugged into power receptacles wherever lights or other devices are to be switched. They are also capable of controlling many satellite Micro-Irrigation Systems. These inexpensive units are installed where needed around the office suite.

In those situations where tubing runs are not practical, freestanding self-watering containers can frequently be used to supplement the main irrigation system. The self-watering containers are most useful in these settings as floor-standing planters.

Hanging planters are frequently used around offices and can be easily serviced by Micro-Irrigation Systems installed overhead through suspended ceiling plenums. The precise control of this technology prevents overflow problems common to manual watering service. These concepts will be discussed more fully in Chapter 13, covering installation techniques.

Hotels There are profound changes taking over the hotel segment of the hospitality industry. Recent industry studies uncovered some of the following trends nationwide. The list[3] has been edited for relevance to this discussion.

- A significant increase in the variety and diversity of facilities, including all-suite hotels, new mid-priced developments, concierge floors, and convention and airport hotels.
- More impressive public spaces than in the past—hotels are incorporating atriums, expansive spaces, and upgrading furniture to create a stronger image.
- Variety in room design with different kinds of amenities to focus on the needs of business travelers, 25 percent of whom are now women.
- Emphasis on high-touch in rooms, meaning basic physical comfort supplemented with visual comfort and good taste.

Live planters are a part of these trends, and automated plant-care techniques are inevitably following to help provide more high-touch in guest rooms and public spaces. Some in the industry like to look at it as a serene oasis from the frenetic environment outside.

In many ways, hotels are similar to office buildings with respect to the application of Micro-Irrigation Systems. All hotels have a lobby or at least a foyer in which to greet visitors. Many modern hotels are designed with quite expansive lobbies, and in many, a major atrium serves the multipurposes of lobby, elevator landings, food service areas, lounges, and art and commercial showcases, as well as access passageways to shops and guest rooms. Most lobbies, large and small, are planted with live foliage, some very profusely. Combinations of in situ (built-in) planters and freestanding containerized planters are normally used. Large landscaped areas demand the built-in planters. If properly designed, they will have been fitted with automatic irrigation systems of some type, generally sprinkler or drip technology. Fully automated plant care stops there.

Other areas of the building are normally scattered with freestanding potted plants to break up the starkness of plain, long hallways or elevator landings. They are used quite extensively in hotel restaurants and lounges to create the relaxing environment guests desire (and seek out). When one thinks of these things, the image that slips into mind is the old Victorian hotels decorated with the ubiquitous potted palms in the lobby and lounge. Even then, the hoteliers recognized the value of live plants to soften the environment and make the place seem more hospitable. In fact, the hospitality industry was the first to make extensive use of live plants in their interior environments. Other commercial industries have seen similar benefits and followed their lead.

The average guest room, however, was seldom graced with live plants in spite of their homey image, until more recently that is. With the advent of the so-called all-suite hotel concept, where the buildings are designed with mostly guest suites rather than just single rooms, the use of live potted plants has gained more popularity. The reason seems to be that the suite more closely represents one's home away from home, and the live greenery adds an element of relaxation and homelike decor. One of the reasons for the slow application of plants in guest rooms has been the difficulty of gaining access to the plants for routine watering and maintenance. Hotel guests cannot be disturbed by

[3]Ann Nydele, "Future Trends in Hotel Design," *Buildings* (March, 1988), p. 60.

plant maintenance technicians—hotel policy does not allow it for fear of upsetting customers. It must be remembered that most commercial plant-care services are provided by independent contractors using manual irrigation practices. They have a regular schedule of visits for any given client. It is highly unlikely that most rooms would be vacant during those visits, and it is conceivable that weeks or possibly months could go by before some of the plants are watered. That kind of irregular care is, of course, intolerable for most decorative plants. Many interiorscapers have minimized the problem by installing self-watering planters. The problems associated with their use has prevented widespread application, but they have served a purpose. Micro-Irrigation Systems can now be designed into the building as integral facilities management systems, or they can be retrofit into individual suites to serve the same purposes. As an integral system, the control center and water distribution tubing would be installed and laid out much the same as in office buildings or apartment buildings. Special irrigation receptacles mounted on the walls of the rooms provide easy access to irrigation water, no matter where in the room (or rooms) the planters are located.

Inexpensive small-area systems can frequently be retrofit to take care of individual rooms or suites. Control centers can usually be installed under bathroom sinks (in the vanity cabinets) where they are connected to the water supply. Tubing is distributed under carpeting around the perimeters of the room.

Freestanding planters in hallways, entryways, elevator landings, covered walkways, snack centers, restaurants, lounges, shops, at the front desk, and in administrative offices can be serviced by branches of the central system (with irrigation receptacles) when properly designed for them.

Long, narrow planter boxes are common decorative elements in the large atrium hotels common to the Hyatt chain as well as other chains. They are used extensively to line the edges of atrium balconies on each floor. Their vinelike plants trail foliage over the edges of the balcony facades, cascading into the atrium and giving the appearance of hanging gardens (Figure 11.7). This is one of the images that architect John Portman has stamped indelibly onto the American commercial landscape.

Planters of that type require precise irrigation to prevent overflow onto balcony carpeting and to the facade and atrium floor below. Because of the sensitivity of this type of planter box, manual plant care is much more demanding and expensive. Common automated irrigation technologies are much too coarse and indiscriminate in their application of water to use with most of these demanding settings. Micro-Irrigation Systems are precise and controlled enough to service such planters effectively and can easily be designed into them, so long as it is done in the early drawing-board stages of the project.

There are outdoor installations of potted plants around hotels as well. These are usually around the pool sun decks, tiki bars, etc. Whenever potted plants are near a sprinklered lawn, a branch can be run to the deck and fitted with a drip system element to service the plants. When this is not possible, Micro-Irrigation Systems, run from a nearby hose bibb or other water line connection, would be a better choice.

Resort Complexes Resort complexes are basically hotels with greatly expanded recreational facilities and other amenities. The hotel portions are generally decorated and planted like any other hotel and have similar maintenance problems. Many resorts also have outbuildings, such as cabins, cottages, and clubhouses that use potted plantings extensively. Cabins and cottages particularly are very comfortable and homelike and can benefit from the natural presence of live plants. Facilities

Figure 11.7 Atrium of a modern hotel decorated with cascading plants.

managers and interiorscapers, however, are faced with the same access problems they run into in conventional hotel rooms and, for those reasons, frequently ignore interior plantscaping in cottages. Micro- Irrigation Systems can easily be designed and installed into small buildings of that type to provide fully automatic irrigation service, freeing the managers of those access problems.

Clubhouses are generally furnished in a manner similar to the hotels themselves, with furnished, carpeted areas in the lobbies, offices, lounges, restaurants, game rooms, locker rooms, fitness rooms, and pro shops. Freestanding planters are frequently found in these locations and can be effectively serviced by the automatic systems we have been describing.

Tiki bars are frequently decorated with greenery (hanging and freestanding), and because of the brighter environment outdoors, the plants require more frequent watering—usually by manual means. Simple inexpensive precision plant-care units can be easily installed to care for these plants automatically instead.

Figure 11.8 Typical Micro-Irrigation System installation in a hotel lobby. (Courtesy of Aqua/Trends.)

Restaurants and Lounges

The appeal of live plants in restaurant and lounge settings is long-standing. For a great many years, restaurateurs have been placing and hanging potted planters around the dining rooms of their establishments. We are hard put to find well-decorated eateries that have no plants—from fast-food chains to five-star establishments. The appeal is the natural, comfortable, and relaxing environment live foliage lends to an otherwise plain interior setting. Artificial planters have encroached on the territory of live foliage, but discriminating restaurateurs recognize the difference and opt for the "real thing."

One can notice the choices of those eating and drinking establishments catering to the upscale young professionals. Hanging live planters and built-in planter boxes are used in abundance to create the casual, relaxed atmosphere that the management knows will draw customers. Because of the extent and location of plant decor of that type, however, maintenance expenses can be quite high. Until precision irrigation technology was introduced, virtually all of the watering had to be done by hand. That generally means a technician climbing a stool or ladder (in some cases, rolling scaffolds are used) to get to the overhead plants with watering cans, tanks, or machines. Freestanding planters have had the option of self-watering containers, which can be used in select applications. Micro-Irrigation Systems offer a comprehensive, fully automated technique that is capable of minimizing manual labor and, thus, costs while providing the most consistent care possible. They also eliminate the overflow and spillage of manual irrigation, so annoying to management and customers alike. Plant-care contractors service restaurants during off-hours, when customers have not yet arrived. Careless maintenance technicians leave behind unwanted signs of their presence. Proper training of field personnel by competent interiorscape firms can keep that type of annoyance to a minimum. The point is that fully automated irrigation systems, in eliminating the irrigation

Figure 11.9 Typical interiorscaped restaurant.

responsibility from the interiorscape contractor, reduce the number of visits to the restaurant and minimize contractor-related problems.

Micro-Irrigation Systems are installed into restaurants much the same as they are in office buildings, with the control center mounted in a service area of the building, usually in the kitchen or utility room where there is easy access

Figure 11.10 Typical Micro-Irrigation System installation in a restaurant/lounge. (Courtesy of Aqua/Trends.)

to the water lines. From there, tubing is routed throughout the dining area, being hidden as much as possible. Many restaurants are built with suspended ceilings, which provide a convenient overhead space for tubing lines. Other restaurants have high vaulted ceilings with planters hanging from joists and beams. These are the most difficult for the average maintenance technician to service. Ladders are usually required, although manual watering machines with long wands will frequently reach most of them.

Large restaurant chains will be among the major beneficiaries of Micro-Irrigation Systems. Annual plant maintenance expenditures easily run into the millions of dollars for some of them. Because of the suspended ceiling design in most of these facilities, it is an easy matter to retrofit relatively inexpensive Micro-Irrigation Systems into the structure to care for the hanging planters. Planter boxes in room dividers and freestanding containerized foliage can also be retrofit, but usually with more difficulty. The most cost-efficient installations are designed into the building at the drawing-board stage and laid in during the construction phases. Under any circumstance, restaurant and lounge plant-care costs and annoyances can be effectively minimized by incorporating this new technology.

Whenever situations arise that make it impractical to install Micro-Irrigation Systems throughout, the main system can be supplemented with freestanding self-watering containers. In this way, the facility will have full coverage.

Shopping Malls and Arcades

Shopping mall facilities have been built in profusion over the past 15 years or so, and most have been well landscaped with a broad variety of planters. Most planters are built-ins (in situ), well integrated into the interior architectural design. They help to break up the stark lines and building materials and give a natural feeling to the interior, making the whole environment comfortable and inviting. Long ago, retailers also learned of the commercial magic live plants can lend to shopping centers of this type. Interiors with greenery draw people and keep them there longer, meaning they will shop longer as well.

If there is one area of interior landscaping that should be automatically irrigated, it is the shopping mall. The malls are very expensive to maintain, yet fairly easy to fit with automated plant-care systems, so long as it is done during construction. Many malls being constructed today are fitted with automatic irrigation systems. Inadequate technology is one of the reasons that these systems have been ignored in the past, but with the advent of more specialized techniques, there is no longer a reason to perpetuate expensive manual irrigation practices. Large trees in planter pits are still being watered primarily with long hoses, sometimes stretching 100 feet from the local hose connection. To water properly and deeply, long periods of time are required. There is no longer the need for such obsolete methods. One simple, relatively inexpensive irrigation pipe network installed underground from tree pit to tree pit, controlled by a timer, can effectively save hundreds of thousands of dollars over many years of the building's use. Bubbler or soaker emitters should be used around the roots, possibly a subterranean emitter such as laser soaker tubing. These are so much more efficient than sprinkler heads and cause no traffic safety hazards from wet walkways. In this case, a conventional irrigation setup without spray heads would be fine. Trees generally need lots of water, particularly those that get plenty of sun under skylights. Five-minute irrigation cycles can be programmed, sometimes twice daily. The emitters can be adjusted down, if necessary. This is a case where Micro-Irrigation Systems cannot be used to best advantage. The short irrigation cycles are frequently too abbreviated for very large trees growing in a bright environment. The same trees growing in dim light would, however, benefit from the more controlled flow rates of Micro-Irrigation Systems. Models having larger flow rates are available to provide greater capacity.

There are many smaller plants and planters used in malls that can also benefit from automatic precision systems. The network would be established with one or more control centers in the building, depending on the size and complexity of the mall layout. Overall control would be delegated to the facility's computerized control center, if that is used. Tubing would be laid in, under or through concrete slabs to the small planter boxes or to the locations in the building where potted plants would likely reside. Irrigation outlets, such as Aqua/Trends' special irrigation receptacles, can be fitted into the floors, walls, or support columns to access the water. Planter ledges can be fit with irrigation manifolds to service many potted plants on the shelf. Water supply tubing connects to the manifold and carries the short pulses of irrigation flow.

Offices and stores in the mall can be established as branches of the overall master system. Automated plant-care systems can be integrated with the facility's computerized energy management controller to provide a highly responsive installation.

Shopping arcades frequently use potted plants along the promenade to provide a natural, gardenlike atmosphere. Some are on the deck, while others hang from suspended planters. Either are difficult to irrigate by hand. These plants are outside, although usually in the shade. Because they get more light than their indoor relatives, they require more water. Hanging planters must be watered carefully by hand to make sure enough water has been dispensed, yet not enough to cause an overflow condition that might cause slippery pavements. Situations like this can be serviced by drip irrigation systems if they are designed properly. Freestanding containerized plants along the building wall can frequently be serviced from interior Micro-Irrigation Systems through appropriate emitters, such as the fountain type. They are mounted on an outside wall next to a plant location. During the abbreviated irrigation cycle, water is squirted the short distance between it and the plant's soil surface (see Figure 11.11). The appeal of using this type of emitter over the conventional ones lies in its vandal-proof design.

Retail Establishments Shops, department stores, and other retail establishments have gotten their share of interiorscaping in recent years. For the same commercial reasons that attracted restaurateurs, merchants are using plants to attract customers. In most cases, potted plants are used. Department stores scatter them throughout, on partitions, display cases, and on the floor near display cases. Room settings in furniture departments and other specialty showrooms are particularly prone to be decorated with various types of greenery.

Micro-Irrigation Systems are well suited for this type of irrigation, as the furnished interiors of shops demand precise water placement. Control centers can be installed in a rest room or a service room where there is convenient access to cold water lines. Tubing can be installed overhead or through partitions to planter locations. Wherever possible, small emitter tubes can be passed under carpeting to isolated containerized plants. Many stores have suspended ceilings that provide a convenient plenum through which to route tubing. Large department stores require extensive advanced planning; preconstruction systems design can provide the most cost-efficient installation. Small areas of stores can frequently be retrofit with inexpensive individual systems capable of

Figure 11.11 Fountain emitter at an outdoor shopping arcade.

servicing many plants. If cold water lines are inaccessible, low-pressure Micro-Irrigation Systems can be used.

Banks These commercial establishments are also being decorated today in a more inviting manner, using live plants to provide color and natural ambience.

Banker's offices are decorated and plantscaped as any other office might be. Containerized plants sit on desktops, filing cabinets, wall units, bookshelves, room dividers, and workstation modules, as well as on the floor. Automatic irrigation can be applied in the same way as in corporate offices.

The open service areas one generally finds in banks also use strategically placed plantings. These environments can be more difficult to automate. In most cases, the key is to integrate a Micro-Irrigation System into the structure. Tubing can be cast into the concrete slab and accessed through service boxes mounted in the floor and hidden by the carpeting. Other branches of the tubing can be brought up into room divider, service counter or partition planter boxes. Other tubes would circle the perimeter of the open area, either along the baseboard or inside the walls, with irrigation receptacles or other connections providing access to the water supply tubing. Tubing can usually be hidden behind decorative moldings and utility channels.

If the bank is in a technically advanced building, the Micro-Irrigation System can be controlled by the facility's computerized energy management system.

Clubhouses and Recreation Halls

These are found around country clubs, resort hotels, and residential complexes. They are generally designed and decorated as homey, lounge-type environments and are well suited for sports, social, and community use. Live containerized plants are frequently used to complement the decor in foyers, game rooms, fitness rooms, hallways, lounges, locker rooms, pro shops, snack shops, solariums, pool areas, and sun decks. These facilities can benefit greatly from the use of automated plant-care systems. Because these areas are usually well furnished, Micro-Irrigation Systems are the only technology of choice. The exceptions are outdoor pool decks and sun decks, which are not as critical. In those cases, either Micro-Irrigation Systems or drip irrigation can be used.

Hospitals and Medical Offices

Medical buildings of all sorts are being plantscaped with a variety of planter schemes, all trying to provide a more casual, natural, and relaxing feeling to alleviate visitors' anxieties as well as to add decorative value. Live plants are being used in hospital lobbies, waiting rooms, cafeterias, administrative offices, and solariums, as well as in the waiting rooms of doctors' and dentists' offices.

As with any other building, the inconveniences and costs of plant-care services can be minimized by using automated techniques. Most times, large planter areas in hospital lobbies can be effectively irrigated with drip systems or Micro-Irrigation Systems. The advantage in using the latter is that other branches of the same system can service the furnished areas of the building as well. Modern hospitals are being built with sophisticated computer controls for a variety of functions, and Micro-Irrigation Systems can become peripherals to these other systems.

Doctors' and dentists' offices can be installed with systems in ways similar to most other office settings, using special irrigation receptacles if the installation can be integrated into partitions.

Apartment Buildings

Although one might consider these as residential buildings and not commercial, most facilities services having to do with the common areas of those buildings are very commercially oriented. Large apartment buildings and condominiums are usually handled by property management companies who are as concerned about methods and costs as are commercial building managers. For the most part, the products and services used are similar.

Large luxury apartments and condos are frequently decorated with lush plantings in the lobby areas. Built-in planter boxes are usually used, supplemented with freestanding containerized plants. Drip systems and Micro-Irrigation Systems are the prime candidates for automated plant care in the planter boxes, while the latter technology is best suited to care for the freestanding potted plants. Retrofit installations can frequently be made to service all or a part of the lobby area. The project's practicality is determined by the construction and layout.

Residential Buildings

Figure 11.12 Condominium lobby interiorscape serviced by a Micro-Irrigation System. (Courtesy of Aqua/Trends.)

So far as common areas go, containerized plants are also found in hallways, game rooms, fitness rooms, sales and management offices, and elevator landings. Micro-Irrigation Systems can service all of these locations if designed into the building prior to construction. Outdoors, potted plants can be found around pool areas, sun decks, walkways, snack centers, and tiki bars.

In a totally fitted building of this type, the Micro-Irrigation System is integrated not only into the common areas but also into the residential part of the structure to service the apartments. This will be dealt with in more detail in the following discussions.

Residential Buildings   Live plants have been used in our residences since early times. We enjoy them, and they help fulfill our spirit. Psychologists say this is due to a deep-seated need that has to do with our evolution as human beings. Now we are finding more practical reasons for including plants in our living spaces. If we listen to NASA's environmental researchers, plants should be used to clean up indoor air pollution—we can make our homes healthier by using lots of them. Of course, many people do this anyway as a means of decorating and hobbying, but the less appreciative part of the population will eventually learn of the more practical benefits as well, and will find that plant care is easier than ever.

Almost all houseplants are taken care of manually. There are some exceptions, wherever self-watering planters, sprinklers, drip systems, and Micro-Irrigation Systems have been installed. Other homeowners use the little single-plant watering devices that appear on the market from time to time. Except in upscale homes, plant care is done by the homeowner. Unfortunately, many lack the knowledge or inclination to care for them properly. Many are too busy with work or other problems to worry about plant care. Plants can create problems for those that travel a lot on business, take frequent vacations, or use a residence as a second home, being away from it for long periods. Many are prized specimens, and many become objects of close emotional attachment—"green pets," as some refer to them. For those that can afford it, plant-care services can be hired. Many don't like the idea of strangers in their homes while they are away, however, so they impose on friends and neighbors to come in from time to time and give the greenery a "drink." The temporary help generally doesn't know how much water the plants require, so they tend to overwater. This results in barely adequate plant care.

Micro-Irrigation Systems now make all of this easy and dissolve most of the problems associated with the care of houseplants. They are capable of automatically watering plants week after week, month after month. Other attention should be given to the plants as well, but the critical, most demanding irrigation step is taken care of and in many cases, integrated with other plant-care tasks. For example, plant nutrients, pesticides and other chemicals can sometimes be applied systematically, carried by irrigation water.

Single Family Homes   Architects and builders are learning to accommodate their client's use of plants by designing skylights, greenhouse windows, atriums, planter shelves, boxes and pits, high-placed windows, supplementary lighting, etc., into new homes and apartments. In many, the design is very much plant oriented.

As previously discussed, prior fully automatic technology is unsuitable for housing interiors (with a few exceptions). Exceptions involve the occasional use of sprinkler system branches from outdoor systems to water built-in planter boxes and pits. These are in large, expensive homes where the planter is far enough from the furnished areas that there is no real concern about accidental water damage. Drip emitters are sometimes used in these planters; they are actually the wiser choice over spray heads. As for the rest of the house, however, these technologies are seldom suitable. Self-watering planters have been used by some to provide a measure of convenience. Micro-Irrigation Systems can provide comprehensive, fully automatic plant-watering service in smart (technically advanced) homes.

Micro-Irrigation Systems can be integrated into the structure of the house during the building process, or they can be retrofit in existing homes. They can be used for full-house service, where all of the plants in the home are irrigated by the system, or for smaller installations that cover only a section of the house. Outdoors, containerized patio plants can be watered by independent systems installed to cover the patio, pool deck, or entry areas of the house. Other partial systems can be installed in atriums and planter beds. Some installations require low-pressure versions of the technology, while high-pressure systems are used in most situations and sourced from the home's water supply. It is technically feasible to service automatically every planter in a home, and when the system can be incorporated into the structure, the feasibility becomes a reality.

Integrated high-pressure systems are installed with the control center in a laundry room or garage, and the water distribution tubing network is routed from there throughout the house, passing mostly through partitions, floor structures, ceiling and attic spaces, and basements. Irrigation receptacles are mounted on the walls of rooms, generally spaced with one on each wall in appropriate places. Emitter tubes are, of course, plugged into them and passed under the carpet edge to wherever freestanding planters might happen to be. There is a great deal of flexibility in that regard, as the emitter tubing can be routed for long distances, if necessary. Whenever plants are moved, it is a simple matter to reinstall the emitter tubes.

Planter shelves and box planters are fitted with special manifolds to dis-

Figure 11.13 Integrated Micro-Irrigation System installed in a residence, showing an irrigation receptacle servicing a planter. (Courtesy of Aqua/Trends.)

tribute water to a number of plants at that location. There is a trend toward high-tech homes, with various systems and services that encompass everything from security to energy management. If the home is fitted with a computerized energy-control system, the Micro-Irrigation System can be interfaced with it; otherwise, the plant-care system would be self-controlled.

Retrofit installations can run the gamut from single, inexpensive units caring for a couple of plants to full-house coverage. Obviously, it is more difficult to install tubing networks in fully constructed, fully furnished homes, making the retrofit installation also more costly. Tubing runs in these instances are mostly through attics, basements, and partitions, as well as under carpeting and behind baseboards around the perimeters of rooms. When tiled floors are encountered, special problems are created that require other simple solutions. The most practical retrofit installations involve inexpensive yet sophisticated carrier-frequency remote-control devices, such as the X-10 PowerHouse System. These are installed to control small satellite plant-care systems scattered wherever necessary throughout the home. Each of these inexpensive low-pressure units is capable of watering up to about 10 nearby plants. *Nearby* in this case can mean up to about 35 feet away. The complexity of individual unit control is made simple by the carrier-frequency controller that injects control intelligence onto the home's electrical wiring to be picked up and reacted to by receiving modules plugged into power receptacles wherever they are needed. In this way, plant-care units can be placed anywhere in the home (or patio/pool areas) to take care of any number of zones, and all can be controlled from a central, convenient point. Whenever they can be used in retrofit situations, multiunit, high-pressure systems can be controlled the same way. A secondary benefit is that a very versatile energy-control system is also in place, able to control lighting (plant lighting as well as general lighting), appliances, heating/air-conditioning, security systems, etc. The combination makes for a very smart house, at low cost.

Many homes have extensive patios around which potted plants can be found in abundance, hanging from trellises, arbors, and covered walkways or resting on retaining walls, steps, and patio decks. Many times, offshoots of the sprinkler system are not adequate for the job—actually, drip irrigation is a better choice than sprinkler technology, but either can be installed from the lawn irrigation system. Because of the superior control, Micro-Irrigation Systems can water these plants without overflow.

Apartment Suites Micro-Irrigation Systems can be incorporated into individual apartment suites either during construction or later as a retrofit installation. High-pressure versions are the systems of choice in most integral installations, and low-pressure versions in retrofit situations. Although it is possible to set up a single control center for a string of apartments, the most practical design permits complete control in each apartment.

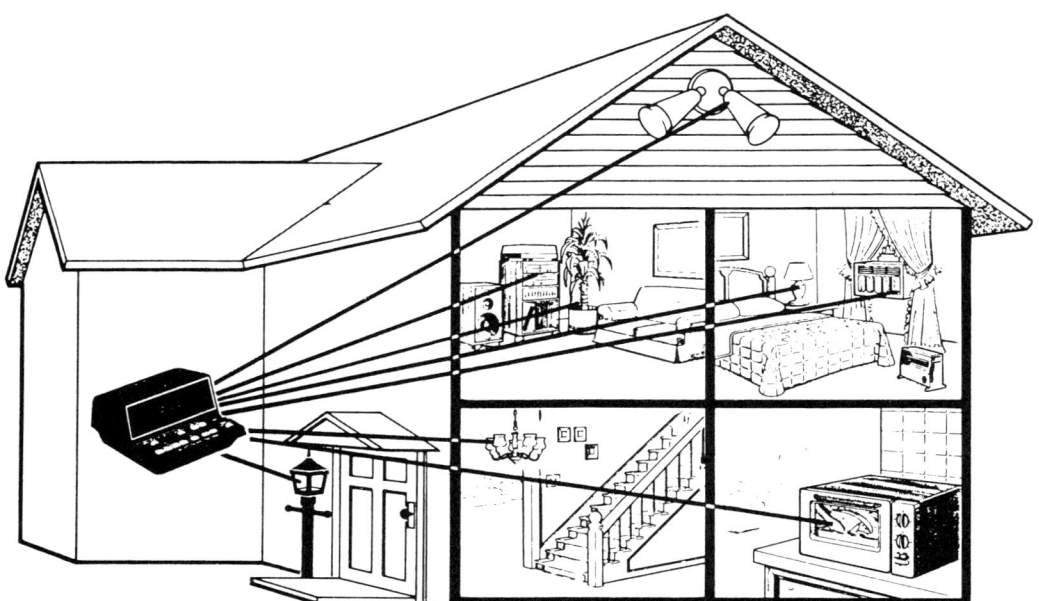

Figure 11.14 Home installed with carrier-frequency controlling plant-care units.

Figure 11.15 X-10 controlled Micro-Irrigation System (light-duty) installed in a home. The pump/reservoir module is hidden by furniture at the left of the photo.

An apartment's system is much like those systems used in hotel suites. The control center would best be mounted in the laundry center (most modern apartments have them) or under bathroom or kitchen sinks. From there, the tubing network is routed throughout the partitions to various rooms of the apartment, with access provided by irrigation receptacles mounted in the walls. Emitter tubes are, of course, plugged into them and laid under the carpeting to the planter locations. Control of each apartment system can be provided by a central computer in the management office of the building, providing timing or system shutoff if the tenant does not want the service. In this way, the building owner can provide a fee-paid amenity for the tenant population. In other situations, the timer/controller is completely controlled by the apartment owner or tenant.

Many potted plants can be found outdoors on the balconies or terraces of apartments. There is generally no way to water these except by hand. Outdoor plants require frequent waterings, which become a bother and a problem to apartment dwellers, particularly when they must travel. Many building codes prevent the inclusion of hose bibbs on the balconies of multistoried buildings. The reason has to do with the annoyance and water damage possibilities of overflow from upper floors cascading down to lower levels.

Low-pressure Micro-Irrigation Systems have self-contained reservoirs to function as the water source. Pumps built into the reservoir module provide flow to the tubing network servicing the plants on the balcony. The timer/controller can be mounted outside on a wall (requiring weatherproof equipment) or inside the apartment with the wiring extending outside to the pump connec-

Figure 11.16 Apartment balcony serviced by an outdoor Micro-Irrigation System.

tions. These systems are ideal for freestanding potted plants as well as hanging planters. Containerized vegetable gardening can easily be accommodated, as can speciality gardening, such as bonsai culture and orchid growing. Apartment dwellers need not give up an enriching hobby just because they choose to live in a multi-family dwelling.

Marine Applications   Luxury yachts are being decorated like homes, apartments, or offices—with style and the best appointments, including the latest in high technology. Live plants are a part of this, and Micro-Irrigation Systems are a welcome addition to the yachtsman's complement of on-board amenities. These systems can easily care for the cabin plants. Tubing can be threaded through bulkheads and partitions with the same ease that they are installed in land-based structures. Irrigation receptacles can be used, as can direct connections. In most cases, these systems must be low-pressure versions operating from the yacht's water tanks or self-contained pump/reservoir modules.

Now that the many areas of application have been discussed, we will concentrate on the design and installation of Micro-Irrigation Systems into buildings. That detail will be found in chapters 12 and 13.

12 THE DESIGN AND COSTING OF MICRO-IRRIGATION SYSTEMS

Overview Previous chapters have discussed precision, Micro-Irrigation™ Systems in general terms. Now comes the practical side, the detail that allows one to translate ideas and concepts into a layout design that can make the project a reality. For those that would like to learn the techniques of Micro-Irrigation System design, we offer the following knowledge in the hope that it will foster further study.

Each and every project is unique in its requirements, making the design tasks different in each case. Yet there are threads common to all, so in conceptual terms, one would be on familiar ground no matter what type of project was under consideration. Integrated systems have their own set of problems and retrofit designs have their unique problems. The challenges of a large office building are much more demanding than a small home, but many of the considerations are the same. In this chapter, we will try to sort it out sufficiently and give one enough expertise to be able to handle most Micro-Irrigation System design projects.

Defining the Design Problem Before anything else can be done, the design problem must be defined. One must (1) know what is to be accomplished by an installed system, (2) have a general feel for the physical attributes of the building under consideration, and (3) have some knowledge of the possible hazards and difficulties. Later, these will be combined with knowledge about the systems themselves to evolve a design plan.

A quick design stage is usually required to get a handle on the nature of the installation and the approximate costs. First, a cursory survey must be made of the building plan. If the building is in place or under construction, a visit to the site is in order to look over the design, structural details, interior finish details, locations of built-in planters, locations of water sources, possible future locations for freestanding planters, etc. Notes should be taken, sketches of the layout made, and at least rough dimensions taken. If building blueprints are available, they should accompany the designer. If any of the building developer's people are available, they should act as guides in pointing out building features that may be relevant and in answering questions. If the building has not yet been built or is in a remote location inaccessible to the designer, this initial study must be made from building plans and other information furnished by the client. The designer should ask for a complete set of plans, including **floor plans** of the building sections under scrutiny, **elevations** of same, and **plumbing and electrical layouts.** The clients should always be questioned as to what they expect out of an automated plant-care system in the interior of their building. All this input is necessary to get a clear picture as to the nature and objectives of the project, potential installation problems, and approximate costs, as well as what must be done to satisfy the client.

A couple of factors have just been mentioned that highlight the need for the designer to have enough expertise in building design and construction to be able to read blueprints, understand some of the terminology used, understand some of the problems associated with building construction, know something about plumbing systems and building codes, and know something about interior design. One does not have to have the knowledge of a civil engineer, an architect, an interior designer, a construction supervisor, or a contractor, but some depth is necessary. If that is lacking, it calls for study into those fields.

The Sample Layout With the information gathered, the designer should make a quick layout of the system to be recommended. That may come after many hours of study. The quick preliminary design should include the following factors:

- The general type of system required—low-pressure or high-pressure versions.
- The recommended control equipment—appropriate models of Micro-Irrigation System timers and controllers should be specified.
- The size of the main water distribution tubing lines.
- The approximate routing of the water distribution tubing network.
- The approximate type and number of tubing fittings required.
- The approximate number of irrigation receptacles required.
- Other details if they seem relevant (installation, labor, etc.).
- The approximate installation cost calculated from the information available.

A report to the client in the form of an installation proposal can then be generated from the design study. Many complex projects require so much preliminary study and design work to determine feasibility that the client must be charged for that service. Other projects are simple to lay out and install, and the design process can be accomplished on-site at the time of installation, *providing it is done with expertise and care.* Most building projects require something between the two extremes. When the installation proposal has been accepted, then a detailed set of working plans must be generated for use by the architect and contractors involved as well as for the owner and property or facility manager. Discussions of that process follow.

Technical Considerations There are many technical details that must be dealt with in systems design. Some have to do with engineering data; others are not as involved. Technical drawing methods are used to graphically describe a Micro-Irrigation System layout. Not all installations require a formal design plan, but whenever the system is to be integrated into the structure of a building, a formal design plan will be required (just as electrical or plumbing design plans are required), so the architect and installation contractors know exactly what the system is all about, where it goes, and how it is installed. In a manner of speaking, the automated plant-care system might be considered a sort of secondary plumbing system, although its purpose, control, and appliances bear no similarity. Most integrated systems will, in fact, be installed by plumbing contractors, and a "road map" must be furnished for them to follow. That is the purpose of generating a technical working drawing. It is called the *interior irrigation system plan* and is drawn with symbolism. Because the expertise is similar to plumbing, the language of plumbing is used to graphically describe these special layouts. Figure 12.1 illustrates a few of the important basic plumbing symbols used in Micro-Irrigation System design.

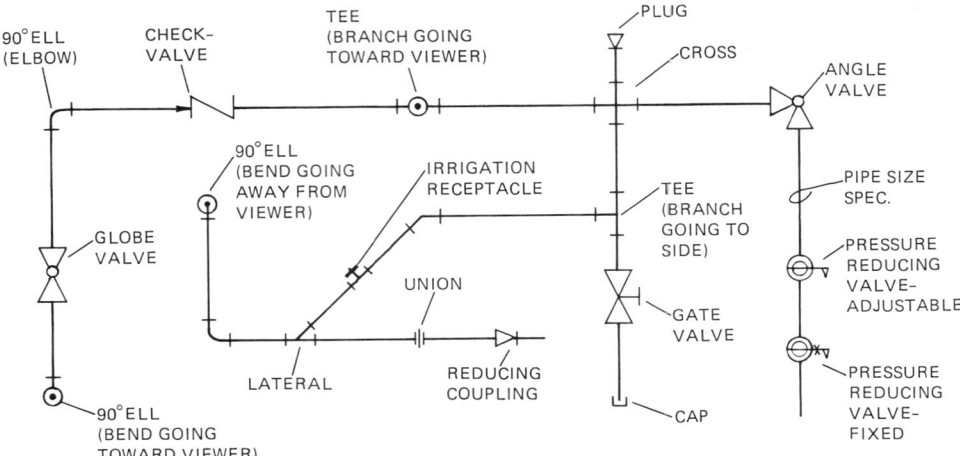

Figure 12.1 Plumbing symbols common to interior irrigation systems.

Solenoid Valve and Pump/Reservoir Ratings

One of the keys in specifying a pump or solenoid valve in a Micro-Irrigation System is the output rating of the device. Put in other terms, it defines the "duty" to which it can be put, or how much of an irrigation load it can carry.

Small pump/reservoir modules are flow rated in terms of ounces/10 seconds (ounces of water delivered during a 10-second operating cycle). Ratings for these light-duty pumps will be in the range of 7 to 12 ounces/10 seconds. Larger units for use outdoors are flow rated in the same terms, but output will be much higher, usually about 120 ounces/10 seconds. These ratings are for ground-level service. Flow ratings at an 8-foot elevation should also be available from the manufacturer. They represent the lift capabilities of the pump and tell the user whether the pump is useful for watering hanging planters or possibly potted plants on the floor above or on high-mounted planter shelves. The output port size of a pump tells the user what it can be connected to—the tubing size and the type of connection.

The size of the reservoir is another important consideration, for it determines how long the unit will service the plants on a system without needing a refill. Reservoirs are supplied in sizes from 2-1/2 gallons to 32 gallons, although almost any size can be made available for special applications. In the light-duty systems, a 2-1/2 gallon reservoir, watering 6 or 8 medium-size, indoor plants, would hold enough water to last 7 to 9 weeks without a refill. This is highly variable because so many factors determine plants' moisture requirements (see Chapter 6). In designing the system, an attempt should be made to provide it with enough reservoir capacity for at least 6 weeks of continuous operation. Keep in mind that low-pressure versions of Micro-Irrigation Systems are automatic only to the extent of their reservoir capacity, then they require attention (refilling). There are ways to link two reservoirs in order to multiply capacity. These alternatives can be considered when necessary.

Solenoid valves are commonly flow rated in terms of the C_v *factor* (valve flow coefficient) or valve capacity index, which indicates the rate of water flow in gallons per minute under certain standard conditions. They give the designer a general idea as to the flow capacity of the valve. The C_v factor is determined by the inlet and outlet port sizes and the internal architecture of the valve. A more practical flow rating for solenoid valves is the flow rate at various water pressure levels. The flow rate of any solenoid valve varies with the water pressure applied to it . . . the higher the pressure, the greater the flow rate, and vice versa. Flow rates of solenoid valves used in Micro-Irrigation Systems range from 0.2 gallons per minute to 35 gallons per minute. The size is chosen according to the volume of water required by the particular installation. Most interior systems require only very low volume. Table 12.1 shows the relationship between C_v factor and flow rates.

In small installations where only a few plants are involved mini flow rates of only 0.2 gallons per minute can be used. This is common when only one or two rooms of a hotel (guest rooms), home, or condo must be serviced. It is better, however, to choose a higher rated solenoid valve so that the system coverage can be expanded at a later time, if required. Temporary reductions can be made at the valve to keep flow rates manageable.

TABLE 12.1

Relationship between Solenoid Valve Rating and Flow Rates (in Gallons per Minute) at Various Water Pressures

Solenoid Valve C_V	20 psi	30 psi	40 psi	50 psi	60 psi	65 psi
0.34	1.6	1.9	2.5	3.0	3.4	3.5
1.4	6	7.5	8.5	9.5	11	11.5
2.8	12	15	17	19	22	23
3.4	15	18	21	23	27	28
3.6	17	20	23	26	28	30
5.5	24	29	34	37	40	43

Sources: Dayton Electric Manufacturing Co., "Solenoid Valves," form 551589 (Chicago, May, 1978) and Aqua/Trends.

Water Flow Variables

Water flow from the output of a pump or solenoid valve is affected by the tubes or pipes, fittings, and other devices that it must pass through on its way to the plants. The direction of flow also has an effect. Water pressure and flow drops as it goes along, and the extent of these changes in a system determines the size of the pump or solenoid valve required by the installation. Rather than guessing what is required or trying to determine it by trial and error, rough calculations can be made to zero in on the appropriate equipment. These calculations must be made with a variety of input data—factors that will be discussed in the following sections.

Flow Characteristics of Tubing

The flow characteristics and diameter of the tubing (or piping) are important. Materials differ in the way they permit or restrict the flow of water, or any other liquid for that matter. The interior surfaces of a tube or pipe create a drag on

TABLE 12.2
Water Pressure Loss in Plastic Tubing at Various Flow Rates and Tubing Sizes

Water Supply Volume (GPM)	Friction Loss in psi/100 ft. of Plastic Tubing (psi)			
	3/8" OD	3/8" ID	1/2" ID	3/4" ID
2.5	200	27	6.4	1.2
5.0	700	100	25	4
10.0	—	360	90	15

Source: Adapted from LCP Chemicals and Plastics, "Friction Characteristics of Water Flow through Rigid Plastic Pipe," monograph (Carrollton, Ohio, September, 1984.)

the water as it rushes through, this because of frictional contact between the two materials. The tubing diameter has the same effect. The smaller the tube, the greater the restriction to flow. The distance that the water has to travel through a tube has the same effect. The greater the distance, the slower the flow rate and the lower the water pressure downstream. This relationship is specified by the industry as the pressure drop per 100 feet of tubing (or pipe). Table 12.2 contains data that display that relationship for some of the tubing and piping used in Micro-Irrigation System installations. It will be seen that for a given flow rate as the diameter of the tubing decreases, the pressure losses due to internal friction increase dramatically. For example, many Micro-Irrigation System installations require input flow rates of 5 gallons per minute. From the data in Table 12.2, we see that if we use small-diameter tubing in the order of 3/8" ID, the pressure drop over 100 feet of tubing is so high as to permit little or no water through. At least 1/2" ID tubing would have to be used for the main water distribution line if the system is to irrigate a large number of plants. Usually, the main line is much shorter than 100 feet, so losses of this magnitude are not prohibitive. Branches can be smaller, as they carry less water.

The Effect of Gravity

Another factor that tends to restrict flow is gravity. Whenever tubing is installed in a vertical fashion (a riser), from floor to floor for example, the weight of the water within the tube creates a drag on its upward motion. Consequently, pressure is reduced in the process. This is called *dynamic discharge head*. It takes 0.434 psi to push water up 1 foot. Pushing the water up 20 feet would require 8.68 psi (20 × 0.434 psi) for that flow pattern alone. Therefore, in designing these system installations, the number of feet that tubing must rise over the water source should be taken into consideration. In large buildings particularly, the effects of static head can be significant.

The Effect of Tubing Fittings

Other things in the path of water flow also tend to restrict its movement. Tubing and pipe fittings, as well as regulators, filters, backflow preventers, etc., all reduce the water pressure as it passes through. The amount of friction loss depends on the design of the device, its material, and the sizes of the orifices in it, as well as the rate of water flow and the pressure acting on it. That's a complex menu of factors, and we shouldn't get into detail here; but suffice it to say that if many of these devices are in the tubing line, the pressure loss can be significant and should be adjusted for it. The data in Table 12.3 provide a guide. Tables related to fittings are specified in terms of friction loss equivalent to a given length of tube (or pipe) of the same size. For example, if a certain 1/2" fitting had a friction loss equivalent to 10 feet of 1/2" tubing, it would mean that 10 feet should be added to the total footage of tubing when calculating the sys-

TABLE 12.3

Allowances for Friction Loss through Fittings (in Units of Equivalent Length of Similar Size Tubing)*

Tubing Fitting	Friction Loss—Tubing Equivalent		
	3/8" ID	1/2" ID	3/4" ID
90° Ell (elbow)	0.5 ft.	1.0 ft.	1.3 ft.
45° Ell (elbow)	0.3 ft.	0.6 ft.	0.8 ft.
Tee-run (straight through)	0.2 ft.	0.3 ft.	0.4 ft.
Tee-branch (90°)	0.8 ft.	1.5 ft.	2 ft.
Globe valve	4 ft.	7.5 ft.	10 ft.
Gate valve	0.1 ft.	0.2 ft.	0.3 ft.
Check-valve	—	30 ft.	—

Source: Adapted from Frank R. Dagostino, *Mechanical and Electrical Systems in Building*, (Reston, VA: Reston Publishing/Prentice-Hall, Inc., 1982), p. 43.

*Varies with the type of coupling and other factors.

tem's friction losses. This is done for each such device in the line. A common practice is to simply add an additional 50 percent to the length of tubing used in the design to compensate for losses through fittings.[1]

The Effect of Emitters

Another type of fitting is the emitter. Its relationship to irrigation flow rates will be discussed separately. Various kinds of emitters are used in Micro-Irrigation Systems. Each has its own flow characteristics and would thus affect the design in a slightly different way. The flow rates from emitters are important in determining the size of water distribution tubing and of pumps and solenoid valves used in the design. The total demand for water in an installation depends on the aggregate amount that is likely to flow at one time from all emitters on the system (if they were all in service). Adjustable mini-valve/emitters are the most common in this technology. The data presented in Figure 12.2 show the maximum rates of flow from that style of emitter at various pressure levels. These variables are most important, for the systems are designed to operate at different pressures to accommodate different needs.

Another emitter type that gets some use in Micro-Irrigation Systems is the Laser Soaker Line. It is, of course, much different than the adjustable valve types and has a different purpose—that of providing a continuous, linear soaking pattern on or in the ground. As pointed out in previous chapters, the laser-drilled emitter holes are spaced at 6" and 12" intervals. These tubes have different flow rates at various pressure levels. That relationship can be seen in Figure 12.3. These data will be useful in designing planter box installations and circular emitter patterns in containerized tree planters.

The Location of Control Centers

An important design consideration common to all projects is the location (or locations) of the control center. This is where electrical and plumbing connections are made to the Micro-Irrigation System's timer/controller, which is the source and "brains" of the system. The first factor that must be dealt with is convenience to a water source.

In the case of low-pressure versions of the technology, the pump/reservoir module is the water source, so it can be located in almost any practical place.

---

[1] Frank R. Dagostino, *Mechanical and Electrical Systems in Building*, (Reston, VA: Reston Publishing/Prentice-Hall Inc., 1982).

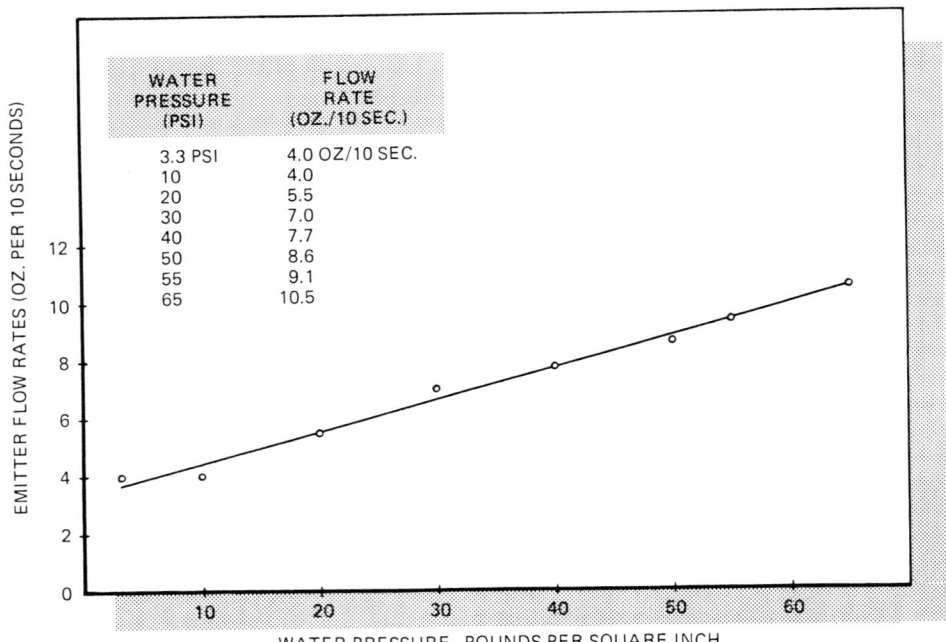

Figure 12.2 Flow rates of the adjustable mini-valve/emitter (maximum flow at various pressures). (Courtesy of Aqua/Trends.)

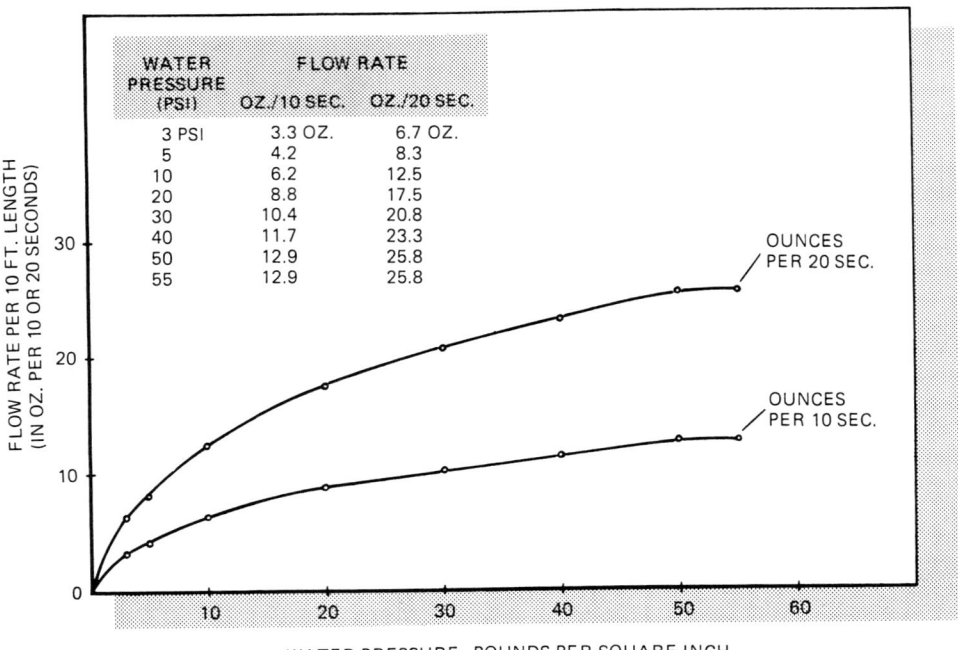

Figure 12.3 Flow rates (per 10-foot lengths) from Laser Soaker Line—6" hole spacing (at various water pressures). (Courtesy of Aqua/Trends.)

However, to make a neat-looking, unobtrusive installation, these units should be in out-of-the-way places. Smaller systems can be hidden behind furniture, mounted inside of cabinets, molded architectural furniture and planter boxes, etc. Larger units must be installed in closets, utility or service rooms, garages, etc. Outdoors, they can be located in little-used corners of patios, balconies, and pool decks. They should not be too far from a faucet so that hauling water to them from time to time will not be too much of a burden. Hoses or tubes can be connected to faucets to refill reservoirs so long as the distance is not prohibitive.

High-pressure control centers are located convenient to a cold water line in out-of-the-way areas of a building. The timers and controllers are surface mounted so that plumbing connections can be made externally. The supply line is connected to a nearby cold water pipe. In residential buildings, the most practical locations for control centers are in laundry rooms, garages, and inside of kitchen and vanity cabinets. In commercial buildings, the control centers should be in a core area location, such as a utility or janitor's room, mechanical room, or rest room. Systems meant to service smaller building areas can frequently be centered inside of bathroom vanity cabinets, in office kitchenettes, etc. Outdoor systems can be connected to hose bibbs.

In all cases, there must be an electrical connection that provides 24-hour service to the system . . . automation cannot be accomplished without continuous power. Since most available Micro-Irrigation System controls are simple plug-in types for ease of installation, power must be available through a receptacle. In those versions that are hard-wired into an electrical junction box, it must be provided at the control center location. Although very little power is required by these systems, some prefer to have the automatic irrigation system on an independent power branch with its own circuit breaker. In most cases, that is not necessary. Wall outlets can be fitted with mini-breakers if desired. In high-tech homes and commercial buildings having a computerized energy management system, it would control the switching of power to the electrical box servicing the Micro-Irrigation System.

Self-powered controllers are available for special applications. They are battery powered and can be installed in locations not serviced by building electricity.

The Routing of Water Distribution Networks

Another general design consideration is the way tubing networks are laid out in an installation. Each situation will be different because of varying building design and construction factors, as well as interior design and plant location considerations. There are rules of thumb for all designs, however.

The first consideration must be to route the tubing in such a way that it is virtually unseen. Completely integrated systems are designed with tubing threaded through the building structure to be terminated at scattered irrigation receptacles, emitter tube connections, or in planter boxes. The only tubing seen outside the walls are the small-diameter clear plastic emitter-tube assemblies. If these are hidden under carpeting and behind cabinets, furniture, and planters, they too are unobtrusive. Retrofit installations are a little more trouble to design; sometimes carpeting is not available to hide the tubing, and other methods must be sought. Commercial buildings frequently have suspended ceilings and the spaces above provide a good area through which to route tubing. The same is true for raised floors in some of the newer office buildings. The basements and attics of homes also provide easy access for tubing installation. When these conveniences are lacking, surface-mounted moldings and channels

Important Flow Controls

can be used to hide portions of the tubing run. More will be said about these techniques when we reach the subject of installation in Chapter 13.

There are times when available building water pressure is too much for a particular installation or when flow rates are high for other reasons. Full pressure and/or flow are required only (1) when great numbers of plants must be serviced by the system, (2) when very long tubing runs are necessary to reach remote planter locations, (3) when vertical rises are extensive, or (4) when there are combinations of all these factors. All tend to naturally reduce system water pressure and flow rates. Most installations operate under more moderate conditions, and some sort of flow reduction is necessary. One method is to use flow-reducer fittings in the solenoid valve input line. Most of these are simple, inexpensive devices that are pressure compensating, keeping roughly the same flow rate through a wide range of pressure variations. The other way is to use either a variable or fixed water pressure regulator. While the variable type is more expensive, it permits easy adjustment to the levels required by the installation. The fixed type provides only one output pressure level.

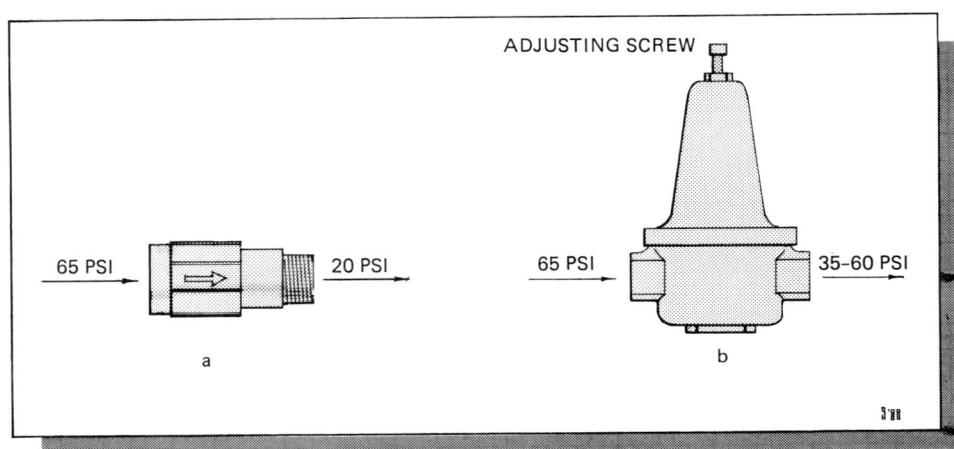

Figure 12.4 Water pressure regulators.

Pressure reduction is desirable, particularly when there are unsecured (not clamped) tubing connections to barb fittings. High pressure can sometimes blow these junctions apart, causing leaks. Under other circumstances, high pressure can strain even compression fittings and sometimes cause leaks in them as well. There are times when only certain branches of a water distribution network require pressure or flow modification, leaving the main part of the network at full operating capacity.

Another important flow control device used in Micro-Irrigation System design is the check-valve. These are used in main water distribution tubing lines (particularly risers) and, at times, in emitter tubes. The theory of check-valve use has already been discussed in Chapter 10. Check-valves are important in this technology for several reasons and must be used with skill.

The Designing and Costing of a System Installation

The procedure for designing and costing a system installation is as follows:

1. Obtain a set of plans for the building area requiring installation. Floor plans are sometimes adequate but elevations, plumbing, electrical,

space, and furniture plans are also desirable because they provide an insight into available crawl spaces, overhead plenums, basements, etc., that can be used to route tubing lines. They also help uncover obstacles to an installation (location of other services) as well as destinations (planter locations). Architectural working drawings are best, but floor plans from sales promotion brochures can often serve the purpose in simpler projects. If plans are not available, make a rough sketch of the area with detailed dimensions in floor plan form.

2. Check building codes for local requirements.
3. Locate a practical and convenient place for the control center on the floor plan. Draw a small rectangle to represent the control center location, and draw symbolic notes for other control center elements (pressure regulators, filters, backflow preventers, shutoff valves, etc.).
4. Where appropriate, mark the location of irrigation receptacles along the room walls. If receptacles are not to be used, mark anticipated future locations of potted plants (emitter locations). Some of these are choices based on logic, others on firm knowledge (fixed planter boxes and pits). It is usually a good idea to consult with the interior designer or landscaper working on the job. In the early design stages, that may not be possible. It is not necessary to know the exact future location of plants ... being close is good enough. Keep in mind that with Micro-Irrigation Systems, plants need not be placed at the receptacles. Flexible emitter tubes will bridge the gap.
Here is an example of the logic which should be used:

Containerized decorative plants are generally placed in or near the corners of rooms, next to large windows, on pieces of furniture along walls, or in front of blank wall space.

They are seldom placed in walkways, doorways, in dimly lit locations, or near emergency exits where they might impede traffic.

Most retrofit installations do not use irrigation receptacles ... other fittings are used to connect emitter tubes to the plants. Manifolds are used in cases where a number of plants are grouped. Mark the locations of these manifolds and connecting fittings as well.

5. Count the number of irrigation receptacles or other output fittings planned, and record the figure.
6. Sketch in the best routes for tubing lines from the control center to the irrigation receptacles or other anticipated planter locations. Tubing should be routed so that it can be hidden in interior partitions, basements, ceiling joists and beams, suspended ceiling plenums, crawl spaces, plenums below raised floors, within cabinets, etc. Use the most direct routes. Main tubing lines should be installed overhead or cast into floor slabs wherever possible (new construction and renovation). These are larger in size than branch lines. Lay out branch lines as extensions from the main tubing lines. If overhead, bring them down through partitions (drops) to baseboard level, and from there branch out to receptacles or emitter locations, routing the tubing through partitions, etc. Try to minimize the number of vertical tubing lines.

To clarify terms, *drops* are vertical tubing lines having downhill water flow, and *risers* are vertical tubing lines having uphill water flow. Both require check-valves at their bases, but the one-way flow of these devices will, of course, be opposite. The term "riser" is usually used generically to denote a vertical tubing line.

In most retrofit installations, tubing is hidden beneath carpeting around the perimeter of rooms, or in cases where there is no carpeting, tubing is installed behind special baseboard moldings.

7. Count the number of vertical tubing lines, and record the figure. This will result in the number of main check-valves required. There will be a check-valve installed at the base of each riser or drop. Mini check-valves may be installed as well in emitter tubes to control localized flow.

8. With a rule, measure the length of main tubing lines (trunk lines) on the scaled floor plan. Use the scale of the layout to estimate the length in feet. Record the figure.

9. Calculate the length of risers required as follows:

Number of risers in the layout × estimated riser length (in feet) = total length of tubing in risers

Floor-to-ceiling distances can be estimated at about 8 feet for residential buildings. Commercial buildings generally have about 12 feet between floors.

10. With a rule, measure the length of horizontal branch lines. Using the layout scale, estimate the length in feet. Record the figure.

11. Calculate the total demand of the system as follows:

Number of emitters in the system × 0.09 GPM avg. demand/emitter = total demand of system in gallons per minute (GPM)

12. Calculate the flow rate required at the input of the system (pump or solenoid valve output) as follows:

Total demand of system (GPM) × 2.0 loss factor = total estimated flow rate required at input

Note that there are situations when tubing runs are short, and only a few fittings and pieces of auxiliary equipment are used in the line. Friction losses are therefore low, and the 2.0 loss factor can be reduced considerably.

13. Choose a solenoid valve controller or pump/reservoir module that can provide the total estimated flow rate. Choose a controller with a built-in solenoid valve having enough flow capacity to provide the required flow rate (consult Table 12.1) at the building's average water pressure or at some reduced pressure level if that is appropriate. Valve C_v ratings should be listed in the equipment manufacturer's spec sheet.

14. Choose a pump controller if low-pressure versions are involved to meet the power and duty demands of the pump/reservoir module chosen in Step 13. That information should be found in the manufacturer's spec sheets.

15. Choose the size of tubing to be used for the main water distribution lines. The inside tubing diameter should be the same as the inlet/outlet ports of the pump or controller's solenoid valve going into service. Adapter fittings can be used to provide compatibility if necessary.

16. Choose the size of tubing to be used for risers and horizontal branch lines. In most cases, 1/2" ID tubing is used for vertical drops or risers,

and 3/8" ID or 3/8" OD tubing for horizontal branch lines. The 3/8" OD tube size is adequate for branches servicing up to about 35 plants.

17. Total the lengths of each type and size of tubing based on the tentative layout (from Steps 7, 8, and 9). Record the figures.
18. Count the number of each type and size of tubing fittings and accessories as required by the tentative layout. Record these figures.
19. Calculate the material costs as follows:

$$\text{Number of units of each item} \times \text{unit costs} = \text{estimated material costs}$$

Total all material cost estimates.

20. Estimate labor costs based on the installer's labor rates in the area. An installation quote can be obtained from a contractor if outside services are to be used (get several quotes). Factors to be taken into consideration are the extent of the installer's training and experience and the estimated number of callbacks for final tubing hookups, adjustments, and service. Plumbing installation figures can also be obtained from various publications written about construction cost estimating.
21. Total the estimated material and installation labor costs to arrive at a grand total expense.
22. Add reasonable overhead and profit margins to arrive at the estimated total installation cost.

These figures will enable the designer to reasonably estimate the cost of a proposed installation. They can be refined, if required, through further study of the job.

The selection of Micro-Irrigation System components can be made easier by using one of the available installation kits where all controls, tubing, fittings, etc., have been preselected for compatibility.

The final working drawings of the interior irrigation system plan should be made on drafting paper or film as an overlay to the architect's floor plan blueprints. They can be reproduced and included with the other project working drawings.

In the sections that follow, we will look at several examples of system design. There are many other applications of Micro-Irrigation Systems for which we have provided no examples or exercises, but be assured that the principles discussed here are typical of many applications.

Designing for Commercial Buildings

It has been mentioned in previous sections that most commercial buildings are built on speculation, and low-cost installations are required in vacant space so as not to pose a burden during the initial period of disuse. *Skeleton* installations of Micro-Irrigation Systems are designed to meet that need. They involve the "bare bones" of a system and can be installed very economically. They prefit the building space with the necessities for hookup and use at a later time. Meanwhile, the building owner can advertise the inclusion of high-tech amenities without having made a large capital investment. The skeleton installation on each floor consists of a control center hooked up to the building's cold water supply and a main tubing line (usually 3/4" ID copper) that is installed overhead, down the middle of the vacant tenant areas. Connections about every 20 feet for 1/2" MPT adapters are plugged and ready for hookup to branches going to

214   Chap. 12 / The Design and Costing of Micro-Irrigation Systems

suites as the space is leased (refer back to Figure 11.3). Design work involves the selection and location of control centers and the layout of main tubing runs. Costing is simplified by the fact that few parts are involved.

Example of System Design for an Office Building

In Figures 12.5 and 12.6 and the discussion that follows, we will take the reader step-by-step through a demonstration design of a skeleton installation. The floor plan is of a modern office building having approximately 33,900 square feet of tenant space on the floor under consideration.

Because of the complex design of the building floor, with major tenant areas split into two sections (or zones) and separated by an elevator landing/bridge, it was decided to design two independent high-pressure irrigation systems, one for each zone. Although the material cost of this approach is slightly higher, the installation is simpler and labor costs would be slightly lower. There is also more system flexibility in having independent zones. The floor plan tells us that Zone A is approximately 18,300 square feet, and Zone B approximately 15,600 square feet. A prior inspection of the building disclosed that the overhead space on the floor was fitted with 20″ deep steel joists spaced every 4 feet. These are adequate for suspending the 3/4″ ID main tubing line over the vacant tenant space (building codes permitting). It was also seen that the mechanical room in each zone is next to the rest rooms, providing convenient sources of water. Therefore, they were chosen as locations for the control centers. The control equipment would be mounted on a wall, and 3/4″ ID input tubing installed up the wall to connect with a cold water line in the overhead space (or other close location). Shutoff valves, pressure regulators, and filters are de-

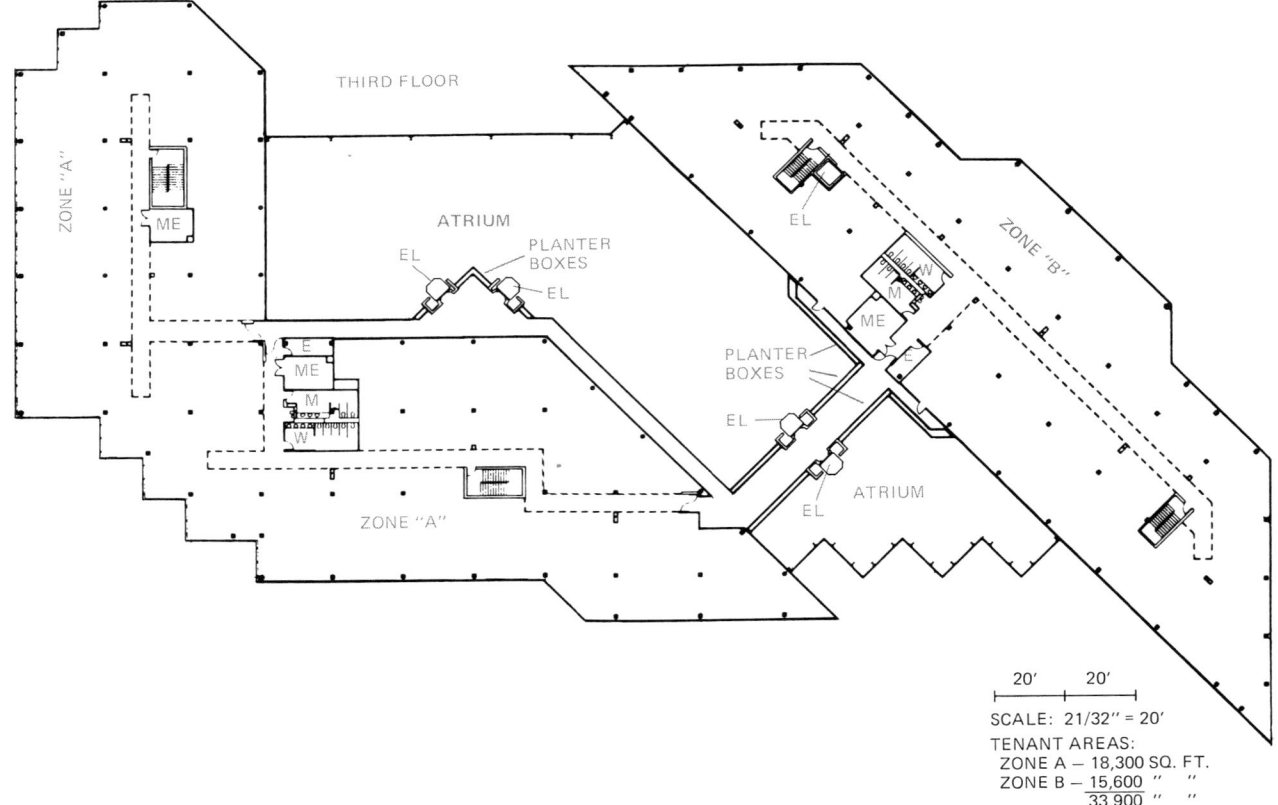

Figure 12.5 Floor plan of a typical modern office building.

Example of System Design for an Office Building

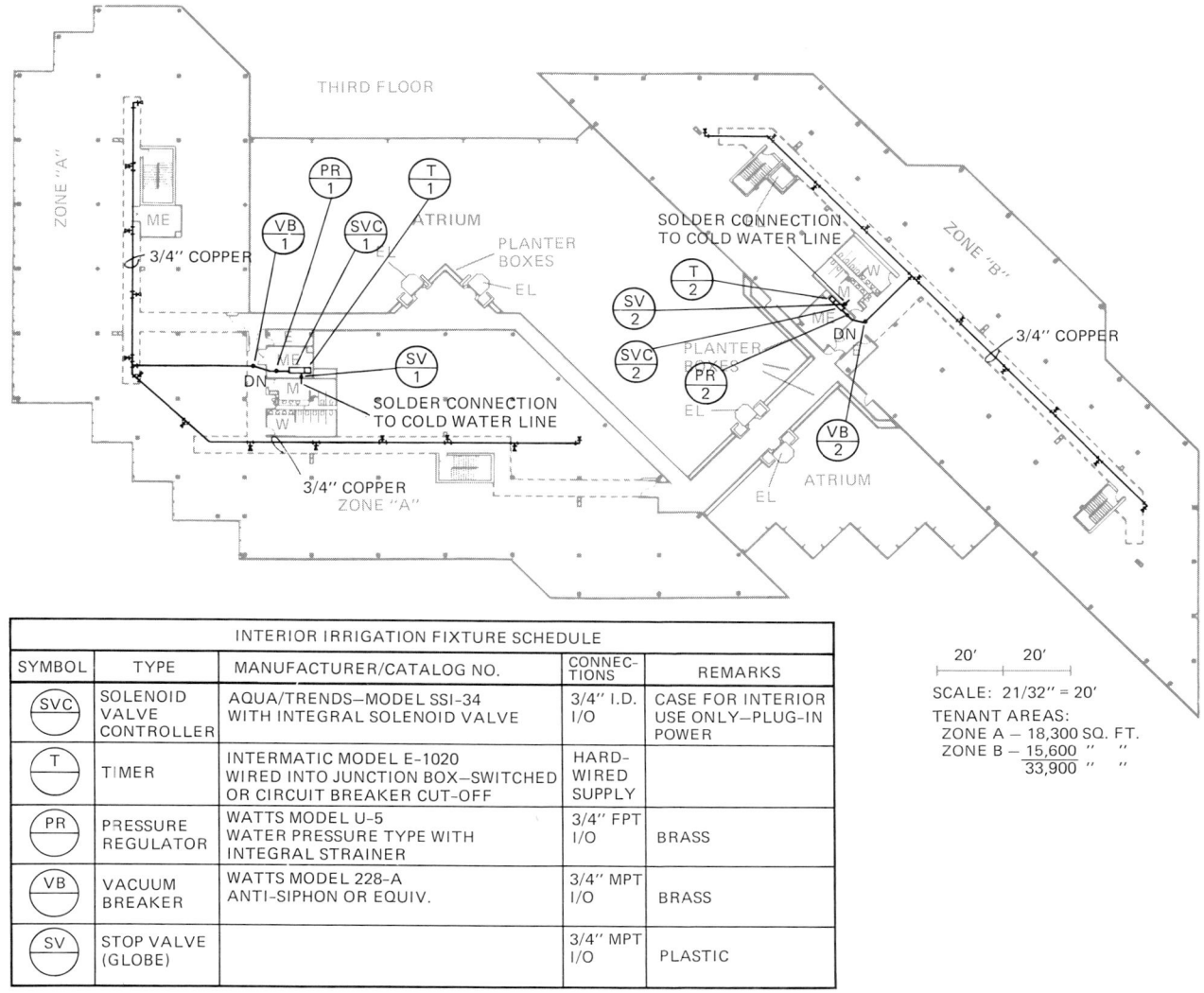

Figure 12.6 Interior irrigation system plan for a skeleton installation in a modern office building. (Courtesy of Ayers, Snyder & Locke.)

signed into the input lines, as well as a vacuum-breaker into the immediate output line. The 3/4" ID output tubes from the controller also rise into the overhead plenum and connect to a "spine," the 3/4" ID trunk line tubing that runs down the center of the tenant space. That is to be suspended from the joists or pipe hangers. Every 20 feet, a fitting is installed to provide for future connections to office suites. Be sure access will be available. That design concept is drawn over the building floor plan (Figure 12.5) either directly or on a tracing paper overlay (Figure 12.6). It then becomes the *interior irrigation system plan*. Symbols for each of the fittings and accessories should be included. In this case, they consist mostly of 3/4" ID (sweat) × 3/4" ID (sweat) × 1/2" FPT branch tees, fitted with 1/2" MPT plugs. A *fixture schedule* is drawn on or attached to the interior irrigation system plan. It is a table of the control equipment and fixtures required but does not include the detail of tubing types and fittings, etc. In this exercise, the interior irrigation fixture schedule (in Figure 12.6) includes the solenoid valve controller, timer, pressure regulator, and vacuum breaker.

After tubing lines are integrated into the floor plan, *a cost estimating work sheet* is filled out. It is a detailed schedule of the materials required, their costs, and the installation labor required. The design is carefully inspected and each of the fittings, fixtures, and equipment is recorded on the costing sheet. The work sheet for this example is shown in Figure 12.7.

Each zone of the floor plan is costed separately. The cost per unit for each part is obtained from the company data base of most recent materials costs either from suppliers' current price lists or from an up-to-date quote obtained from the supplier. The length of each type of tubing is estimated from the floor plan using the scale of the drawing. A rule or architect's scale is laid on the

Cost Estimates — Work Sheet
Office Building — Floor #3 — 33,900 Sq. Ft.
Skeleton Installation

Materials	Cost Unit	Zone A No. Req'd	Zone A Total Cost	Zone B No. Req'd	Zone B Total Cost
Controller-Model SSI-34	$235	1	$235	1	$235
Timer-Model E-1020	18	1	18	1	18
Pressure Regulator-adjustable with strainer	34	1	34	1	34
Backflow Preventer-vacuum-breaker	14	1	14	1	14
Stop Valve-globe	6	1	6	1	6
Connector—for water line	1	1	1	1	1
Tubing/Pipe/Hose					
¾" ID-Rigid-Type L-copper	$0.56/ft.	290 ft.	$162	240 ft.	$135
Fittings					
¾" ID Ells-Sweat-Brass	$0.80	2	$ 2	2	$2
¾" ID × ¾" ID × ½" FPT-Brass Branch Tees	1.10	10	11	8	9
¾" ID Ells-Sweat × ½" FPT	0.85	2	2	2	2
½" MPT Plugs	0.55	12	7	10	6
Other (Misc.)			$45		$45
Total Materials Expense	$1,044		$537		$507
Total Labor Expense	440*				
Installation 390*		(1 Plumber-8 hrs. @ $27/hr.; 1 Helper-8 hrs. @ $20.25/hr.)			
Other (electrical) 50*					
Grand Total Expense	$1,484*				$1,484
Profit/Overhead (installing contractor—15% of materials)					157
Grand Total Cost of Installation					$1,641
Cost per Square Foot (at 33,900 sq. ft. of tenant space)					$0.05/sq. ft.

*Includes subcontractor profit and overhead.

Figure 12.7 Costing sheet for office building design.

drawing in continuous tubing segments to measure the approximate length. To this is added the estimated length of tubing used in the risers and a waste factor of about 5 percent to get the total amount of tubing (of each type) required by the installation. Fittings are counted by size and type, and the numbers recorded. While doing this, a great deal of thought has to go into each connection and the type of fitting needed to make that joint. In the current example, the main tubing lines are rigid copper, and the fittings must be soldered in place to join sections and make connections into office suites. Copper ell fittings are used at the end of the main lines. These are 3/4" ID sweat × 1/2" FPT to make connection with the 1/2" branch tubing at a later time. Plugs are temporarily used in the 1/2" FPT legs of these fittings. Hardware and miscellaneous supplies are added to the list. These items include tubing clamps and/or pipe hangers, screws, masonry nails, wood support blocks, solder, flux, etc. Subtotals are calculated for each item on the list, and material totals are derived from these. All cost figures should be rounded to the next highest dollar amount for simplicity.

Labor costs for the installation must then be estimated. The basis for these costs are quotes from installation contractors, past experience, or from data published in recent editions of *Means Mechanical Cost Data Book* published annually by R. S. Means Company, Kingston, MA. This valuable reference lists installation labor costs by job category. In getting quotes from potential installation contractors, be sure they have all of the facts before making a bid. Labor costs will vary from one location of the country to another, and even widely within an area. Quotes should include the subcontractor's profit and overhead on labor. Installation labor costs will be different if a developer has in-house plumbers, as opposed to using an independent contractor.

Once labor costs are calculated, they are added to the total material costs to obtain a grand total expense. A factor for installer's overhead and profits must then be added to this. A 15 percent subcontractor markup on materials is considered a fair figure for estimates. That figure is added to the grand total expense figure to arrive at an estimated grand total cost of the installation. In the present example, the total for that floor came to $1,641, or $0.05 per square foot. General contractors will frequently add another 10 percent to the totals.

A recommendation could then be made to the client to extend the branches of the system to service the planter boxes flanking the elevators in the common area. It might cost an additional $0.01 per square foot or so for that floor, but the $350 spent during construction would save many thousands of dollars in plant-care costs over the life of the building.

Example of System Design for an Office Suite

This example of Micro-Irrigation System design will deal with connections from the building's skeleton installation to service a hypothetical corporate office suite. The irrigation system design work would be done just after the architects and interior designers have completed their layouts, and an office floor plan, or, better, finished working drawings, are available. In the example at hand (see Figure 12.8 and 12.9), the office suite was designed with approximately 5,000 square feet of space.

The interior design calls for the use of about 109 containerized foliage plants. They are to be placed on filing cabinets, desks, credenzas, bookcases, and tables; there will be freestanding planters on the floor as well. The numbered circles and polygons on the floor plan are tentative locations for the plants. Actually, that fact has little relevance in designing this installation. Irrigation receptacles must be designed into the walls at logical yet convenient locations, so plants can be serviced easily, regardless of where they are eventu-

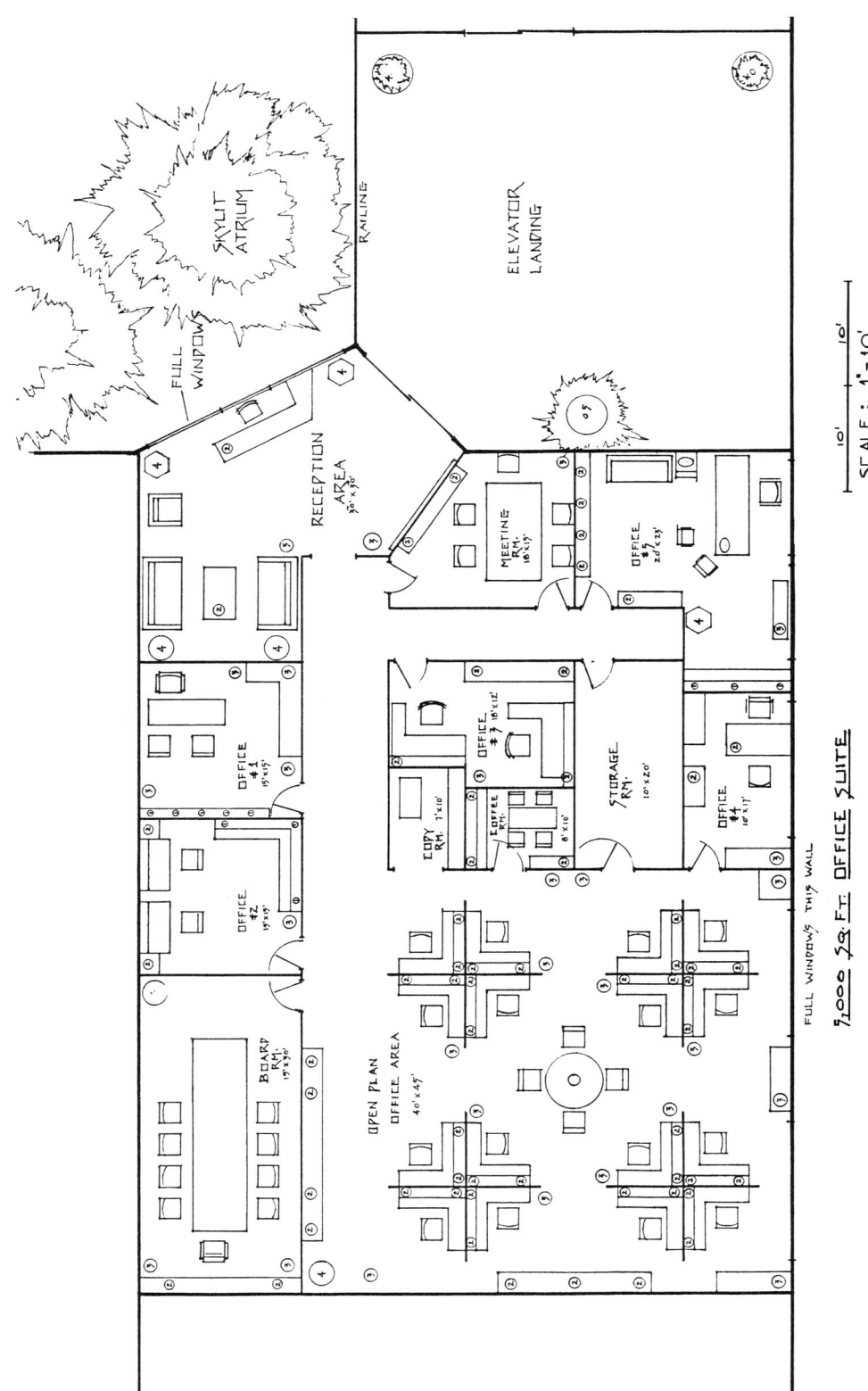

Figure 12.8 Floor plan for a typical office suite.

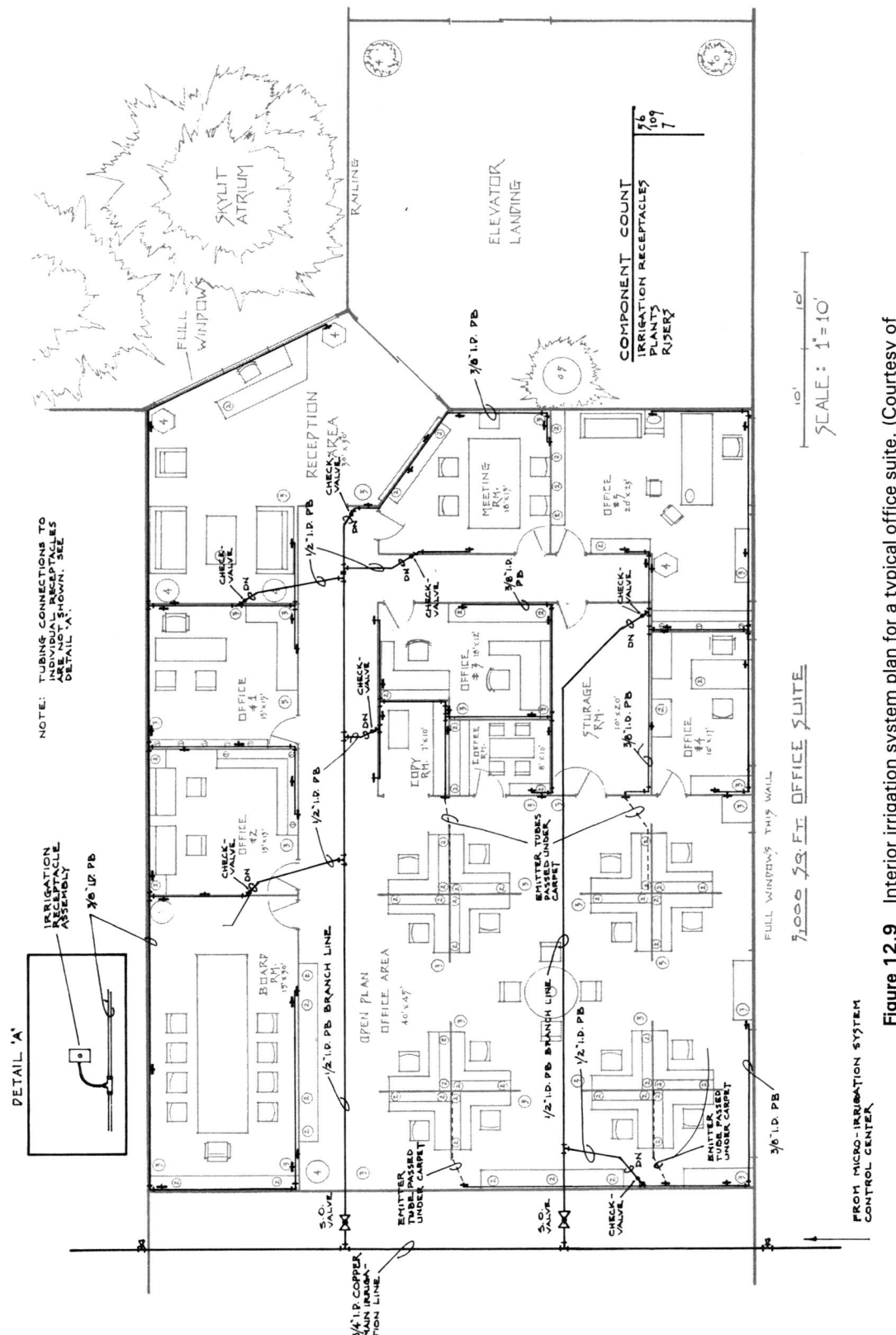

Figure 12.9 Interior irrigation system plan for a typical office suite. (Courtesy of Ayers, Snyder & Locke.)

ally placed (or moved to at a later time). The receptacle locations were chosen according to the logic discussed earlier in this chapter. When all the receptacles were located on the plan, they were counted, and the total was found to be 56. If 109 potted plants are eventually serviced, there would be that many emitters used in this branch of the building's system. The demand by this branch for irrigation water is calculated as follows:

Number of emitters × 0.09 GPM average demand/emitter = total demand
109 × 0.09 = 9.8 GPM

Because most of the plants in this particular suite will be small, each using less than 0.09 GPM of flow rate, it is not necessary to double the calculated total demand figure with a loss factor. The main irrigation line servicing this floor is 3/4" ID, and the solenoid valve controller used would have an output of at least 40 GPM to accommodate all the suites on the system. Only full-pressure output would be considered; however, if only one or two suites were on the system during the early leasing stages, pressure levels would be reduced. Another way of doing this is to design a fixed pressure regulator into each suite's input line, so that no matter what the general system pressure is at the time, the suite's pressure would be held at a fixed level. This is the preferred method.

To be sure that the flow is well distributed to the suite without too much demand on any one branch input line, it was decided to feed water through two independent 1/2" ID polybutylene branch lines instead of just one. That means making connections overhead to two branch tees in the 3/4" ID main irrigation trunk line. Some building codes do not permit plastic tubing or piping in overhead areas of commercial buildings; thus, copper or another acceptable tubing material can be used instead. Local codes must always be checked in the early stages of design work. The 1/2" ID branch lines are also routed over the office suite, with lateral subbranches taken from them and dropped down through partitions at six strategic locations. These are marked "DN" for "down" on the plan. The locations for these drops, or risers are chosen so that as many receptacles or emitters as possible can be serviced from each. The tubing material can be 1/2" ID polybutylene in this case, without violating most building codes. **A check-valve must be designed into each vertical line at its base.** The smaller lines branching off from the bases of the risers are 3/8" ID polybutylene; 3/8" OD PB tubing can generally be used as well. They are routed through partitions at baseboard level to irrigation receptacles and other emitter locations. It may be noticed in Figure 12.9 that an extra drop was made into the reception area partition. This was done as a safety measure to assure that area of enough water to accommodate a number of larger plants, such as trees. A common interior design practice is to place more prominent plants in reception areas.

The open plan office area must be serviced in a different way. Assuming the floor of the office suite is not raised (as is done in many modern office buildings), one of two courses of design can be taken. Each workstation island, or grouping, can be serviced with a low-pressure, Micro-Irrigation System, with the control center and pump/reservoir module installed in an empty cell of a desk or cabinet. Small-diameter tubing lines would be routed from there through the workstation partitions to the planter locations (including those free-standing on the floor). The neatest, most professional installation would be made with irrigation receptacles mounted on the surface of cabinets, desks, or partitions to provide easy plug-in access to the water lines. The designer can decide to use this more refined method, but must take the extra material and installation cost into consideration.

The second method that can be used in open plan office areas is using emitter tubes from the suite's high-pressure irrigation system to service the workstation islands. An emitter-tube assembly is simply plugged into a nearby irrigation receptacle and routed under the carpeting to a workstation. This works well when the office carpeting is installed with a thick foam underpad. PVC emitter tubing is small in diameter and soft; it will sink easily into the soft underpad and not be seen from the surface. Other methods of installation can make the tubing runs under walkway areas even less obtrusive, virtually invisible. It is the walkways that are usually the problem. Wherever furniture covers the flooring, a slight ridge from below makes little difference, but one across an open floor would be considered unsightly. The same problem exists when trying to retrofit electrical wiring to a workstation. It must also be passed under the carpeting and creates the same situation. Tubing, being smaller and more flexible than commercial wiring, is less of a problem. Keep in mind that thin-walled tubing of this type tends to flatten down somewhat; it does not become completely flat but takes on an oval configuration. That reduces its propensity to show a ridge on the surface. At the same time, water flow is not restricted, because so long as the tube is not flattened completely, the interior area remains the same and carries the same amount of water as when the tube is rounded. Obviously, furniture and other objects cannot be placed over the tubing as that would cause a flow restriction. More and more residential-type carpets with foam underpads are being used in commercial settings. Aside from the fact that there is a more decorative selection, interior designers find that sound absorption qualities are better, that they are more comfortable to walk on, and that the underpad helps the longevity of the carpeting because it cushions pedestrian pressure and reduces friction on pile fibers. A difficult situation occurs when the office carpet is of a commercial type. These types are usually thinner and are cemented directly to the concrete slab floor without the use of supplementary foam underpads; they are generally foam backed. As a result, it is difficult to pass electrical wiring, tubing, or anything else under these carpets without a ridge showing. Sometimes electrical wiring is passed across the top surface of the carpet and covered with a thin plastic strip called a cord cover. The same technique can be used with small tubing. In any case, the various methods under consideration should be discussed with the client and a final decision made as to how the workstation islands are to be serviced with automated plant care. In this case, we will assume that the decision was to plug into irrigation receptacles and pass the tubing the short distance under the carpeting to the workstation islands, then route it through the workstation structure to the planter locations, as with low-pressure systems. Shallow channels can be cut into the concrete floor slab to accommodate this short tubing run.

For the sake of clarity, the interior irrigation system plan for this example has been drawn on an overlay sheet (see Figure 12.9). In costing this project, it must be remembered that no control center or main water distribution lines are necessary because that service has already been provided by the building's system. Figure 12.10 illustrates a cost estimating work sheet like the one we did when designing the skeleton installation.

Each type and size of tubing is listed and measured off for approximate length. Each type and size of fitting is identified on the plan and listed. Each accessory and piece of hardware that is likely to be required is also listed. Start with the largest tubing lines first and work down to the smallest. Add some extra to all figures for waste. The amount of actual waste will vary with the skill

Cost Estimates—Work Sheet Office Suite—5,000 Sq. Ft.			
Materials	Cost/ Unit	No. Req'd	Total Cost
Stop Valve-globe	$6	2	$12
Pressure Regulator—Fixed—15psi	10	1	10
Tubing/Pipe/Hose			
½" ID PB	$0.33/ft.	280 ft.	$93
⅜" ID PB	0.29/ft.	460 ft.	133
¼" OD PVC	0.09/ft.	110 ft.	10
Fittings			
½" ID Tees-Acetal-Barb type	$0.60	5	$3
½" ID Check-Valves-Acetal-Barb type	4.50	7	32
½" ID × ⅜" ID Ells-Acetal-Barb type	0.62	2	2
⅜" ID Tees-Barb type	0.53	5	3
⅜" ID Ells-Barb type	0.44	12	6
⅜" ID × ⅜" ID × ½" ID Branch Tees- Acetal-Barb type	0.53	5	3
Emitters			
Adjustable, Mini-Valve-Plastic-Barb	$0.53	130	$69
Accessories			
Irrigation Receptacles (with accessories) ¼" OD	$23	56	$1,288
Emitter Hold-Downs	0.10	130	13
Tubing Clamps-½" ID	0.20	20	4
Hardware			
Crimp Rings-½" ID	0.14	58	$9
-⅜" ID	0.11	180	20
Misc.			$50
Total Materials Expense			$1,760
Total Labor Expense* (2 Plumbers-8 hrs. ea. @ $27/hr. 1 Helper-8 hrs. @ $20.25/hr.)			594
Grand Total Expense*			$2,354
Profit/Overhead (installing contractor- 15% of materials)			264
Grand Total Cost of Installation			$2,618
Cost per Square Foot (at 5,000 sq. ft. of office space)			$0.52/ sq. ft.
*Includes subcontractor profit and overhead on labor.			

Figure 12.10 Costing sheet for office suite design.

of the installers. A 5 to 10 percent waste factor (on tubing, fittings, and hardware only) is usually normal. There would not be an expectation of waste when considering electronic equipment, irrigation receptacles, and the like. The total material expense for this design is about $1,760. There is no such thing as an absolute price when it comes to making cost estimates on any project. The cost of materials and labor varies with so many factors that unless firm quotes have been obtained and the project is carried out within the period that the quotes are valid, only close estimates can be stated in design and cost studies. This is particularly true of quick, early-stage investigations. Estimated costs can usually be guaranteed, however, within 10 or 15 percent. Installation labor costs vary widely by geographical location, type of craft, experience levels of the workers, union affiliation, and use of in-house labor as opposed to hiring independent contractors. In the case at hand, installation is predicated on the use of an independent contractor. It is estimated by the contractor that two plumbers and one plumber's helper would be required for 8 hours. Wage rates were quoted at $27.00 per hour for each of the plumbers, and $20.25 per hour for the helper. Therefore, total labor expense comes to $594 for the project, making for a grand total expense of $2,354. Another 15 percent of the material expense is added for subcontractor profit and overhead charges. That brings the total to $2,618. This is what it would cost the building owner in this example to extend an existing Micro-Irrigation System into a new tenant suite. On a square footage basis, the total represents $0.52 per square foot. The building owner might mark up his cost and charge the installation to the tenant at a higher rate, either as a onetime charge or amortized over the period of the lease.

Example of System Design for a Single Family Home

This section will demonstrate the design steps of a residential Micro-Irrigation System. Assume that a home builder requested the cost of installing a fully integrated version (Aqua/Trends' Mirage III Systems, for example) into a luxury single family residence of 3,200 square feet while under construction. The floor plan furnished by the builder (see Figures 12.11 and 12.12) shows a three-bedroom, two-and-a-half bath house with family room and den.

Further investigation discloses that above the living room, dining room, and foyer is a high vaulted ceiling. That fact is relevant here, because it means there is no overhead space through which to route tubing in those areas. The large garage is chosen as the most appropriate location for the control center. Not only is it out of the way, but it is next to the laundry room where there is convenient access to a cold water line. A stop valve is designed into the input water line to shut off the entire branch when necessary; an adjustable pressure regulator that includes a built-in screen filter (strainer) is also designed into the input water line.

Next, irrigation receptacles are located at appropriate places along the walls of the rooms. Three are planned for the foyer as potted plants are frequently placed there to provide a casual look to the entrance. All the other rooms are fitted with these special outlets as well, including the kitchen. There two receptacles are designed into the splashboard above the countertop. Many homeowners like to have plants on kitchen counters and the convenience of having automated plant care there as well as in the other rooms is a welcome amenity. Modern bathrooms are fitted this way as well, particularly those having skylights or windows. This is not the case with our current example, however. The number of irrigation receptacles included in the plan are counted and recorded (28 for this example).

Now knowing how many outlets are involved, we can calculate the total system demand. Assuming only one emitter per irrigation outlet on average, there would be a maximum of 28 emitters in operation at any given time; there will usually be less. The total demand of this system is calculated as follows:

Number of emitters on the system × 0.09 GPM average demand/emitter
= total demand
28 × 0.09 = 2.5 GPM

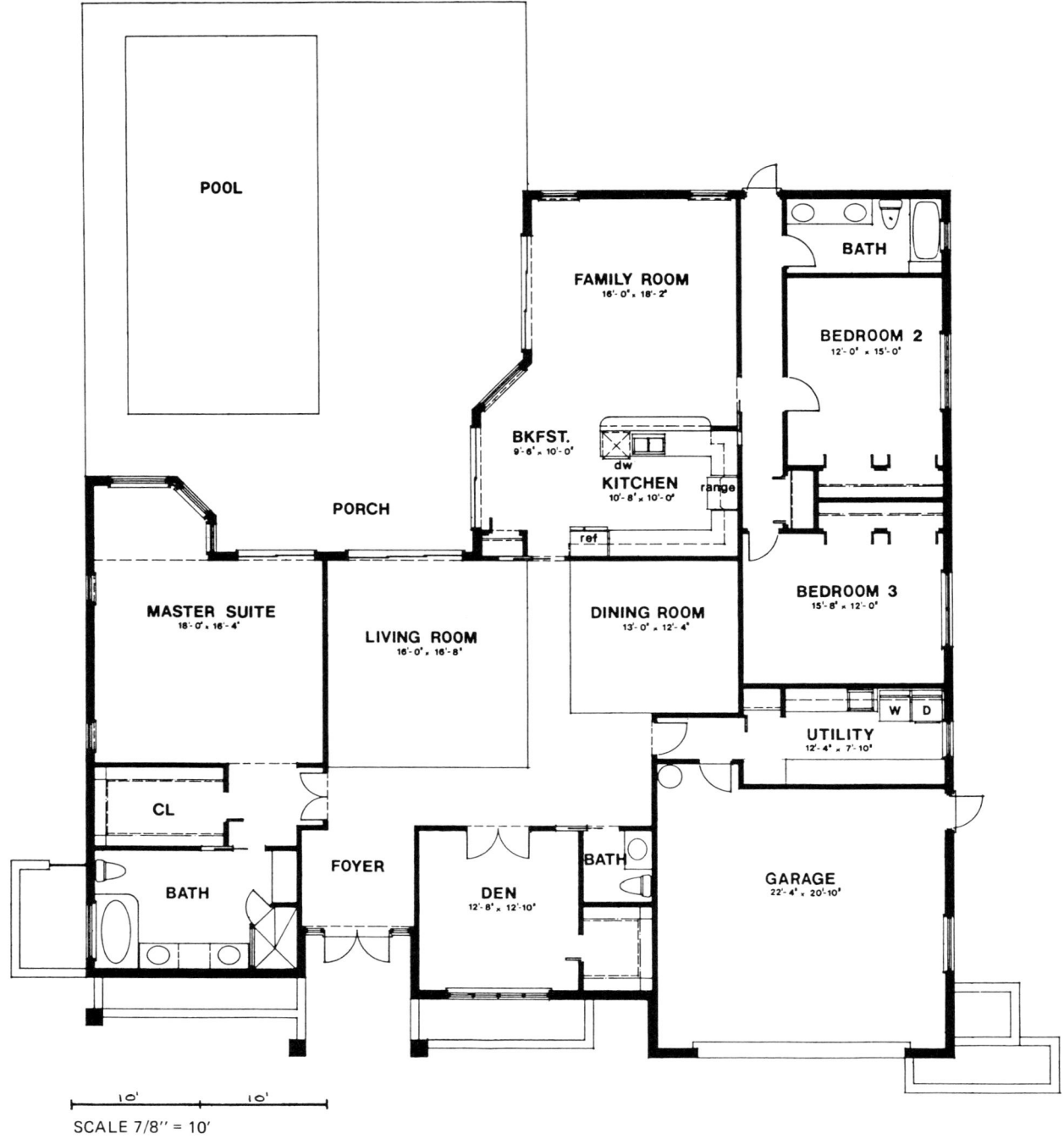

Figure 12.11 Floor plan of a single family home.

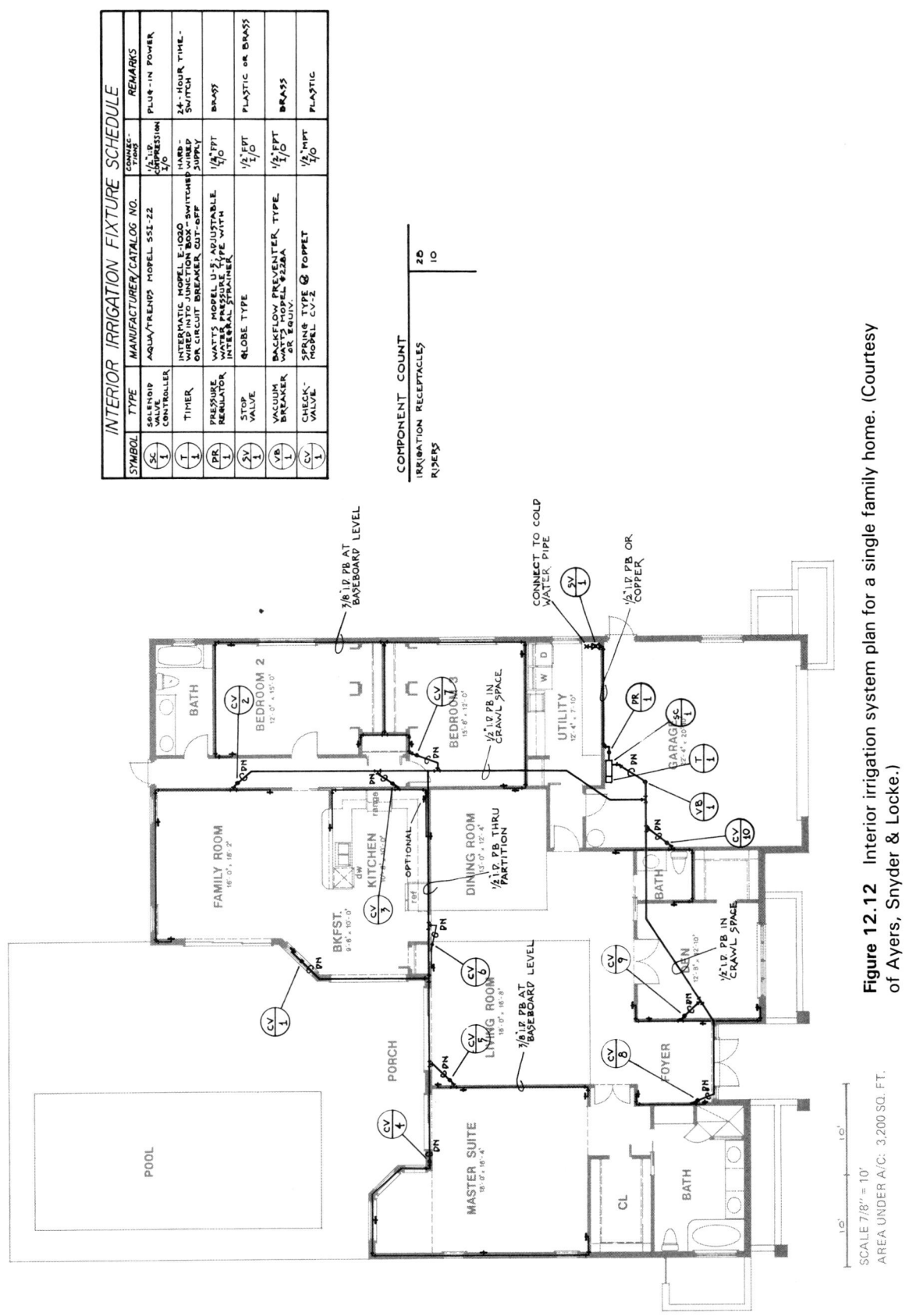

Figure 12.12 Interior irrigation system plan for a single family home. (Courtesy of Ayers, Snyder & Locke.)

From this, the flow rate required at the irrigation system input can be calculated as follows:

Total demand (GPM) × 2.0 loss factor = total flow rate required
2.5 × 2.0 = 5 GPM

A solenoid valve controller with an output of at least 5 GPM must be chosen to control system operation. Referring to Table 12.1, we see that a solenoid valve with a C_v of at least 1.4 is required. The better choice would be one having a C_v of 3.4 or 3.6. For this example, we have chosen Aqua/Trends' Model SSI-22 as it has enough capacity to service the plan under consideration as well as other auxiliary branches, if required, indoors or out. It has an input/output tubing size of 1/2″ ID.

Next, we must find the best routing for the tubing lines. They must be integrated into the framework of the building so as not to be seen. One of the problems we will encounter, however, is the vaulted ceiling area. These rooms generally require that tubing, wiring, etc., be installed around their perimeter, as space is seldom available overhead. Other problems to be encountered have to do with the many interior and exterior doorways. There are many doorways in homes, and subbranches running along the baseboard usually do not traverse an open doorway—although there are practical ways to do it. The line must either be terminated there or routed up and around the door frame to the other side. The former choice is preferred. As most rooms in this house have conventional 8-foot ceilings, there is space above the ceiling joists to route tubing lines. From the output of the solenoid valve controller, 1/2″ ID polybutylene tubing is routed to the garage ceiling plenum. A vacuum-breaker type of backflow preventer is installed in the line at its highest point. Its output is branched off in two directions, one toward the building entrance and one toward the back of the house. Once past the vaulted ceiling area, another branch is connected into it to be run through partitions for service to the master suite, living room, and family room. The other main branch is passed over the den and over the front door frame to service the receptacles on the far side of the foyer. Drops down through partitions are made at appropriate places from these 1/2″ overhead branches. At the base of each drop (riser) is positioned a check-valve and connections to 3/8″ ID polybutylene tubing lines that traverse the room perimeters at baseboard level. To these are connected the irrigation receptacle input tubing. The interior irrigation system plan should be laid out on vellum tracing paper or drafting film if the floor plan furnished by the client is small. The drawing becomes too confused with all of the elements involved if it is laid out directly on the floor plan. This example has been drawn on an overlay (see Figure 12.12).

The costing of this installation follows the same procedures we used in previous examples. Control center elements had previously been chosen while designing the system. They should be listed, along with their latest prices on the cost estimating work sheet (see Figure 12.13). The length of tubing runs is measured with a ruler or scale and estimated from the drawing scale. A little extra length is added to compensate for waste. Estimates for 1/2″ ID and 3/8″ ID tubing are recorded, as are the counts for each fitting size and type, accessories, and hardware. The latest prices for each are extracted from supplier price lists or recent quotes. Labor costs should be obtained from several installation contractors as job quotes. Totals should then be calculated for materials and labor. In this case, grand total expense was estimated at $1,882 or $0.59 per square foot, based on the 3,200 square-foot area of the house under air-conditioning. Garage area is not counted, as it isn't considered part of the living

Example of a System Design for a Single Family Home

Cost Estimates—Work Sheet Single Family Home—3,200 Sq. Ft.			
Materials	Cost/ Unit	No. Req'd	Total Cost
Controller-Model SSI-22	$224	1	$224
Timer-Model E-1020	18	1	18
Pressure Regulator-adjustable, with strainer	34	1	34
Backflow Preventer	14	1	14
Stop Valve-globe	6	1	6
Connector-for water line	1	1	1
Tubing/Pipe/Hose			
½" ID PB	$33/ 100 ft.	240 ft.	$80
⅜" ID PB	30/ 100 ft.	240 ft.	72
Fittings			
½" ID Tees-Acetal-Barb type	$0.60	8	$5
½" ID Ells-Acetal-Barb type	0.48	3	2
½" ID Check-Valves-Acetal-Barb Type	4.50	10	45
⅜" ID Tees-Acetal-Barb type	0.82	2	2
⅜" ID Ells-Acetal-Barb type	0.72	18	13
⅜" ID × ⅜" ID × ½" ID Branch Tees-Acetal Barb type	0.53	7	4
Accessories			
Irrigation Receptacles (with accessories)	$23	28	$644
Tubing Clamps ½" ID	0.18	24	5
⅜" ID	0.10	38	4
Hardware			
Crimp Rings-½" ID	0.07	40	$3
-⅜" ID	0.06	64	4
Misc.			40
Total Materials Expense			$1,220
Total Labor Expense* (1 Plumber-14 hrs. @ $27/hr. 1 Helper-14 hrs. @ $20.25/hr.)			662
Grand Total Expense			$1,882
Profit/Overhead (builder @ 50% of expense)			941
Grand Total Cost to Homeowner			$2,823
Cost per Square Foot (at 3,200 sq. ft.)-Expense (Profit/Overhead Excluded)			$0.59/ sq. ft.
*Includes subcontractor profit and overhead on labor.			

Figure 12.13 Costing sheet for single family home design.

area. Many home builders typically add on another 50 percent to cover their profit and overhead, making the cost to the homeowner $2,823 for this system. Because capacity is large enough in this case, a suggestion would be made to the homeowner that the Micro-Irrigation System be extended outdoors to the pool/patio area. This could be accomplished easily and without much additional cost and would conveniently service freestanding and hanging containerized patio plants.

The house used for this example is larger than average but was selected because it contained most of the challenges normally faced in designing residential systems. Smaller homes and apartments are typically less difficult to design systems for than this home.

Computer-Aided Design and Costing Modern computer-aided design (CAD) methods are well suited to projects of this type. They can significantly reduce the time and effort required to design and cost Micro-Irrigation System installations, particularly when dealing with complex configurations. Computer software programs will soon be available for practitioners.

13 THE INSTALLATION OF MICRO-IRRIGATION SYSTEMS

Overview In many respects, installing the elements of precision, Micro-Irrigation Systems™ is much the same as installing electric power wiring, plumbing system piping, security system or intercom system wiring, or central vacuum cleaner system ductwork. In each case, the distribution network must be routed through the building in such a way as to be operational and safe, yet unobtrusive. Those are the main objectives of Micro-Irrigation System installations as well. There are many techniques unique to this concept that have been developed over a period of time, mainly through firsthand installation experiences. In the limited context of this chapter, we will pass on to the reader most of what must be learned to be knowledgeable, skillful, and versatile in this specialized field.

General Layout Considerations As we have seen in Chapter 12, there are many impediments to the easy application of this technology (or any other for that matter) in a building interior. The design of the building is a major factor, as are its construction materials. If done with foresight and skill, the designer should have taken all or most of the installation problems into consideration and provided a plan that would make the installer's job much easier. The interior irrigation system plan with its complementary fixture schedule and cost estimating work sheet can provide a "road map" for the installation. It shows the equipment and parts involved, the general layout of the control center, the general routing for the water distribution network, the types of tubing connections that must be made, the installation points for flow control devices, and the approximate locations of emitters as well as decorative plants. The installer, in looking at this material, should be able to get a quick idea as to what is required of them. If the installer was involved in the bidding process, the plans and job requirements would already be familiar.

Probably the most difficult part of an installer's work is the routing of tubing throughout the structure. There are many decisions that must be made as the job progresses. This is particularly true of retrofit projects, in which the techniques have mostly to do with being able to hide tubing lines from view. When rooms are empty and are fitted out with residential-type carpeting with adequate space beneath for tubing lines, then the installer's job is relatively easy—but somehow things are never that simple. Problems are created by furniture and room layout, potted plant locations, partition structures, and other such details. Many times, they cannot be foreseen in the design stages, and appropriate action must be decided on the spot. An installer's training and experience will be important during these times.

The integrated versions of Micro-Irrigation Systems also run into a wide variety of situations that must be dealt with in specific ways. Probably the least difficult are the skeleton installations in commercial buildings because the floors to which they are applied are vacant and wide open. There are few physical impediments in roughing these systems in. Probably the most difficult situations involve tubing installations at inappropriate times, for example, after much of the interior partition drywall has been installed. Situations like this are usually accidental; however, competent project management should minimize their occurrences.

Installing Control Centers The location of control centers has been discussed in previous sections, and out-of-the-way spots are preferred. In dealing with high-pressure Micro-Irrigation Systems, control centers are generally located in utility rooms of some sort in the core of commercial buildings, or garages or laundry rooms in residential structures. Water source connections are made to the building's cold water supply in a couple of ways. If no convenient hose bibb is available, a tee fitting is soldered into the copper line and the irrigation branch tubing connected to it, usually with a ½" MPT or ¾" MPT fitting. A stop valve (usually globe type) is installed near the connecting point, followed by a strainer or other filter (see Figure 13.1). If a pressure regulator is to be used in the system (as in the majority of cases), it can be installed next in line. Some adjustable types

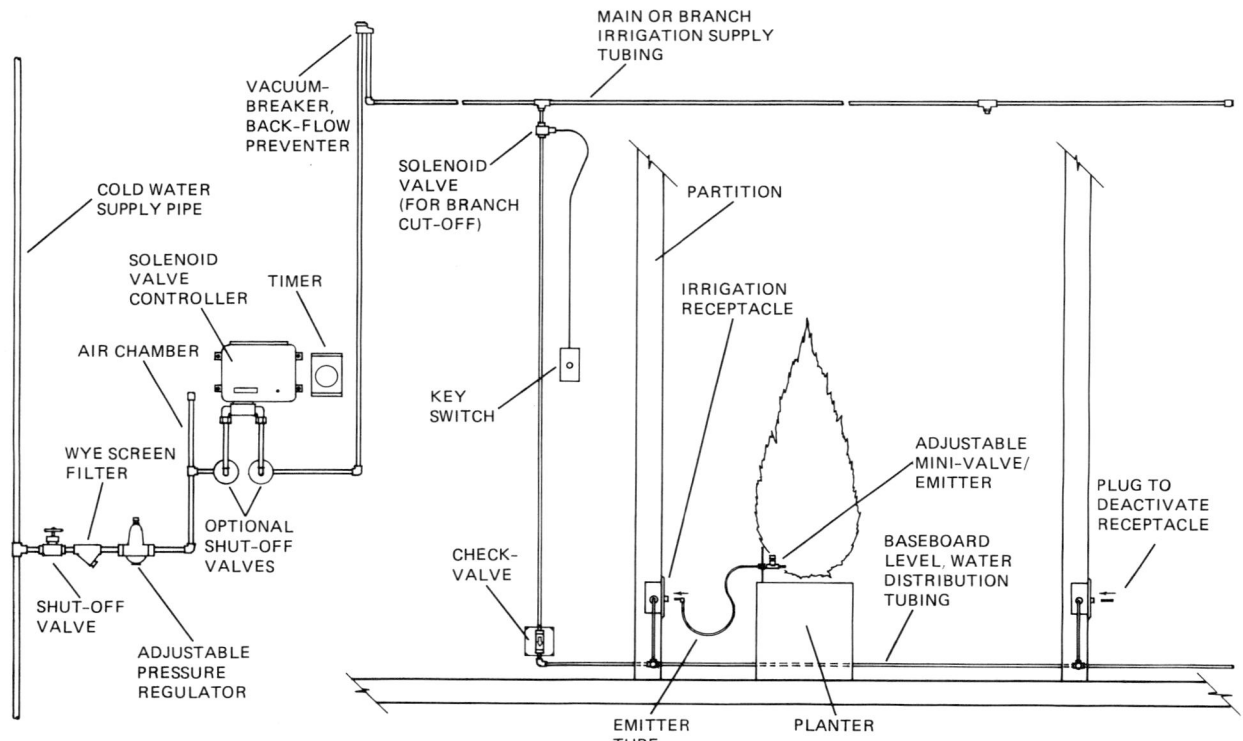

Figure 13.1 Configuration of a high-pressure Micro-Irrigation System integrated into a partitioned structure. (Courtesy of Aqua/Trends.)

have built-in screen filters so as to integrate the two devices. All of these connections can be made with threaded fittings, preferably made of brass. The output side of the pressure regulator is connected to the input side of the solenoid valve. Part of this connection should include an anti-water-hammer air chamber. It is simply a capped bypass tube that absorbs the shock of high-volume water flow stopping abruptly. It can save wear on fittings and the solenoid valve itself.

All solenoid valve controllers are surface mounted. The built-in solenoid valve accepts input tubing from water source connections on one side, then sources the water distribution tubing network on its output side. Input tubing can come from inside the wall behind the controller, in which case escutcheons are used to dress up the pass through drywall. When tubing is integrated into a building during construction, it is roughed in, usually some time before the electronic control elements are finally installed. At the control center location, tubing stubs are passed through the drywall and loosely capped, ready for inspection and pressure testing of the irrigation tubing network and final connection to the solenoid valve. The temporary caps are later replaced with a bridge of tubing between the input and output stubs to permit continuous flow through the system during testing. This connection is made to be temporary by the use of compression fittings. If the irrigation branch shutoff valve is not close to the control center, a globe valve should be installed in the solenoid valve's input line for emergency shutoff.

Limited light-duty systems useful in residences, offices, hotel suites, and restaurants can be sourced from plumbing connections under kitchen or bathroom sinks. Control centers are located in the cabinets, and tubing is routed through partitions to irrigation receptacles or other emitter connections.

Low-pressure versions are installed in a convenient, hidden place. When large pump/reservoir modules are involved in interior installations, they are located in a closet or utility room of some type. Pump controllers and timers are mounted on the wall near the reservoir. Most controllers are plug-in types, so a timer having a power outlet must be used. Small low-pressure systems with 2 ½ gallon reservoirs can be easily hidden behind or under furniture, behind large planters, and inside of cabinets or desks. With these as well, the small pump controller is fastened to the wall in close proximity to the reservoir. In each case, the low-voltage output wire from the controller connects to pump terminals or wire leads; the installer should be careful to observe proper DC polarities.

If electromechanical time switches are used in the system, they must be mounted next to the controllers. One type commonly used is a simple plug-in design that is plugged into a power source, with the controller plugged into it. Another common type used in Micro-Irrigation Systems wires into an electrical box with the controller connected to it either through hard wiring in the same electrical junction box or externally plugged into it. Wiring a wall switch to the junction box for power cutoff is not recommended unless a key switch is used. Experience has shown that it is too easy to throw the switch accidentally, and the system can languish inoperative for days or weeks before anyone recognizes the problem. Key switches make such mistakes more difficult. Some designers may want to use a separate electrical circuit with its own circuit breaker for the Micro-Irrigation System. It is not considered necessary to do so as the power required is seldom greater than 50 watts—much less power is required in "cruising mode," which is 99 percent of the time. There are internal fuses protecting the control equipment. As a compromise step, it may be desired to install a localized fuse or circuit breaker to protect just the irrigation branch of the electrical circuit. If in doubt, it would be a good idea to discuss the situation with local building inspectors or electricians. The main requirement is that continuous power be made available to the irrigation system control center.

If a centralized power management system is used in the building, it must be able to selectively control electrical outlets or junction boxes, providing power to the control center. The computer must be programmed to apply power to the control center for at least one minute, two or possibly three times per day. It is a good idea to provide a supplementary power lead, independent of the computer control circuit, to use when the main supply is disfunctional.

Installing Water Distribution Networks

The routing and connection of tubing networks for Micro-Irrigation Systems are a complex subject; there are a variety of materials used in a variety of ways. The commonality is that the tubing must carry water to planter locations, must be leak free, must provide local control over water flow, and must be sufficiently hidden to make the installation neat and unobtrusive. Each of these requirements will be dealt with in turn, as will the special requirements of various types of installations.

Routing Tubing Lines for Integral Systems

When Micro-Irrigation Systems are integrated into the structure of buildings, such as with Aqua/Trends' Mirage III Systems™, most of the tubing network is installed while the interior partition work is in progress, at the same time as the plumbing and water supply are roughed in. In those instances where skeleton installations are made in vacant commercial floor space, the task is easiest because no partitions have yet been constructed to impede progress. Competent plumbing contractors should be employed to do this work. Small secondary installation jobs can be handled by handymen, but keep in mind that the tubing system should be pressure tested and inspected before it is put in use. Local building codes require this in most instances. The point is that the installer should be knowledgeable and skillful about tubing connections. After partitions are enclosed, leaks are difficult to service. Of course, the problem is no different than with the primary plumbing system; as a matter of fact, it is much less severe for several reasons. If installers are competent in handling that work, chances are good that a little further training will make them competent in Micro-Irrigation System work as well.

Tubing networks in skeleton installations branch out from the solenoid valve controller to overhead trunk lines. This is usually copper tubing work. The vertical line (riser) from the controller is installed to about 12" higher than the highest tubing line and fitted at that point with a vacuum-breaker backflow preventer (as shown in Figure 13.1). From there, the trunk line is routed down the middle of the vacant tenant space and fastened to suitable supports or pipe hangers. Every 20 feet, a copper tee fitting is soldered into place. (For further descriptions, see Chapter 12.) A copper ell fitting is soldered to each end of the trunk line. All threaded outlets are closed off with ½" MPT plugs.

Tubing networks routed into rooms and suites sometimes require that their water supply lines be cast into the concrete floor slabs and then brought up into the partitions. Local building codes differ on these points, but usually the tubing must be sleeved with outer tubing as it is installed in the concrete form. They also require that the tubing be continuous, with no fittings or other joints cast into the concrete. Because of that requirement, tubing runs must sometimes be brought above ground for branch or continuation connections. This can be done most conveniently in a fixed planter box or, if necessary, in a large junction box mounted in a partition. The two ends of the tubing lines are simply connected with a fitting bridge between them (see Figure 13.2).

Routing Tubing Lines for Integral Systems

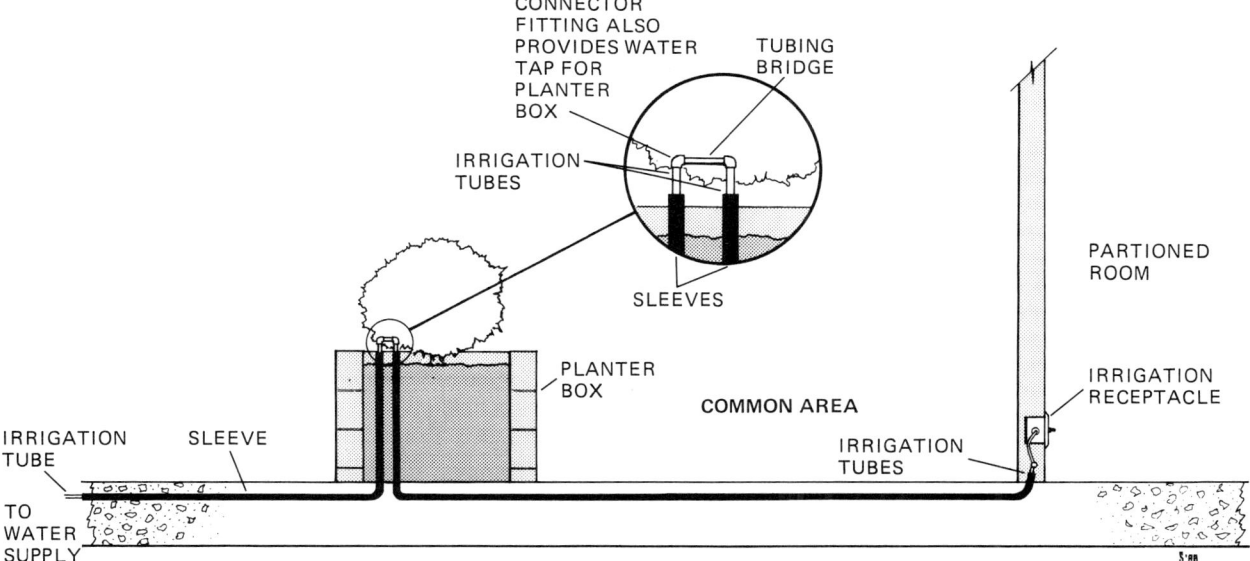

Figure 13.2 Junction technique for irrigation tubing lines cast in concrete slabs.

In most instances, water supply tubing for Micro-Irrigation Systems is introduced into room partitions from overhead connections. When the branch tubes are dropped down into the open partitions, it is important that a spring-loaded check-valve be installed in-line with its one-way flow path pointing toward the baseboard. The check-valve must be chosen for a specific location. Its cracking pressure must be higher than the static pressure in the riser above it. Polybutylene tubing lines (½" ID) are normally used in these runs, and the connections to and from the check-valve are made with barb fittings and secured with copper or aluminum crimp rings. The installer will find that slipping barb fittings into polybutylene tubing is made easier if the tubing end is first softened with heat, either from very hot water or a heat gun. The latter method is more effective, so long as the temperature used is not excessive. This technique works as well with other plastic tubing. Building codes generally require that there be convenient access to flow-control devices, such as check-valves, so they can be removed easily and repaired in case of problems. Access can be provided with these systems by having an access plate mounted in the drywall (gypsum board). An electrical "mud ring" is then attached to the opening in the drywall to provide a mount for a 4" x 4" blank electrical plate that provides the access when necessary. It can simply be unscrewed to gain access to the check-valve in the partition behind it (see Figure 13.3).

Just below the check-valve, connections are made to branch off further with ⅜" ID or ⅜" OD tubing, which circles the room within the partitions at baseboard level. A reduction is necessary from the ½" ID input tubing size to ⅜" ID or ⅜" OD. Tee fittings are available with ½" ID branches that make the transition easy. These, too, are barb-type fittings and must be secured with crimp rings. As a matter of fact, all barb connections of plastic tubing that are to be sealed up within partitions must use crimp rings. Many building codes require it.

The smaller tubing lines that run around the room at baseboard level must be passed through partition studs. Holes are drilled in wooden studs about 4 inches above the floor. In the case of metal studs, they are frequently made with

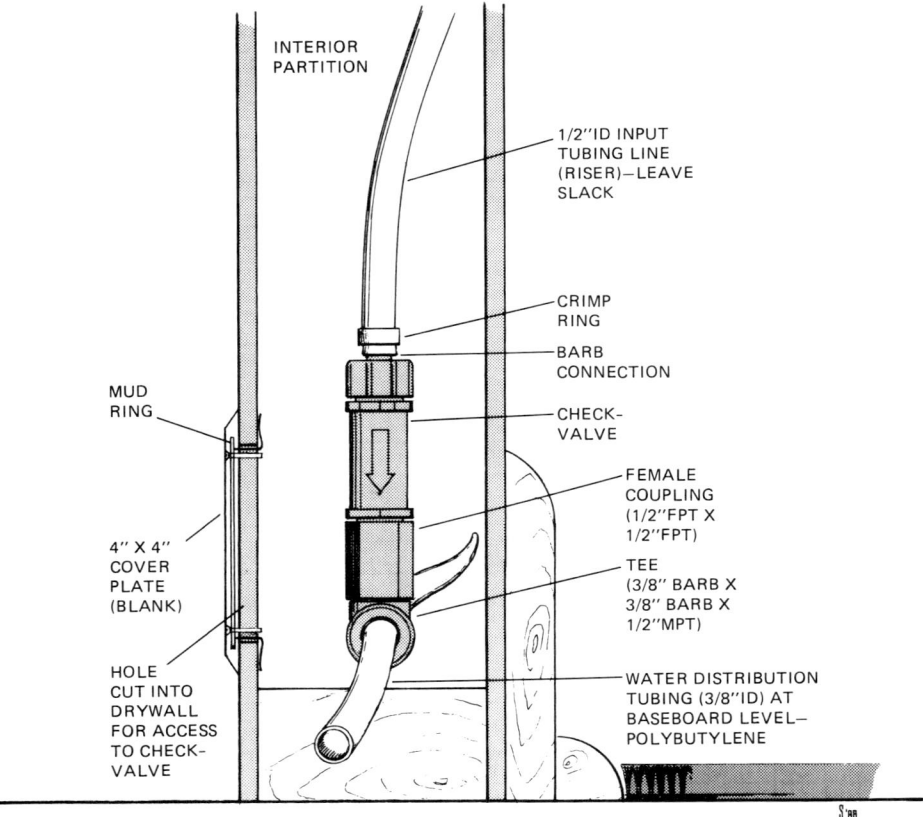

Figure 13.3 Detail of a typical check-valve connection for installations in interior partitions.

holes already punched into them for electrical and plumbing service lines. They can be used for interior irrigation tubing as well. Protective plastic inserts should be used in those holes supporting irrigation tubes. If properly located service holes are not already available, they must be punched out of the metal, stud by stud. This is a routine procedure for plumbing contractors, so the installer will normally know what to do. IT IS IMPORTANT IN INSTALLING MICRO-IRRIGATION SYSTEMS THAT THESE LATERAL BRANCH TUBING LINES BE AT LEAST 6 INCHES BELOW THE IRRIGATION RECEPTACLES AND LOCATED SO AS TO AVOID PUNCTURE BY NAILS AND DRYWALL SCREWS.

During the rough-in stages of installation, the small lateral branches are fitted with feeder lines for irrigation receptacles to be installed in the room. A special tee fitting, with $\frac{3}{8}$" OD feeder tube attached, comes with the irrigation receptacle kit. It is a barb-type fitting that installs into the baseboard tubing line about 9 inches to the side of where the irrigation receptacle will be installed; it is secured by crimp rings (see Figure 13.4).

A junction box also comes with the irrigation receptacle installation kit. It should be mounted on a stud at the desired location, with the box centered about 12 inches above the floor (see Figure 13.5). The irrigation receptacle feeder tube should be passed through the side and then the front of the box. A tag on the tube end should be noted with instructions for the drywall contractor. It should read something like: "ATTENTION DRYWALL CONTRACTOR. Interior irrigation tubing. Cut drywall around box opening."

The routing of tubing lines in a home is not much different than installing

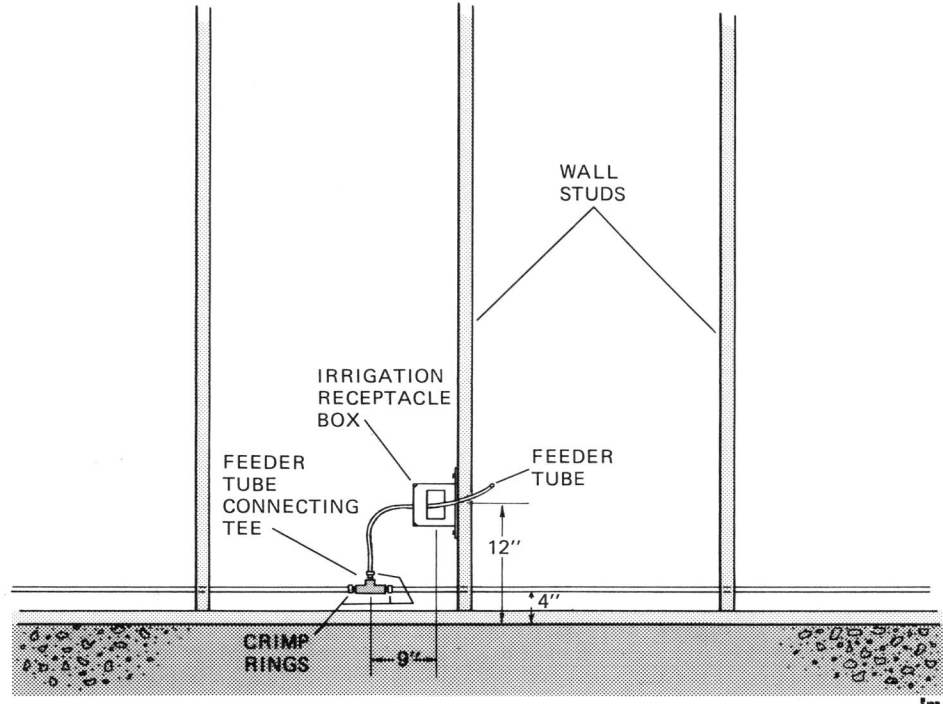

Figure 13.4 Detail of an installation showing irrigation receptacle assembly.

a system in a commercial building. The tubing is passed through beams, studs, joists, and partitions in the most practical way available. The residential design example in Chapter 12 gave the reader an idea of the problems likely to be encountered in this type of installation. If there is an overhead attic, crawl space, or basement, then the job is much simpler. As much tubing as possible should pass through these out-of-sight locations. Any tubing easily accessed in these plenums can be connected with the appropriate compression fittings instead of the barb types. The compression fittings are easier to install but are more costly. In cold weather locations, care should be taken to prevent the pipes and tubes from freezing. This can be done by routing the lines through well-insulated spaces or providing pipe insulation. When routing irrigation lines through basements, attics, or floor joists, it becomes an easy matter to bring an irrigation receptacle feeder line into a partition to service one or more outlets (see Figure 13.6).

One of the advantages in using polybutylene tubing is that a flexible type is available and comes in 100-foot coils. This makes it simple to snake the tubing through tight places, and bends can be used wherever possible to replace elbow fittings. The installation will go faster, and the costs of labor and parts are reduced. In places where tubing lines are exposed to view it is advantageous to use the rigid form of PB tubing because the rigid tube makes for a neater appearance.

There are times when it is advantageous or necessary to route irrigation feeder lines outside the walls instead of within. In these cases, no irrigation receptacles would be used, just connections for emitter tubes or irrigation manifolds. There are a number of ways this can be handled. In cases where tubing must be routed to the center of a room in order to service a future freestanding planter, a tubing branch is laid in the floor slab form (with sleeving) and termi-

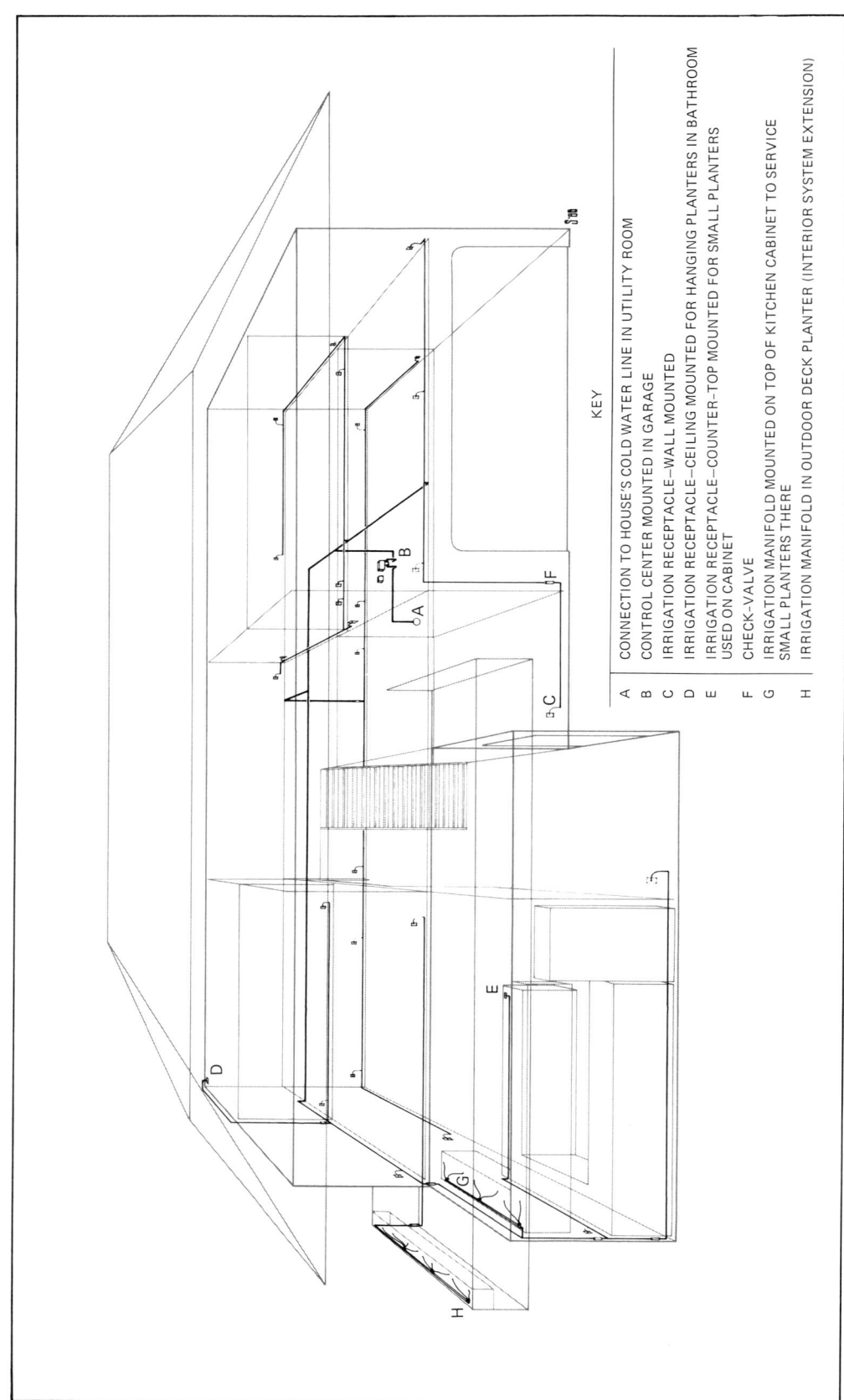

Figure 13.5 Typical home installation of a precision Micro-Irrigation System showing the routing of the tubing within the structure.

Routing Tubing Lines for Integral Systems

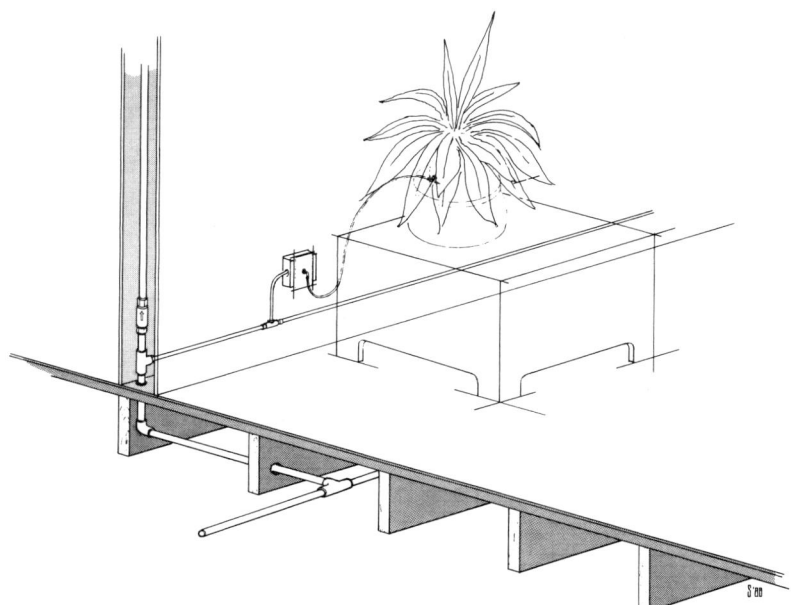

Figure 13.6 Bringing irrigation receptacle feeder lines into a room partition.

nated in a 4" x 4" junction box (with single-gang mud ring) that is also set into the form at the desired location, with the top flush with the floor surface (see Figure 13.7). The irrigation tubing extends through the box and is end capped before the concrete is poured. This tubing would later be connected to an irrigation receptacle plate screwed to the 4" x 4" box. The final result is an irrigation receptacle surface mounted on the floor in the middle of a room to service a freestanding planter to be later installed there.

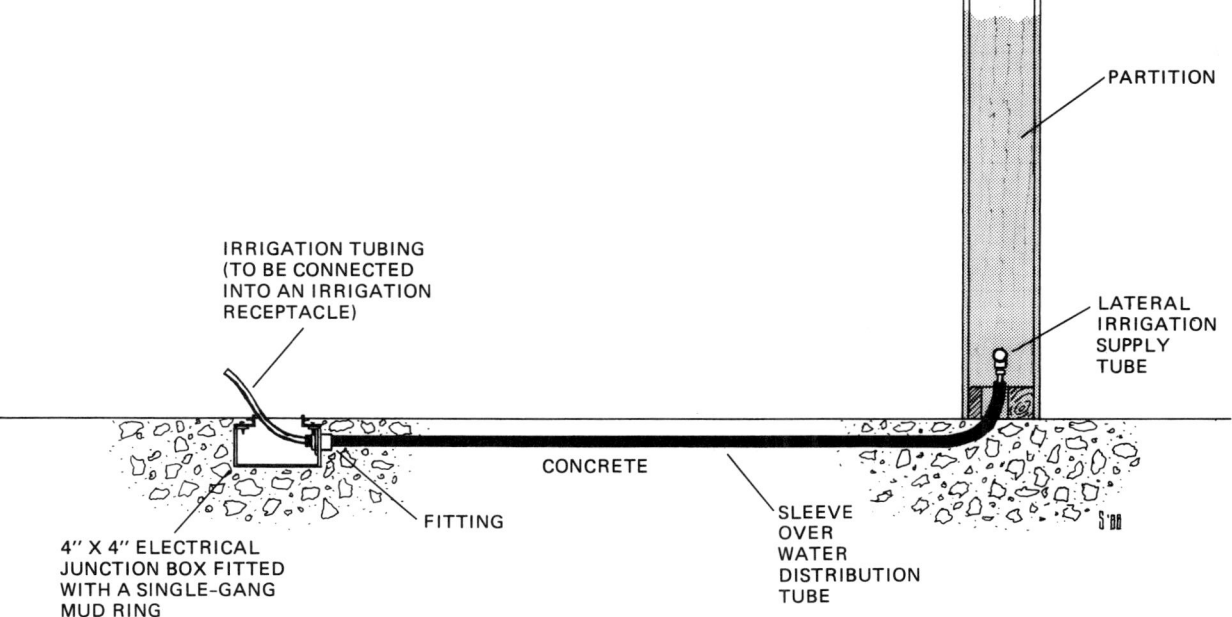

Figure 13.7 Isolated irrigation service box embedded with water distribution tubing into poured concrete floor slab.

Another way of doing this is to use the type of floor box available for electrical service. They are made with decorative faceplates, usually in brass. The junction box itself is recessed into the concrete floor slab, while its decorative faceplate fits flush with the floor surface. A hole can be drilled into the faceplate through which to pass an emitter tube connected to the irrigation supply line within the junction box (see Figure 13.8). Other types have screw-in service ports that provide the same pass-through access.

A similar application utilizes a floor channel. Floor channels are also made to provide electrical or other service to center-room locations, but in a strip pattern so that access to the service can be gained at more than one spot. This concept can be useful with Micro-Irrigation System installations as well. The channel is embedded in the concrete floor slab with an embedded junction box receiving a water distribution tubing line to feed the strip floor fixture. It has a decorative brass cover plate with slots designed into it. Emitter tubes can be connected to irrigation feeder tubes inside the channel and passed through the cover plate to freestanding planters (see Figure 13.9). If vandalism is a concern, emitter tubes of copper, brass, or aluminum, can be used.

Problems are easier to solve along the borders of rooms. When tile or marble flooring is installed, tubing can first be placed at the bases of partitions with small branches to the surface to be used for later connections. They are then sealed in grout when the flooring is installed, as shown in Figure 13.10. Tubing diameters must, of course, be small.

Vertical wall surfaces (partitions, columns, and area dividers) also need special techniques to meet special needs. Irrigation receptacles are more than acceptable for most wall surfaces, either mounted on the surface of drywalled partitions or faced with tile, marble, wooden paneling, etc. Cutouts are made in the decorative facia material to accommodate the special receptacle, as is done with electrical outlets. Abnormal conditions may require a different approach. For example, some public areas are subject to greater than normal vandalism. An owner, fearful that emitter tubing may be destroyed if conventional irrigation receptacles are used to service freestanding planters lined up along building walls (indoors or out), can opt for other types of integral outlets that can overcome this problem. One such special device is the fountain plate, which was mentioned in Chapter 10. It is sourced from an irrigation feed line as are all other outlets and uses a 4" x 4" junction box to enclose connections and mount the exterior plate (see Figure 13.11). A fountain plate must be

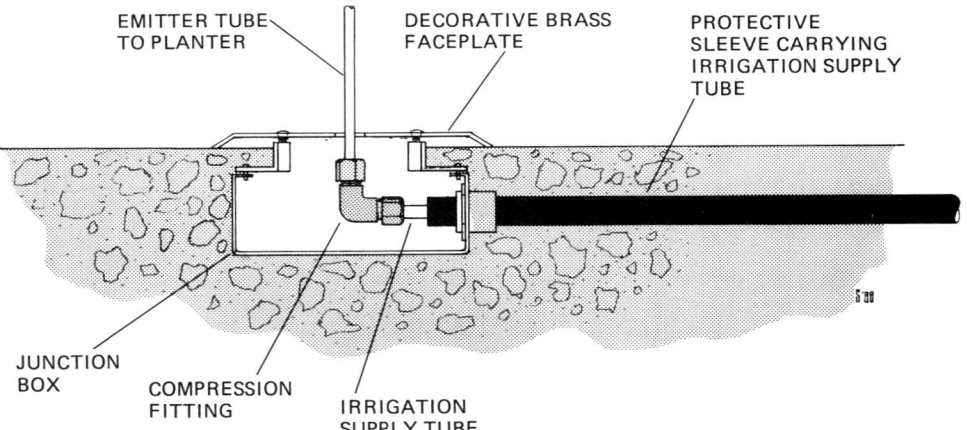

Figure 13.8 Floor box cast in concrete slab and used to access irrigation tubing lines.

Routing Tubing Lines for Integral Systems

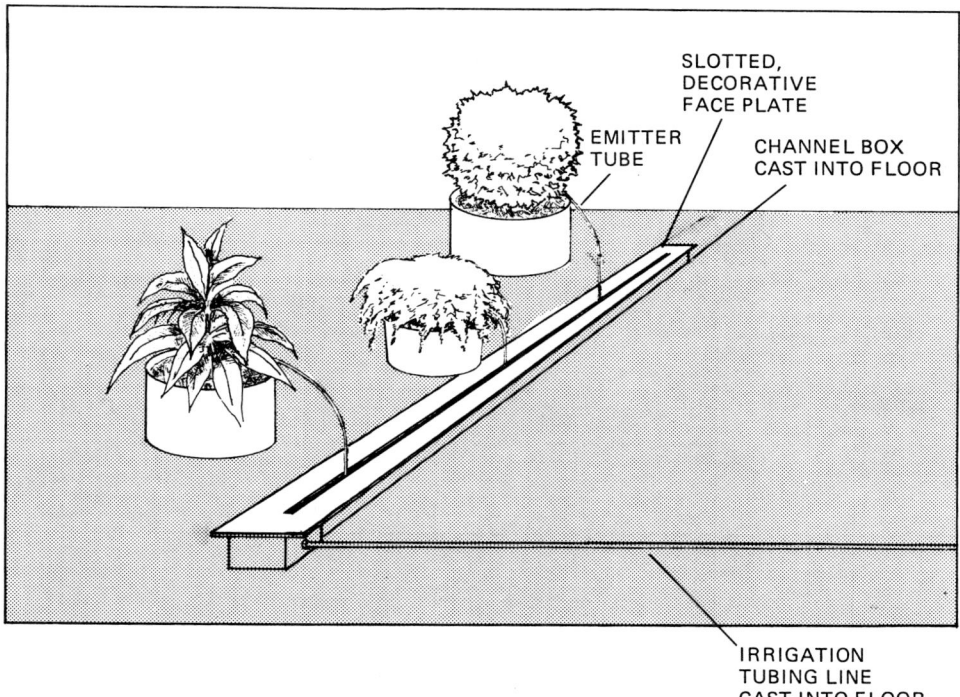

Figure 13.9 Floor channel method of installation used to access irrigation tubing lines.

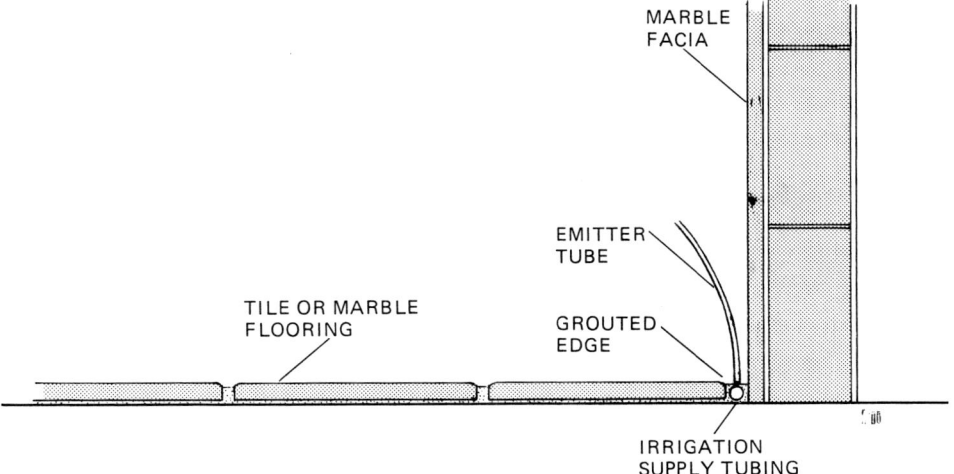

Figure 13.10 Tubing cast into the grouted perimeters of a tile or marble floor.

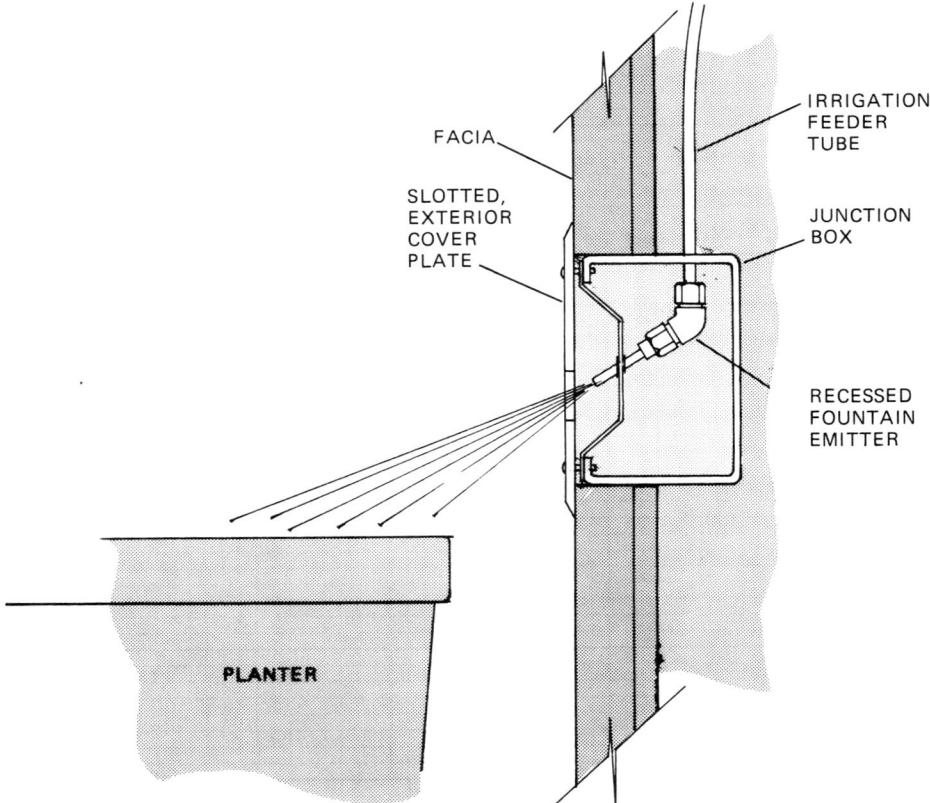

Figure 13.11 Typical mounting of a fountain plate emitter.

mounted at each planter location, and there is generally no flow adjustment. During construction, the junction box is positioned in the wall; irrigation tubing is routed through its top and out the front, ready for further connections to the emitter assembly.

Routing Tubing in Retrofit Installations

Existing buildings are usually more difficult to install than new ones. Small Micro-Irrigation System installations in furnished rooms can be a simple affair, however. The smallest versions are available as kits with all of the equipment, tubing, and fittings packaged together. Tubing lines are not much larger than $\frac{1}{4}$" OD and are usually only about $\frac{3}{16}$" OD. They hide nicely under carpeting. In a case like this, where only 1 to 10 plants are to be serviced in a relatively small area (possibly one or two rooms), a high-pressure control center would be mounted under a kitchen or bathroom sink where plumbing connections are made. If low-pressure versions are more appropriate, then the pump controller and timer would be mounted on a wall behind furniture or planters, or possibly inside furniture or cabinetry, with the pump/reservoir module close by. Tubing would be routed from the control center, under the edge of the carpeting, and around the perimeter of the rooms to the planter locations. Residential as well as some commercial carpeting is laid over thick foam padding with a tack strip holding the edges of the carpet around the walls. There is a gap of $\frac{3}{4}$ of an inch or so left between the tack strip and baseboard molding. This gap makes a convenient channel in which to route the water distribution tubing. The tubing can follow the perimeter of the room, since this is where decorative plants are generally placed (see Figure 13.12). It may be risky to install the tubing before

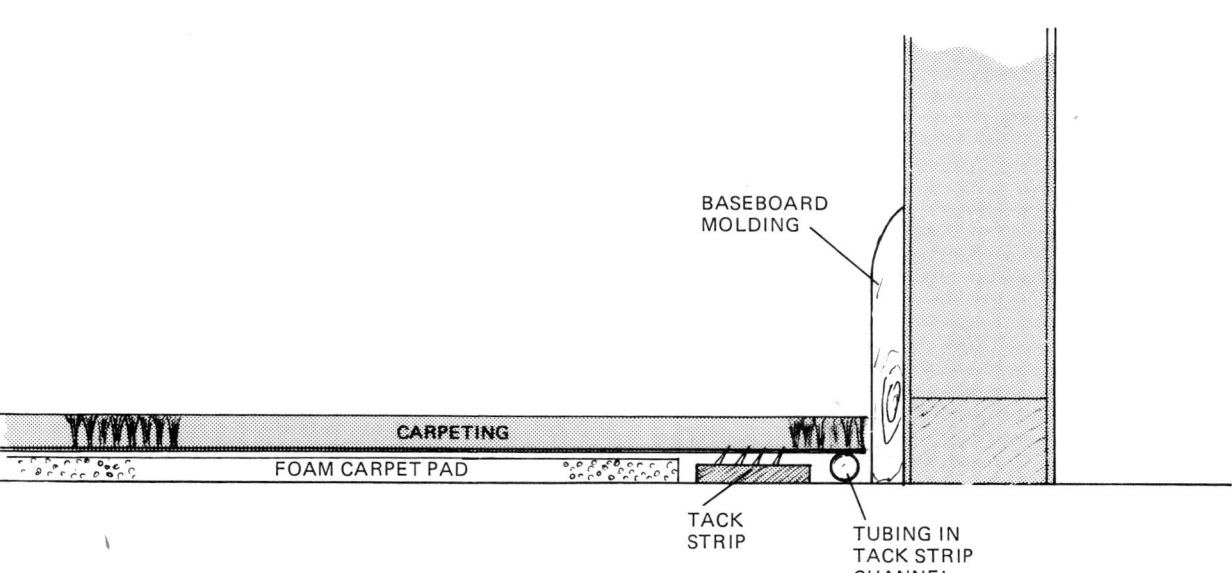

Figure 13.12 Tubing installed under carpet perimeters.

new carpets are laid in renovation projects. Careless edge cutting could damage a tube.

The system illustrated in Figure 13.13 is sourced from the building's water supply. The control center is mounted inside of a kitchen (residential) or kitchenette (office or hotel suite) cabinet, with the tubing distributed around the room under the carpet perimeter. The number of plants that can be serviced and the layout complexity will depend on the tubing size and the duty rating of the solenoid valve controller. Tubing fittings in this case should be a barb type with crimp rings or a compression type. Tubing of $\frac{3}{8}$" OD should be used for the main water distribution lines, if possible, so the system can be extended to other locations or to many more plants, if desired.

The system illustrated in Figure 13.14 is sourced from a low-pressure pump/reservoir module. The assumption is that there is no convenient connection to the building's water supply lines. The control center is located in the corner of the room behind a planter with the pump controller mounted low on the wall so as not to be seen. The small-bore tubing is passed under the carpet edges around the perimeter of the room, and barb-type tee fittings are installed at appropriate points to attach emitter tubes for servicing nearby planters. Because of the low system pressure, it is not necessary to use crimp rings on the fittings; however, the recommended installation practice is to use crimp rings or adhesive bonding to secure the tubing/fitting joints. Fewer plants are serviced by these small low-pressure systems than by the high-pressure versions. The number of plants serviced will depend on the duty rating of the kit chosen.

When tubing must be routed from the perimeter to the center of a room, it can be passed beneath the carpet in a couple of ways. A slit $1\frac{1}{2}$ to 2 inches long can be cut in the carpet backing (between the pile tufts) at the center of the room where the freestanding planter is to be located. A snake, such as one used to draw electrical wiring through partitions, is then passed through the slit and carefully worked under the carpet to the perimeter of the room where a water distribution line is located. A $\frac{3}{16}$" OD PVC tube is connected to the end of the snake and pulled through the slit, with enough exposed length to route it to the planter. The other end is connected into the water distribution line with a barbed tee fitting. The tubing can be run either under the carpet and

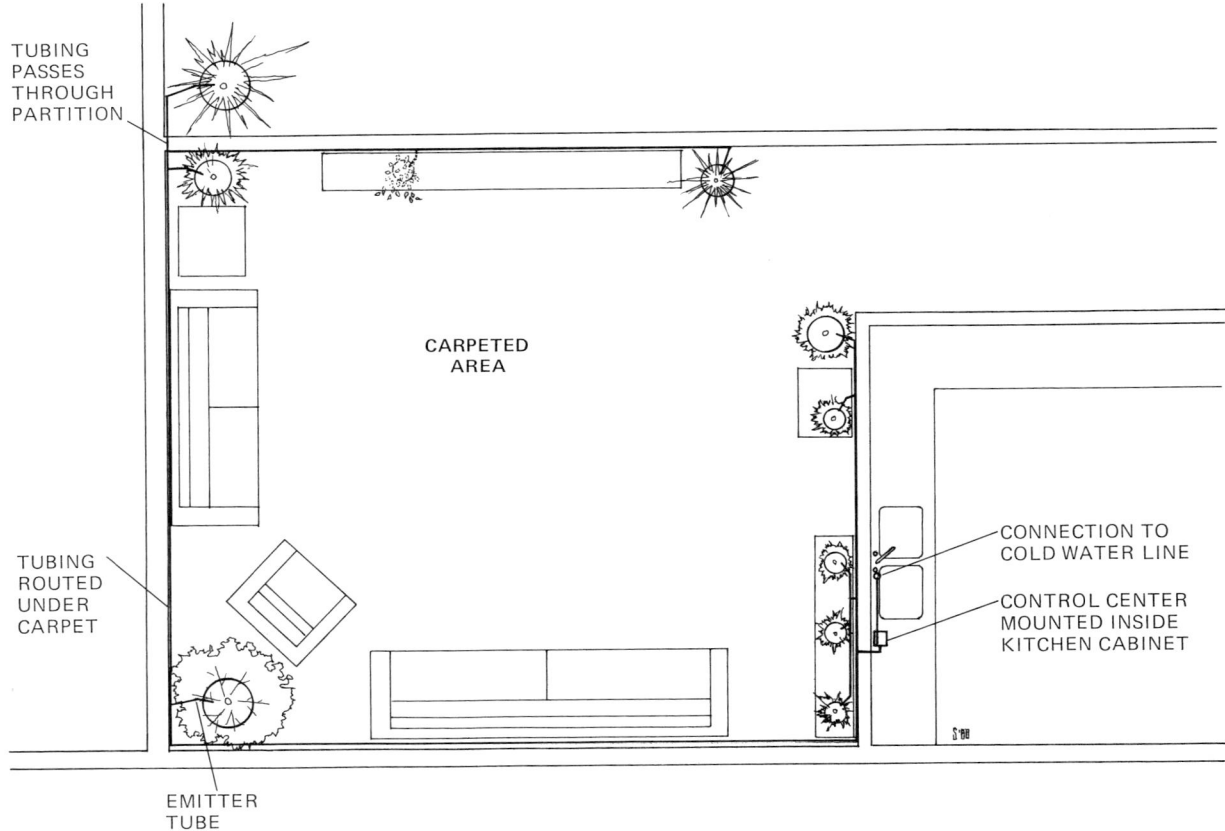

Figure 13.13 Typical limited-service Micro-Irrigation System installation. This is a high-pressure version retrofit into a residential setting.

padding or between them. The latter is preferred because the tubing is then cushioned by the padding beneath it. There is sometimes a slight ridge in the carpet where the tubing passes, but it is seldom objectionable. Flexible plastic tubing installed under carpeting in this manner will tend to flatten somewhat. However, this will cause no problem with the passage of water, *so long as the tube has not completely collapsed.* Elliptically shaped tubing cross sections have the same internal volume as round tubing, and therefore passes the same amount of water (see Figures 13.15(a) and 13.15(b)). Flow restrictions occur when the tubing is flattened too much, as in Figure 13.15(c). The best method of making such installations is to cut the foam padding along the path of the tubing line so as to create a channel for it beneath the carpet (see Figure 13.16). This would normally be done during carpet installation before the carpet itself has been laid.

Where most commercial carpet installations are involved, either there is no underpadding or the padding is part of the carpet, laminated to its underside. The carpets are usually cemented to the floor to prevent slippage. In these cases, it is more difficult to pass tubing beneath the carpeting, but it can be done by cutting a slit first (as previously described); then with a stiff metal or wooden stick passed through the slit, work the carpeting away from the floor, breaking the adhesive bond along a narrow channel all the way to the wall. Plastic emitter tubing can then be fished through this channel, as would be done with an electrical cord. Carpet mastics are made to provide only a semi-permanent bond, so this technique may not be as difficult as it seems. In these

Routing Tubing in Retrofit Installations

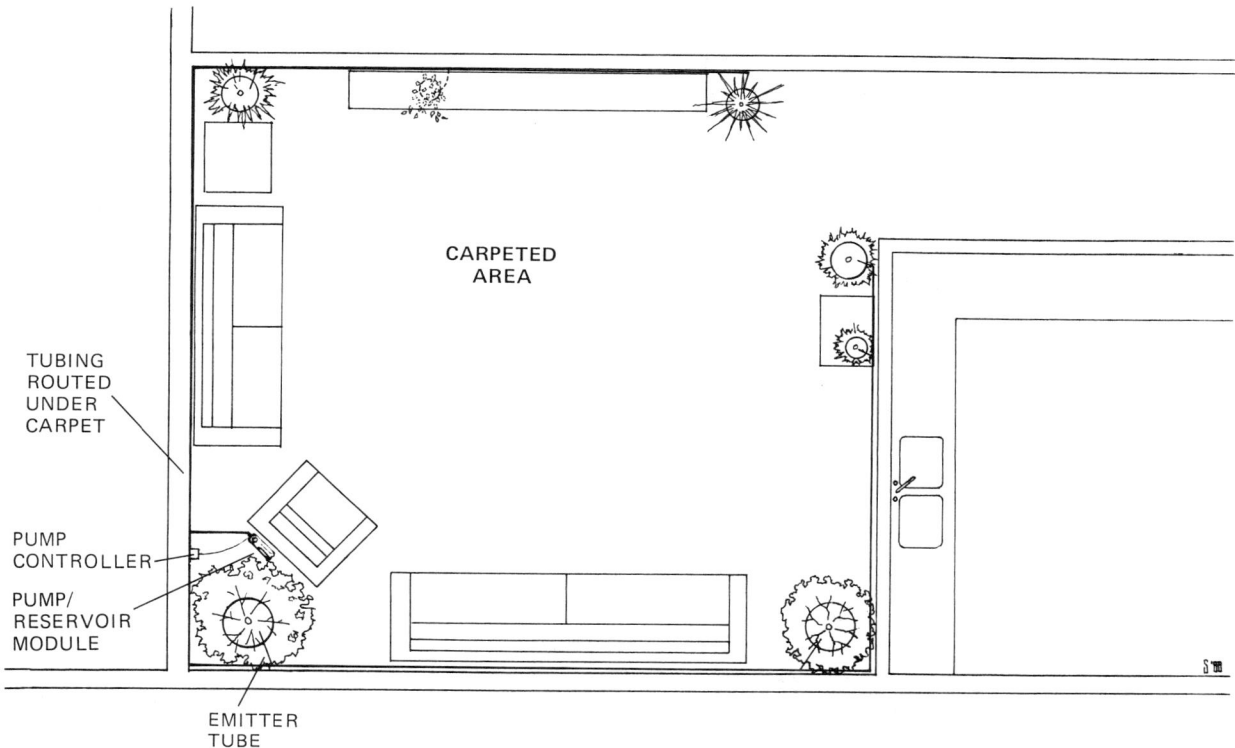

Figure 13.14 Typical limited-service Micro-Irrigation System installation. This is a low-pressure version retrofit into a residential setting.

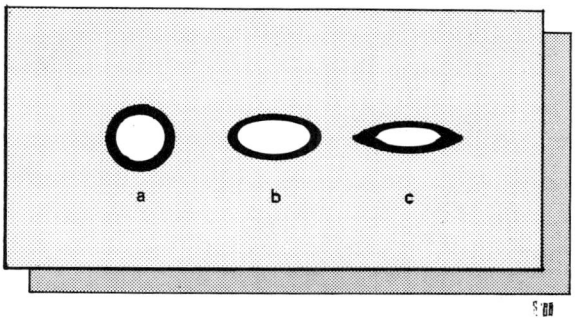

Figure 13.15 Irrigation tubing: (a) round tubing cross section, (b) partially flattened tubing cross section, (c) restricted tubing cross section.

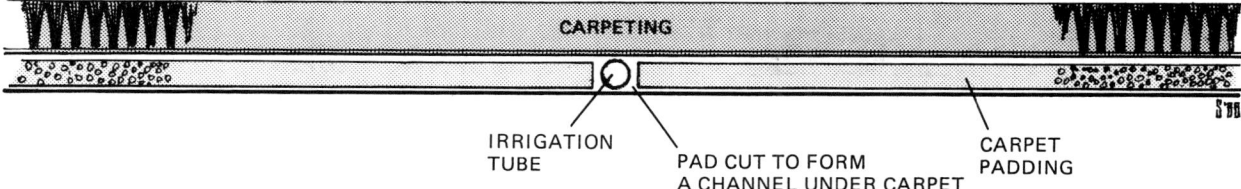

Figure 13.16 Tubing installed under carpeting through a channel of cut carpet padding.

commercial carpet installations, there is no tack strip to hold the edges—thus, no channel exists around the room perimeter. Space can be provided for tubing there, however, by lifting the edges of the carpet ever so slightly and laying the small-bore tubing in the space created. Another way of doing this is to raise baseboard moldings by about ½ of an inch above the bare floor level. This will create a channel in the crook between the molding, drywall, carpet edge, and floor (see Figure 13.17). This crook will be adequate to run ⅜" OD tubing (or smaller) around the room. Yet another technique is to use rubber or vinyl baseboard moldings around the perimeter of the room and hide the small-bore tubing behind the bottom lip (see Figure 13.18).

Another way of accommodating tubing runs is to use channel moldings to hide the installation as shown in Figure 13.19. These can be used not only at the baseboard but up and across wall and ceiling surfaces as well. Several types of channel moldings are available for electrical service installations, and they can also be used for Micro-Irrigation Systems. Among the various types available, some are more decorative than others, some are metal, and others are plastic. The latter types are easier to install. All can be painted to match the decor. Some brands are made with self-adhesive backing to secure the molding to various surfaces. These are recommended for certain circumstances only. The adhesive mounting strip does not work well on many painted surfaces. Loose paint particles will clog the adhesive surface, and it will lose its tack rapidly. Flat wall paint seems to create the worst problem in this regard, as it is heavily filled with pigment, which tends to migrate to the paint surface and then slough off. The best attachment surface for self-adhesive strips are smooth metal surfaces finished with lacquer or hard enamel. The surfaces should be clean and dry before mounting is attempted. Other types of channel moldings are attached to surfaces with screws, either directly or through the use of mounting strips. These types are preferred because of their more secure attachment. They take longer to install, but the callbacks to repair separated moldings

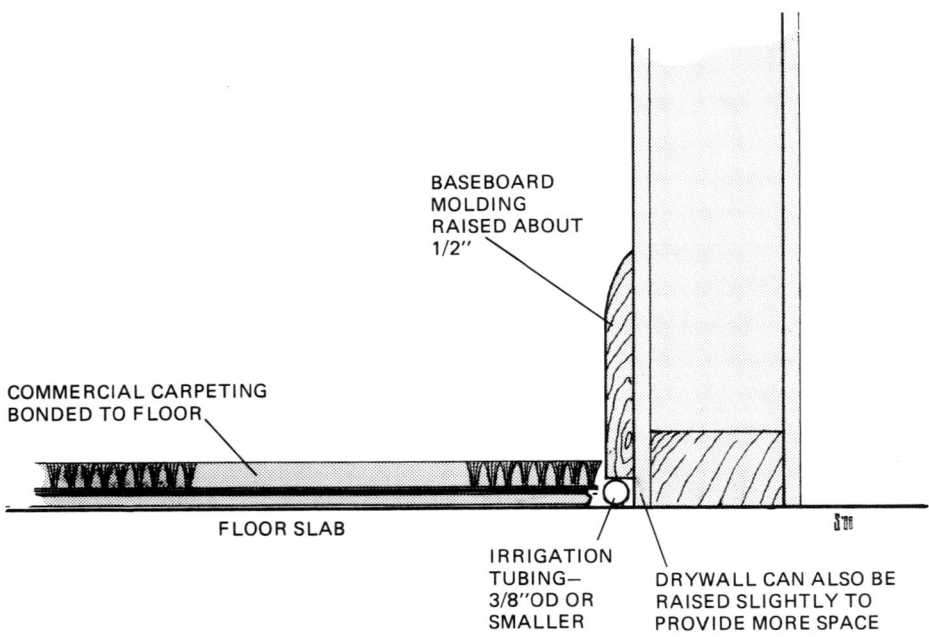

Figure 13.17 Tubing routed under baseboard moldings in a commercial carpet installation.

Routing Tubing in Retrofit Installations 245

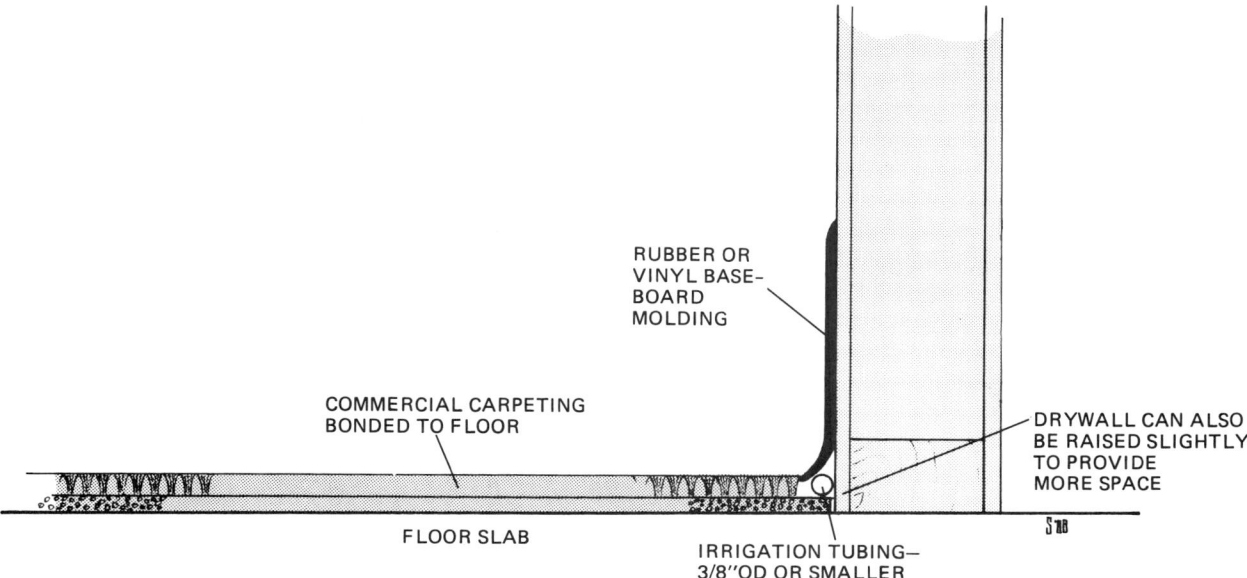

Figure 13.18 Flexible baseboard molding used to hide irrigation tubing lines.

will be fewer. When using any of these channel moldings, connections to the emitter tubes must be made at the appropriate locations (near planters), and cutouts must be made in the molding to allow the tubing to pass through.

Another technique for hiding tubing around baseboards is to modify the existing baseboard moldings with a channel cut into its back side (see Figure

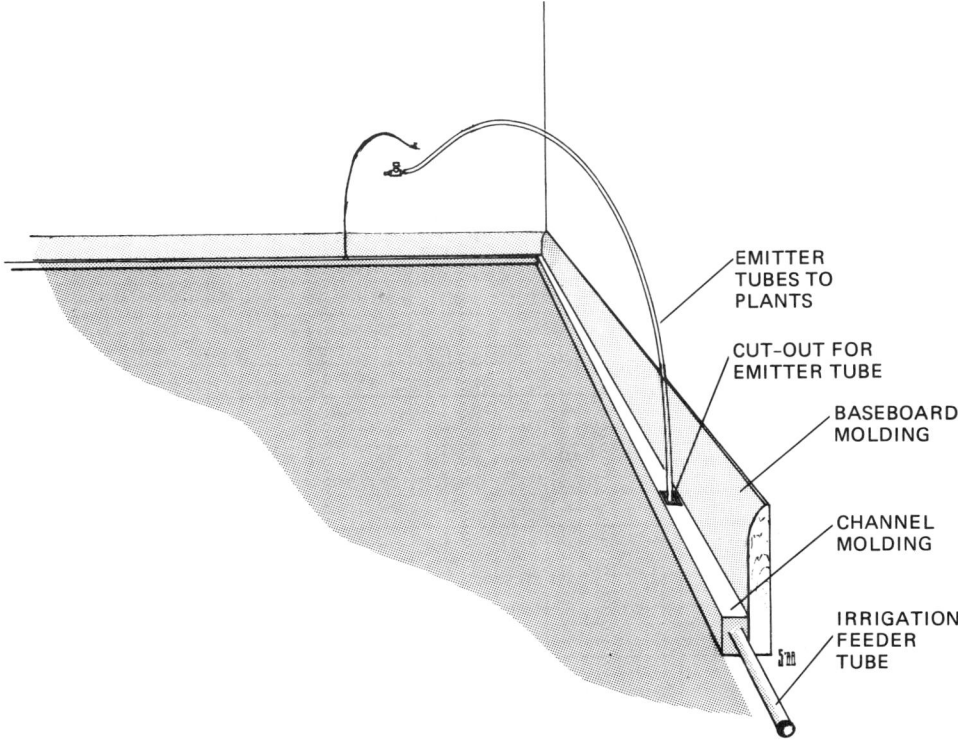

Figure 13.19 Baseboard molding modified with a channel to accommodate irrigation tubing.

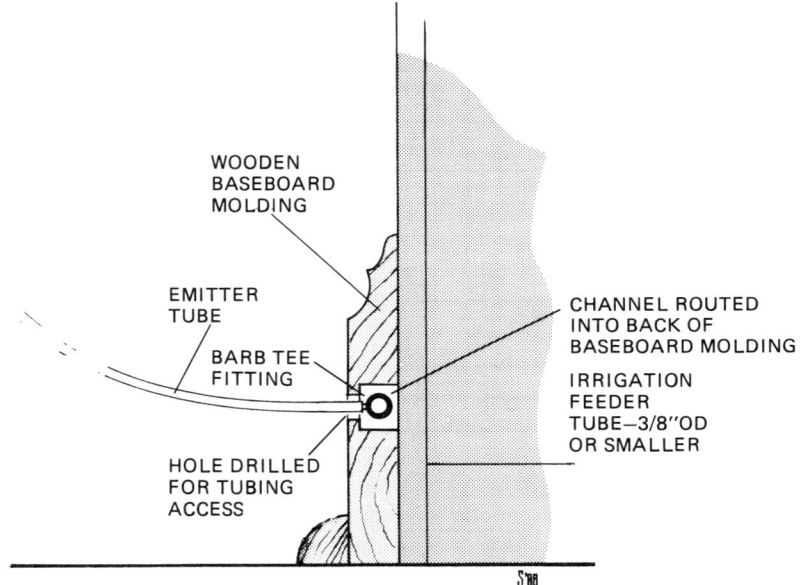

Figure 13.20 Use of channel moldings to hide irrigation tubing.

13.20). That can be done easily with production routing machines or molding cutters. Holes are drilled in the molding at appropriate locations, permitting access to the irrigation feeder line for emitter tubes. Carpenters must be cautioned to keep nails away from the tubing area when installing such moldings.

Yet another technique is available that utilizes common moldings. There are certain brands of small cove moldings that provide a space behind them when mounted. That space can be used to channel irrigation tubing around baseboards (see Figure 13.21). Cutouts or holes must be cut into these as well for emitter tube passage at the appropriate locations.

There will be many situations encountered where no carpeting at all is used on the floors of the rooms to be irrigated—tile, marble, terrazzo, sheet vinyl, or other floor finishes will be used. The techniques just described can be used for these problem situations as well. When large ceramic tiles are used as flooring, tubing can sometimes be installed into the joints between the tiles, then covered with grout.

There are other retrofit situations when it will be necessary to consider cutting a channel into the flooring for passage of the irrigation tubing lines. This is commonly done in construction work, and modern tools make it feasible. After the channel is cut, tubing is laid in and cemented over. This can be done during building renovations without much disruption.

Tubing/Fitting Connections

In all Micro-Irrigation System installations, providing secure tubing connections is paramount. Because of the wide variety of fittings used in the various situations encountered, each requires a different technique.

Barb-type fittings are the most common types used in Micro-Irrigation Systems. They simply slip into tubes and hoses. Low-pressure systems don't usually require any additional joint security as long as there is a natural, friction fit. A more acceptable installation practice, however, dictates that tubes and fittings in low-pressure systems be secured either with an appropriate adhesive,

Tubing/Fitting Connections

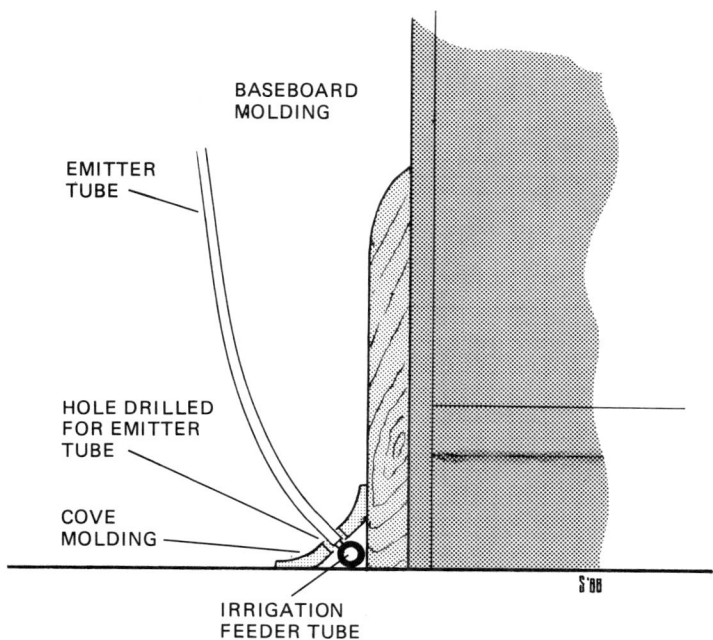

Figure 13.21 Common cove molding used to hide irrigation tubing.

tubing clamps, or with crimp rings. The adhesive types that have shown to work best are latex or neoprene based with volatile solvent thinners, such as solvent-type contact cements.

Crimp rings provide the most secure connections with barb fittings. They are slipped over the tubing end prior to its receiving the fitting and then crimped down over the tubing with a special tool. The crimp should be evenly made around the joint; otherwise, localized tubing distortion, as well as a possible leak, could occur.

Compression fittings are probably the easiest to use. The tubing is simply slipped into the socket provided, and a nut is tightened down until the joint is secure. A gripping collar within the fitting socket crimps down on the tube when the nut is tightened. Plastic compression fittings require not much more than hand-tightening, but an extra $\frac{1}{4}$ turn or so with a wrench makes the joint even more secure. Even though they are easy to use, quality compression fittings carry surprisingly high pressure ratings. Most carry the approval seals of the important organizations in the field—IAPMO (International Association of Plumbing and Mechanical Officials), UPC (Uniform Plumbing Code), NSF (National Sanitation Foundation), and CSA (Canadian Standards Association). Properly installed, they are as safe as any tubing connection. When brass compression fittings are used with softer plastic tubing, particularly PE and vinyl, a brass insert should be used in the tube end to prevent it from collapsing when the fitting is tightened with a wrench.

Threaded fittings may be used but require an adapter of some sort to connect the device to the tubing. For example, most pressure regulators are designed with threaded connections at input and output. Since tubing has no threads, a threaded adapter must be attached to the pressure regulating valve. These fittings are also simple to install, using a wrench for tightening. It is a good idea to use Teflon tape or a paste thread sealer (silicone, etc.) when connecting threaded fittings; they will be more secure and less likely to leak.

Installing Irrigation Receptacles and Performing Other Finishing Tasks

The installation of the tubing lines is only part of the job, particularly in integrated systems. After most of the finish work has been completed on the building interior, the controller and timer are mounted and connected into the rest of the system, and irrigation receptacles and manifolds are connected. These details will be covered in the following sections.

Installing Control Center Equipment

Controllers are surface mounted on a partition wall at the location selected. Tubing lines have been terminated there and are ready to receive the control equipment. Water distribution connections to the solenoid valve are either barb/crimp rings or compression fittings, generally the latter. Depending on the model, electrical connections are made either by hard wiring into a junction box installed behind the controller or by simply plugging into the timer receptacle. The timer itself is installed in a similar way. Some models require hard wiring into a junction box, while others plug into a power receptacle.

Before turning on the water supply, check the operation of the control center. Timers have a manual operating mode that is generally a switch permitting the timer to be bypassed and overridden. Consult the timer instructions. By means of this switch, apply power to the controller; a red indicator on it will light. Listen for the click of the solenoid valve as it opens and closes. Time the interval between its opening and closing. It should coincide roughly with the irrigation cycle rating of the device, which is normally 10 or 15 seconds in duration. After manually turning off the timer's power switch (the controller's red indicator light will go off), wait about 15 seconds and try it again. Run this test at least three times to be sure the equipment comes close to specifications. The operating duration should be within ± 2 seconds of the published specs. With the power OFF, throw the irrigation cycle switch on the controller to the longer cycle position, which is generally 20 or 30 seconds. Perform the same tests as with the shorter cycle to again check for proper operation. The operating duration in this case should be within ± 3 seconds of the published specs. While performing these tests, the plug must be removed from the remote-control jack on the controller. Next, check its operation. Apply power to the controller and let it complete its irrigation cycle. After the click is heard, denoting the closing of the solenoid valve, insert the plug into the remote-control jack. The click of the solenoid valve should be heard again as it opens. It will remain open for as long as the plug is in the jack. **Do not leave it plugged in for more than 1 minute.** If a remote-control cord has been purchased (optional equipment), test the system by plugging it into the remote-control jack at the appropriate time. When first plugged in, nothing will happen. Press the switch on the end of the remote-control cord. It should open the solenoid valve and keep it open for as long as the switch is pressed.

At this point, the timer must be programmed to operate the system at the desired times. Most interior installations of Micro-Irrigation Systems will require operation twice a day. Occasionally, a third cycle is added, and in very large installations having big trees, four or five short operations per day may be needed. If an electromechanical timer is being used, changeable tabs or pins on the clock wheel are positioned to switch the power on or off at the times indicated. If a central computer-controlled timer is used, it must be programmed to switch the power on to the irrigation system controller at the appropriate times and **for at least 1 minute.** This gives the controller a "power window" of enough length to permit it to generate a complete irrigation cycle. The times of day that the system operates are important only for reasons of convenience. In natural environments, rain may water the plants at any time, day or night; thus, the plants have adapted to variation. The irrigation cycles should be spaced as far apart as possible to permit water from the last cycle to diffuse fully into the soil. If two daily cycles are used, space them 12 hours apart; if three daily cycles are used, space them 8 hours apart, and so on. The precision of these schedules is not critical but should be close to provide the most efficient operation. It is a good idea to have at least one daily cycle while someone is there to check on the consistency of operation and occasionally to check that all emitters are flowing as they should. As a suggestion, cycles are practical at 9 o'clock in the morning and 9 o'clock at night. In residences, someone is usually there in the evenings at that time, yet it is not late enough to disturb attempts at sleep. In offices, someone is usually there by 9 A.M. The installer's judgment should be used, or better yet, the irrigation schedule should be discussed with the building occupant as well as the interiorscaper, if one is involved. In large buildings, normally there would be an independent Micro-Irrigation System installed on each floor. However, it would not be desirable to have each system come on at the same time. Even though the flow rates through each would not be large, in aggregate, they would put a strain on the building's water supply. The best irrigation schedule would have the systems operate sequentially, in a staggered manner, at night when the load on the building's water system is least. A practical schedule might be one cycle between 6 and 9 P.M. and another between 6 and 9 A.M. These schedules may be best for the interiorscapers as well, for many try to make their rounds during off-hours. This gives them a chance to observe system operation and make adjustments as necessary.

The installation of low-pressure Micro-Irrigation Systems follows essentially the same pattern just described for high-pressure versions. Controller connections are different, of course. After the pump controller has been mounted on a partition wall, the low-voltage power cord is connected to the pump terminals. In small light-duty models, these terminals are on the pump itself, or in larger systems, there are power leads (wires) from the pump/reservoir module. As these are usually DC pumps, polarity must be carefully observed when making connections.

Installing Irrigation Receptacles

Irrigation receptacles are installed at locations provided for them by making a simple connection to the receptacle feeder tube, which should have been positioned through the front of the receptacle junction box during the rough-in. The tubing connection is either a barb/crimp ring or compression fitting, depending on the model. At this point, with all receptacles connected, it is possible that the building inspector may want to run a pressure test on the system. The receptacle assembly comes fitted with deactivating plugs. Leave them in place during the pressure test. With those procedures out of the way, the irrigation receptacle assembly is attached with screws to the junction box through its decorative plate.

Installing Irrigation Manifolds

Irrigation feeder tubes should have been routed to the areas where manifolds are needed. They would normally be passed through holes in the adjoining partition. The recommended installation procedure calls for mounting an escutcheon over the tube where it exits the wall to make for a neat-looking pass-through (see Figure 13.22).

Connections are made to the manifold with a barb fitting secured by a crimp ring, or with a compression fitting. The manifold will have been fastened to the base or wall of whatever it services during the rough-in stages. Emitter tubes are connected to manifold outlets wherever needed. Other outlets are

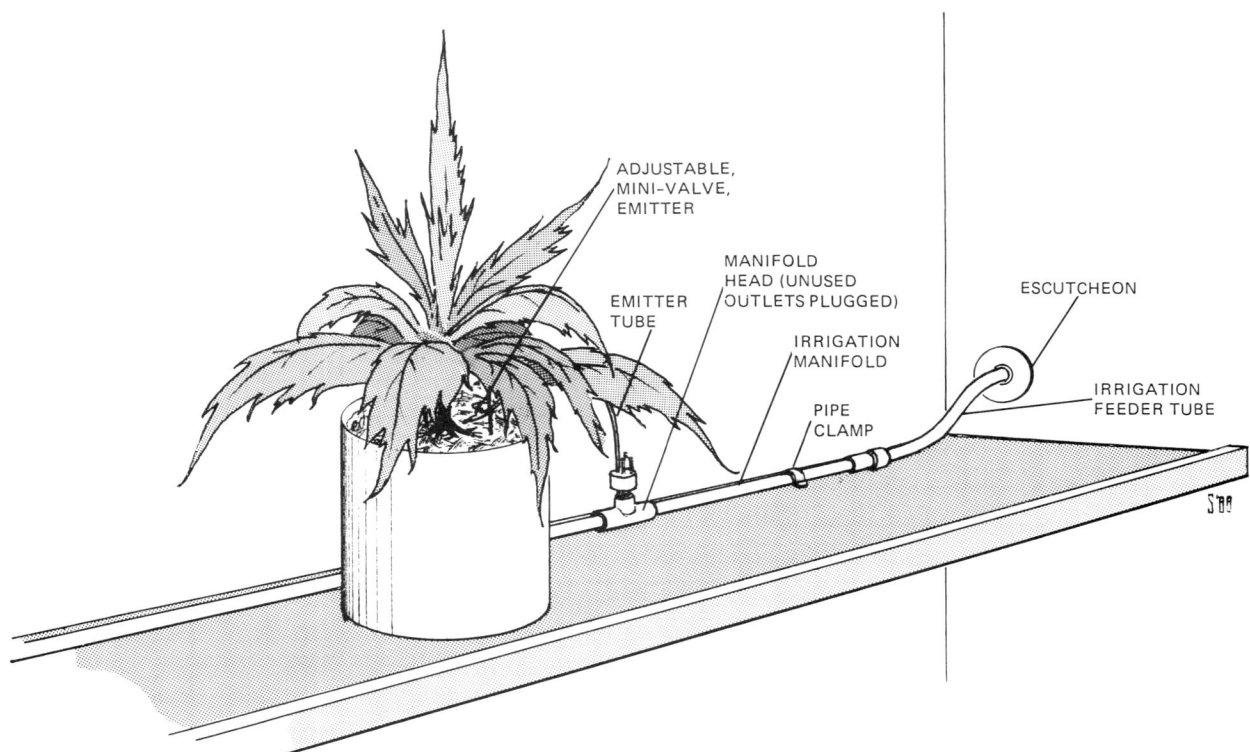

Figure 13.22 Typical irrigation manifold installation servicing a planter shelf.

plugged. Adjustable mini-valve/emitters are connected to the ends of the emitter tubes, and the adjustment screws are opened a bit to permit flow.

Installing Emitter Tubes

Emitter tubes can now be plugged into irrigation receptacles and other fittings to make them ready for duty. The emitter-tube assembly that comes with the unit is simply plugged into the receptacle, after the deactivating plug is removed, of course. Pushing in on the receptacle's release ring will permit removal of the deactivating plug. The tube is routed to the locations of freestanding containerized plants. Each tube should service the closest plants. Whenever required for esthetic reasons, emitter tubes can be placed under the edge of carpeting and brought to the surface in the vicinity of the planters even 20 or 30 feet from the receptacle. More than one plant can be watered from an emitter tube. They can be branched off into two or more subbranches using $\frac{1}{4}$" tees, each with its own adjustable mini-valve/emitter. Whenever a tube must be raised to heights greater than about $1\frac{1}{2}$ feet, it should be fitted with a mini check-valve with the one-way flow direction pointing upwards. This keeps water in the tube after system shutoff and prevents undue air-purging time during the next watering cycle. This is not a hard-and-fast rule; it applies mainly to low-pressure systems or to high-pressure systems where the flow rate has been greatly reduced.

There are times when it is necessary to bring emitter tubes up close to the ceiling in order to water hanging planters or planters on high shelves. Although clear emitter tubes can be used which don't show up very easily against a background, it may be desirable to hide them in the partition. Small holes are first cut in the partition at the top and base, and the tube is fished through the wall in much the same ways as electrical wiring is installed. Connections are made at the top and bottom with supply and emitter fittings. Holes in the wall around the tubes can be patched with spackling or crack filler to make a neater installation.

The emitters themselves are installed by simply plugging them into the end of the tube after it has been cut off to a convenient length, while still leaving a little slack in the tube. It can be difficult to mount fittings in some flexible tubes, such as PE and PB, because of their inherent stiffness. The process can be facilitated by first softening the tube end with heat. The emitter is held in place with an emitter stabilizer, which is no more than a stake and clamp arrangement that is pushed into the planter soil. The small clamp holds the emitter to the stake.

Installing Systems in Planter Boxes

Planter box installations of Micro-Irrigation Systems are different than those for freestanding planters. In many ways, they are less difficult to install, for the esthetic and safety requirements are not as stringent. Irrigation supply tubing is routed to the built-in planter box through partitions or flooring, depending on the situation. The most efficient way when designed into the building is to cast a tubing line into the concrete floor slab between a remote irrigation supply branch and the planter box (see Figure 13.23). The tubing line is connected to an irrigation manifold that is premounted in the planter. Emitter tubes connect to the manifold and service the plants closest to them. Bubbler heads can be used in place of emitter tubes to distribute the water. Mini spray heads can also be used on the manifold when the types of plants are compatible with the technique.

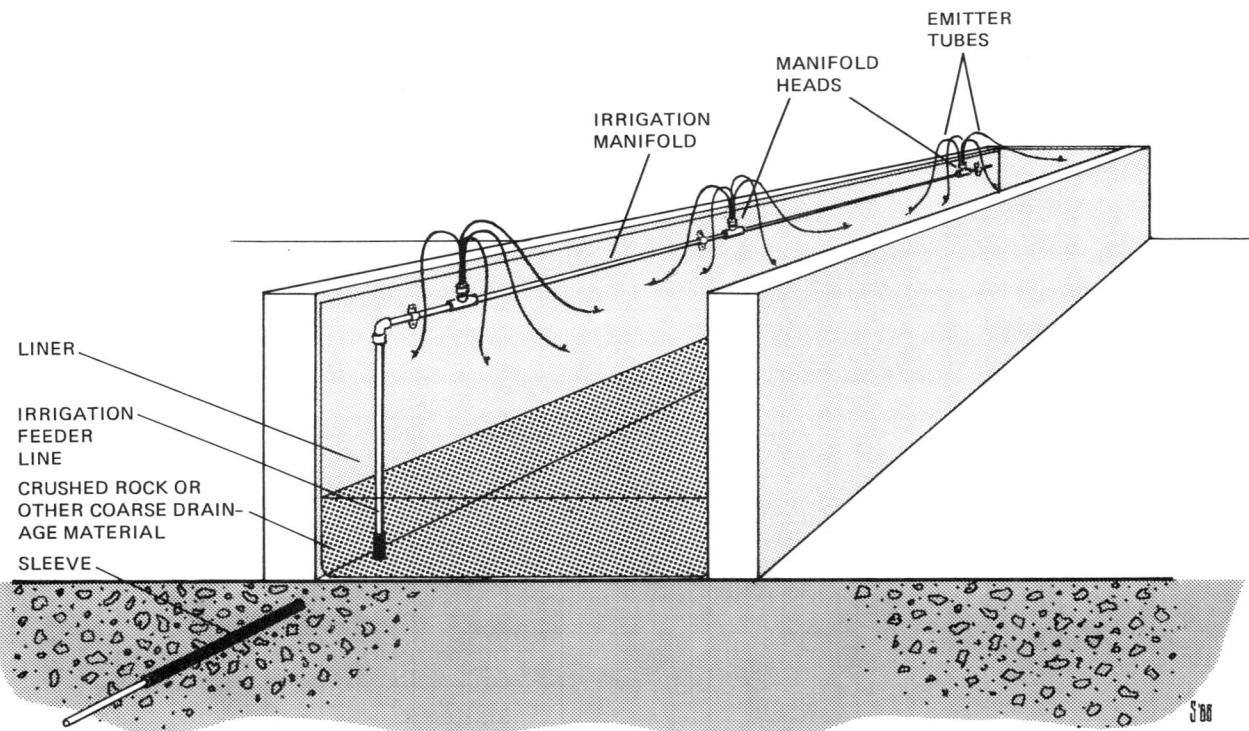

Figure 13.23 Typical planter box installation from a remote water source control.

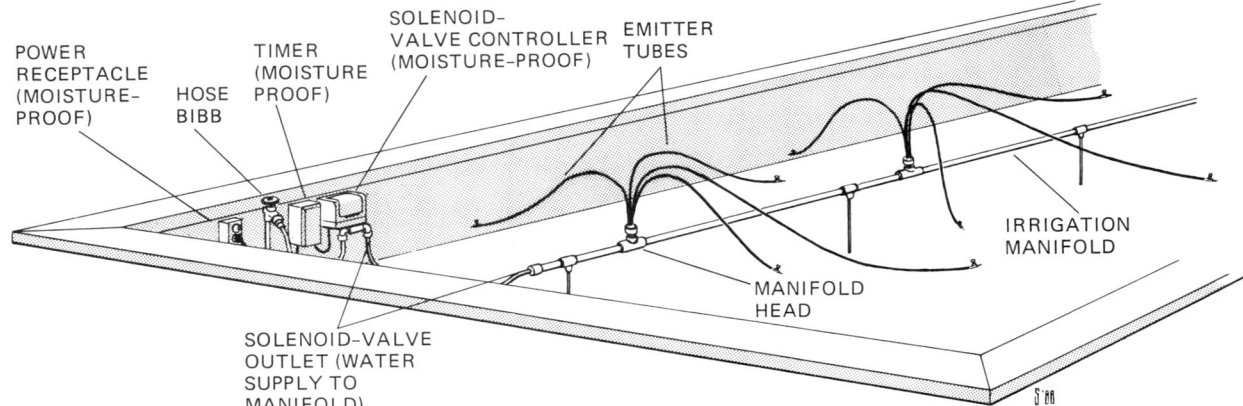

Figure 13.24 Typical planter pit installation with self-contained services (water and electric).

When electrical and water services have been designed into a built-in planter box, a different type of installation is required. Proper design practice will provide hose bibbs and electrical outlets in each planter box or pit. The hose bibb is usually there to permit plant maintenance people to connect a hose, so they can manually water a tree or smaller plants in the same vicinity. The electrical outlet is usually for accent lighting or holiday lights. These services can be used to operate a small independent Micro-Irrigation System established for that planter alone (see Figure 13.24). The type of solenoid valve controller used can be a light- to medium-duty unit, but it must have an all-weather case because of the damp or wet conditions it will encounter. It should be mounted on the wall of the planter near the hose bibb and electrical junction box. The timer, which should also be an all-weather type, mounts next to it. The input side of the solenoid valve is connected to the hose bibb (with a backflow preventer) and the power cord is plugged into the timer's outlet. The output side of the solenoid valve is routed to the water distribution network, which can be as simple as a loop of laser soaker tubing ringing the roots at the base of a tree or as complex as a multibranched irrigation manifold that distributes water throughout a large planter box. If light-duty use is expected, it would be wise to use a fixed pressure reducer to modify the flow rate.

Installing Outdoor Systems

The discussion that follows refers not only to installations of Micro-Irrigation Systems outdoors but also indoors in atrium-type settings; the features and requirements are similar.

Outdoors, Micro-Irrigation Systems provide the same service as sprinkler or drip systems, but in a more precise manner so that planter overwatering and overflow are not a problem. They are very suitable for potted plants around patios, pool decks, terraces, apartment balconies, storefronts, shopping promenades, etc., where not only foliage plants are grown, but flowers, bonsai, and containerized vegetable gardens are grown as well. Both high-pressure and low-pressure systems are used outdoors. Each system must have control center equipment that can withstand the elements, so in choosing equipment, look for all-weather types. Under any circumstance, it is wise to install control centers in sheltered locations whenever possible. This can minimize problems.

Low-pressure system use is common for outdoor installations. Part of the reason has to do with the fact that many building codes prohibit the use of hose bibbs, or any other type of water service, on high-rise apartment balconies. The low-pressure system provides its own water source (pump/reservoir module), and so long as the electrical service is readily available, an installation can be made. The pump/reservoir module is generally large and should be located in an out-of-the-way place (preferably a sheltered location), yet it must be near an electrical power source or where a power line can be extended. The pump controller and timer are mounted next to the module on an exterior wall, and the low-voltage power cord is plugged into the pump leads, with polarity intact (see Figure 13.25). The controller power cord is plugged into the timer's power outlet, and the timer's power cord is plugged into the closest power receptacle that has continuous or controlled service. The pump output tubing is connected to water distribution tubing lines, which are usually a network of ½" Schedule 40 PVC sprinkler system pipes. They can be fitted with manifold outlets, emitter tubes, bubbler heads, or spray heads to accommodate installation needs. Figures 13.26 to 13.28 illustrate several low-pressure systems in outdoor applications.

High-pressure systems are installed in a similar way, except that the water source is generally a hose bibb connection. They are most commonly used around residential and commercial patios and pool decks, storefronts, shopping promenades, and interior atriums (see Figure 13.29). So as not to incapacitate the hose bibb, irrigation system connections should be made to a wye fitting, with shutoff valves in each leg. The leg connected to the Micro-Irrigation System should have its valve open at all times. The other can be used for utility purposes and is always closed when not in use. Remember, the hose bibb valve must be kept open at all times, and it is a good idea to remove the valve handle to prevent accidental shutoff.

Atriums are generally well lit from skylights and oversized windows, so their moisture requirements are high, although usually somewhat less than out-

Figure 13.25 Pump controller in sheltered outdoor terrace installation.

Figure 13.26 Outdoor patio installation of a Micro-Irrigation System.

Figure 13.27 Containerized tomato plant being serviced by an outdoor Micro-Irrigation System.

Installing Outdoor Systems

Figure 13.28 Bonsai planter being watered by an outdoor Micro-Irrigation System. (Courtesy of Aqua/Trends.)

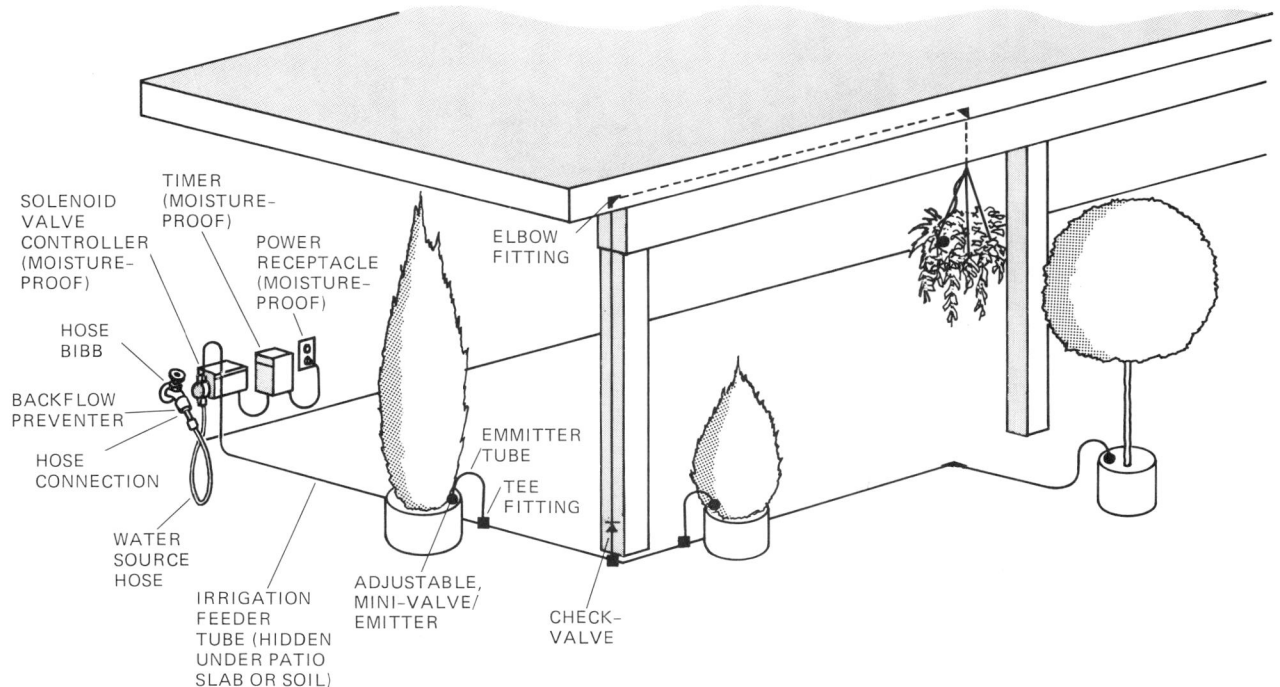

Figure 13.29 Typical high-pressure system outdoor installation servicing a patio area.

door requirements. It should be kept in mind that atriums can become baking hot when the sun is overhead, and a screen of some sort is sometimes used in order to modify lighting conditions. Also keep in mind that atriums do not have as much air circulation as outdoor settings, and moisture can build up to the danger point. Mildew and fungus problems are common. One advantage of Micro-Irrigation Systems is that watering rates are more controlled than with other technologies, and proper moisture can be maintained at the root ball without overdosing—excess moisture is not as much of a problem.

Output tubing from the solenoid valve is connected to the water distribution network, which, again, is usually PVC sprinkler pipe. This can be routed into patio areas, planter boxes, overhead canopies, pool decks, and building overhangs, along walkways, and around vegetable gardens. In cold weather areas of the country, provisions should always be made for drainage prior to frost in order to protect the tubing and fittings. The system should be purged of water by blowing out or draining the tubing lines, and the hose bibb should be closed down.

Interfacing with Other Technologies

It has been mentioned in other chapters that there are other ways to activate Micro-Irrigation Systems. There are many configurations of automated energy management systems. The simplest, of course, is the electromechanical time switch. The most complex are multipurpose computer-based control systems made for large building management. The ones most useful in the application of Micro-Irrigation Systems are fairly simple as modern technical devices go. One of the most versatile of the middle-ground devices is the carrier-frequency energy-control system. The main supplier of this technology, X-10 (U.S.A.), Incorporated, calls it the X-10 Powerhouse System. Its main functions have to do with the computerized, automatic, or manual remote switching of lights,

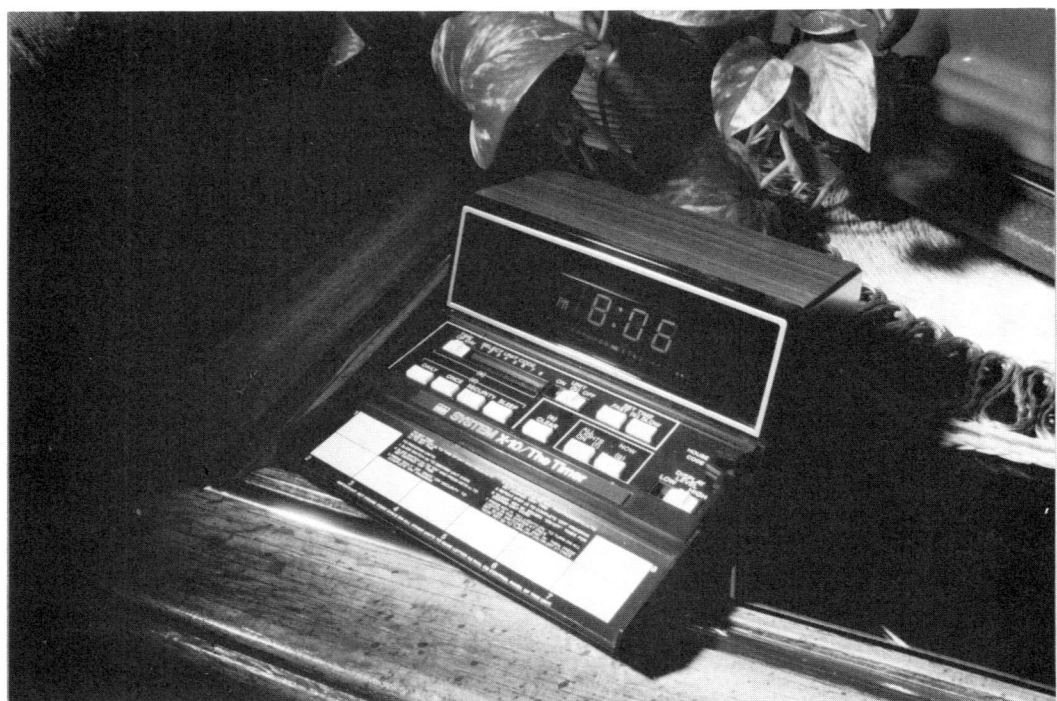

Figure 13.30 Programmable time/controller.

Interfacing with Other Technologies

Figure 13.31 Manual push-button timer/controller. (Courtesy of X-10 (U.S.A.), Incorporated.)

Figure 13.32 Plug-in receiving module. (Courtesy of X-10 (U.S.A.), Incorporated.)

appliances, security devices, heating/cooling equipment, home entertainment systems, etc. The controllers are capable of operating as self-timed automatic switches or as manually triggered remote switches. Manual activation can be by button control panels and radio controllers from somewhere in the same building or from thousands of miles away by telephone. Figures 13.30 and 13.31 illustrate two types of X-10 system controllers.

In all cases, the control signals are superimposed onto the building's electric power wiring and eventually picked up by small receiving modules plugged or wired into power outlets in other parts of the building. Into these modules are plugged the device (or devices) to be controlled. The receiving modules have programmable device codes that correspond to similar codes on the remote controller. When properly coded signals reach the receiver through the power wiring, they cause power to be applied to the devices plugged in. Figures 13.32 and 13.33 illustrate two types of X-10 receiving modules.

The X-10 controllers are capable of orchestrating up to eight power control programs, each switching devices at different times of the day for different periods of ON time. The activation/deactivation cycle can be repeated twice daily. One can see how useful this is in Micro-Irrigation System installations. Small satellite systems can be scattered around a house or office, each plugged into a receiving module. Most (or all) would have the same power application program and would, therefore, be coded into the same control channel. Another control channel can be used for supplementary plant lighting, artificial lights that would come on for several hours during the day or night to supple-

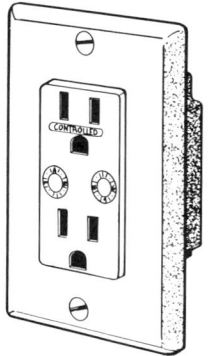

Figure 13.33 Wall receptacle receiving module. (Courtesy of X-10 (U.S.A.), Incorporated.)

ment the natural daylight. Another control channel can be programmed to operate outdoor Micro-Irrigation Systems at other times of the day. Installing these carrier-frequency remote-control systems is an easy matter. In the simplest of cases, all elements of the system are simply plugged into wall outlets at the locations most convenient for them. The timer/controller can be placed in a living room, bedroom, office, or any other practical location. The receiving modules are plugged or wired into junction boxes or outlets wherever a satellite Micro-Irrigation System can be placed (central to the potted plants being serviced). Any number of these satellite systems can be used in the building, and at any location, so long as they are on the same electrical wiring network. The combination of technologies makes a highly versatile system capable of automatically controlling not only the Micro-Irrigation Systems installed in the building but also all lights, appliances, and other electrical devices (see Figure 13.34). Another version of this technology has a timer/controller that wires into an electrical junction box, rather than simply plugging in as with the X-10 system. This makes for a more professional-looking installation.

Other remote energy-control technologies are coming into use. Some are based on a computerized timer/controller panel installed in the home or office, which also automatically switches lighting, appliances, heating/cooling equipment, etc. These systems are simplified versions of the complex building management systems in common use in intelligent buildings. The peripheral devices either are hard wired to the controller or receive control intelligence through carrier-frequency signals. The idea is to bring some of this usefulness to private homes and smaller commercial applications.

Applications of these themes are being introduced in other concepts. One technology just becoming recognized uses local video cable companies to transmit control signals, along with video signals, to their customers. A customer subscribing to the service has special receivers installed that pick the control signal out of the overall reception and distribute it throughout the building for various control chores. The timing and program manipulation in this case is handled at the cable company. A similar concept is being introduced by some telephone companies. In this case, the timing and programming is done at the phone center, with control signals coming through the phone lines. Receiving modules are installed to turn this into a useful system at the remote location.

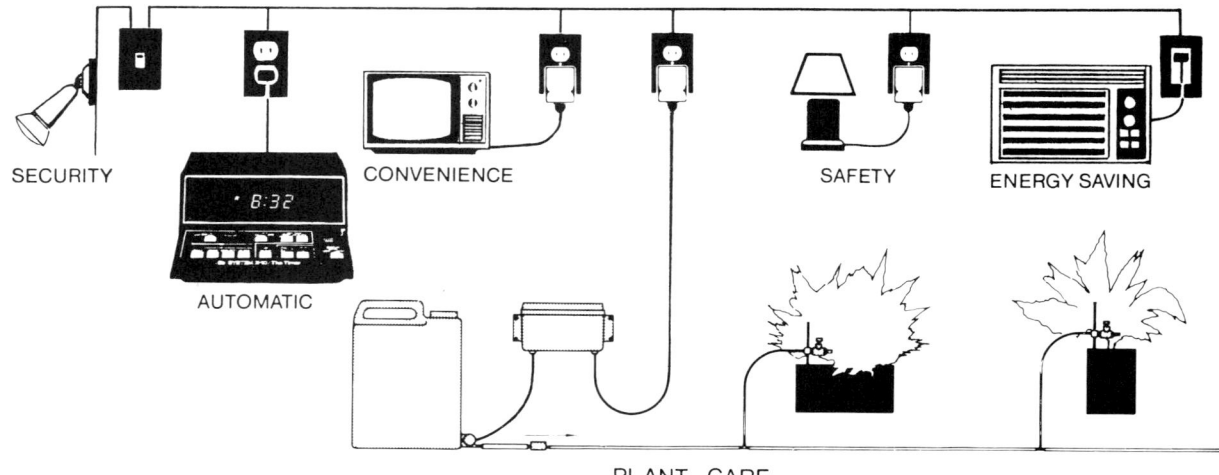

Figure 13.34 X-10 system with various peripherals connected. (Courtesy of X-10 (U.S.A.) Incorporated and Aqua/Trends.)

The point of this discussion is that Micro-Irrigation Systems can be controlled, or rather, power can be applied to turn them on at appropriate times, with all of these energy management concepts. Automatic irrigation becomes a peripheral to the main system, along with the lights, appliances, etc. Installation techniques will depend on the nature of the control technology involved.

14 STARTING AND OPERATING A NEW INSTALLATION

Overview Once the mechanical work of installing precision, Micro-Irrigation™ Systems has been completed, there are established procedures that must be followed to be sure everything has been completed acceptably and to make initial adjustments to the system—things that must be done prior to leaving it confidently as an automatic plant-watering system requiring minimal attention. After that, routine operating and maintenance procedures are instituted. Some of this is done by the installer, some by the installer and plant maintenance person together, and some by plant maintenance person working as the sole custodian of the plants and their electromechanical life-support system. Most procedures are simple, yet many times are complicated by the idiosyncrasies of the project. Techniques have been worked out to cope with most situations, and they will be discussed in detail within the following pages.

Start-Up Procedures The installer is responsible for starting up the Micro-Irrigation System after the mechanical part of the job is finished. The following list outlines the start-up procedures:

1. Check all connections to exposed tubing to be sure they are complete and secure.
2. When the plants are in place, insert an emitter stabilizer stake into the potting soil of each. Adjust the height of the emitter to 2 to 3 inches above the surface of the soil. The emitter outlet should be above the root mass.
3. Open the adjustable mini-valve/emitters about halfway to permit some flow.
4. Be sure an overflow saucer is beneath each freestanding planter.
5. With small light-duty systems, check to see if the mini check-valve is operating properly. This can be done by removing it from the pump outlet, blowing into its input side, and feeling for air coming from the output side. If it seems operable, refit it into the tubing line. If not, replace it.
6. Check to be sure all power connections have been made and electrical service is available to the system (circuit breakers closed, etc.).

Start-Up Procedures

7. Check the timer's irrigation cycle program settings to be sure it is on the proper schedule. If a computer-based energy management controller is used to supply power for the automatic irrigation system, it should be checked by requesting a screen readout of the programmed schedule.

8. Check the controller's irrigation cycle setting. It should be switched to the shortest cycle (normally 10 to 15 seconds).

9. If the water distribution tubing network is multizoned, shut off all zones but the one to be tested and adjusted. Office suite and restaurant installations typically use shutoff valves of some type to deactivate a zone. This is sometimes a manually operated globe valve in overhead irrigation feed lines and sometimes a key-switch-operated solenoid valve in a tubing riser.

10. In large installations, have two people work together to adjust the emitters. It is recommended that the actual adjustment be done by the interiorscape technician assigned to the routine maintenance task. One person should be stationed at the control center, and the other at the planters. There should be some means of communication between them. If they are too far to be heard normally, a walkie-talkie system should be used.

11. Turn the unit on via the timer's manual power switch. The system will begin timing the programmed duration of operation. Water will flow during this 10- or 15-second period. The controller's red indicator will be lit, showing that power is being applied.

12. Insert the shorting plug into the controller's remote-control jack. This causes the solenoid valve or pump to operate, and water to flow. Water will flow for as long as the plug remains in the jack. THIS SHOULD BE NO MORE THAN ABOUT ONE MINUTE AT A TIME. Repeat the process until all tubing lines are filled with water. Some planters could overflow during this operation if they are serviced by tubing near the source, which will readily fill with water. Keep a close watch on them and close down their emitters temporarily, if necessary. If water refuses to flow from an emitter, it is usually because all the air has not yet been purged from its feed tube. Check to see that the emitter adjustment is open enough to allow air to escape.

13. After the tubing has been filled, reinsert the shorting plug for one minute. During that period, emitters can be adjusted one by one. The amount of flow permitted by each emitter will be determined by the type of plant it services, its size, its location, etc. These judgments should be made by a person knowledgeable in **interior** plant-care. Approximate settings can be made without taking measurements, by being judged visually. As a rule of thumb, small-sized plants in 5- to 8-inch containers should have the flow pattern shown in Figure 14.1(a). Remember, if needed for small dry-loving plants, that the minimum flow with these adjustable emitters is a mere **two drops** during a 10-second irrigation cycle. Medium-sized plants in 10- to 14-inch containers should have a flow pattern similar to that shown in Figure 14.1(b). Large-sized plants in 17- to 30-inch containers should have a flow pattern similar to that shown in Figure 14.1(c).

It is recommended that a remote-control cord or electronic remote-control device be used while adjusting emitters. If a remote-control cord is used, it can be extended to the planter locations. Irrigation will continue for as long as the switch is held down, and with

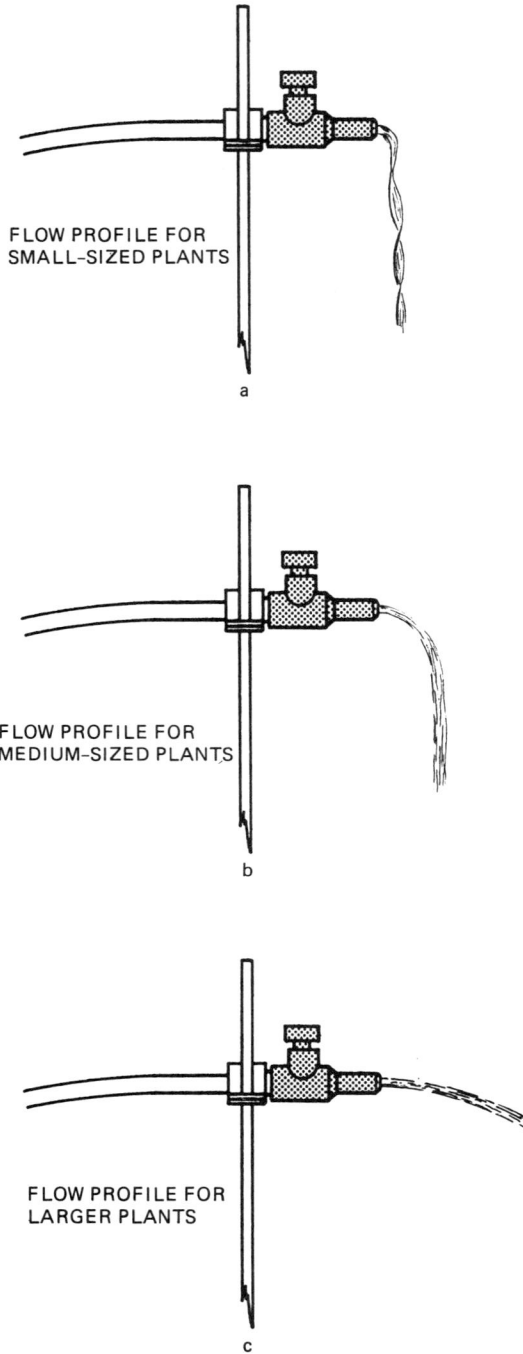

Figure 14.1 Emitter adjustments—recommended initial flow patterns for various plant sizes and moisture requirements (Courtesy of Aqua/Trends.)

this accessory, the flow can be controlled closely so as to prevent excesses. An electronic remote-control switch also permits precise on/off operation of the controller from planter locations, but does it by a more sophisticated switching means. As long as the control button is held down, signals will be received by the controller to keep the flow of water going.

Keep in mind that while individual plant emitters are being adjusted, the other plants are receiving flow as well. When making initial adjustments to large numbers of plants, it is not uncommon to have some overflow from the bottom drain holes. The soil gets saturated and can absorb no more, so it gives up the excess. Plastic plant saucers should be placed under all containers to catch the overflow. Flow should be permitted just long enough to make the adjustment at each plant. If only a short period of time is required to adjust the system, as would be the case with a simple installation, plant soil may not be moist enough. The plants should be hand watered to bring soil moisture up to recommended levels. REMEMBER THAT THE FUNCTION OF THE SYSTEM IS TO CONSTANTLY MAINTAIN AN OPTIMUM LEVEL OF SOIL MOISTURE, NOT TO WET OUT DRY PLANTERS. Don't be too concerned over how precisely or how evenly adjustments are made from plant to plant. The technology is very forgiving, and approximations usually suffice—particularly for the initial adjustments. Fine-tuning comes later, after a week or more of operation.

14. After all plants in a zone have been adjusted individually, check their flow again and readjust where necessary. This has to do with *balancing* the system.

15. Adjust each zone in sequence, shutting down all zones not being worked on. After all zones have gotten their initial adjustment, recheck them. The installer will find that in adjusting many plants on an extensive system, flow to the first plants drops off as the succeeding emitters are opened. In these cases, some water is being taken from the first plants to service the last. Once the system has been roughly balanced, additional adjustments make little difference.

16. Inspect emitter tubes that rise greater than 1½ feet. If they have been fitted with mini check-valves, be sure backflow has been eliminated. If the water level in the tube declines after the system shuts down, the check-valve is leaking and needs to be replaced.

17. If low-pressure systems are being used, dissolve the recommended soluble fertilizer in the reservoir. A weak solution should be prepared. If the fertilizer has color, the solution should be pale.

18. Before leaving the project, hold a meeting with the interiorscaper and the building or facility manager or the homeowner to instruct them on the proper use of the system. It should be made clear that from that point on, the responsibility for operation becomes theirs, and that the installer does not become involved with plant care (unless, of course, the installer is also an interiorscaper). A quantity of spare parts should be left at the site, particularly emitter tubes, adjustable mini-valve/emitters, and emitter stabilizers, as well as ell and tee fittings for emitter tubes.

Operating Procedures

After the mechanical installation has been completed and the Micro-Irrigation System has been checked out and coarsely adjusted, the interior plant maintenance expert (in-house or under contract) then begins to oversee its operation. The following guidelines and procedures should be used:

1. A newly installed system should be monitored weekly for about one month to six weeks, and emitter adjustments made in response to the maintenance person's inspection. Because of the small quantities of

water used daily, it may take a week or two for the soil moisture to reach equilibrium.

2. The use of a good moisture meter is recommended. It reads relative levels of moisture around the root ball, and can penetrate where fingers and gauge sticks can't. In many cases, the soil surface may be dry to the touch, yet the root ball may be properly damp—the meter will discover situations such as these.

3. If a plant still appears dry after an adjustment has been made to the emitter (slightly opened), the soil should be manually watered to bring moisture to the desired level.

4. If all plants seem to be drying out and the emitters are feeding their maximum amounts, then the system should be reprogrammed to water more frequently, for example, three times daily, instead of two times. This is accomplished by rescheduling the timer. Also, for a given irrigation schedule, watering rates can be doubled by means of the switch on the pump and solenoid valve controllers. Most light- and medium-duty interior controllers can be switched between 10 and 20 seconds. Medium- and heavy-duty interior controllers can be switched between 15 and 30 seconds. Controllers made for outdoor use are normally switched between 20 and 40 seconds.

   High-pressure systems have additional adjustments that can be used, if necessary. Water pressure can be increased (if originally reduced with a regulator), or flow rates can be increased (if restricted by a flow-control washer). These coarse adjustments become an important part of seasonal irrigation control. The summer months require a greater water application because of increased soil surface drying and higher plant transpiration rates. As the hotter months approach, soil moisture levels will decline. Heating systems in the winter months also tend to be drying. Appropriate adjustments should be made when necessary.

5. The system can be calibrated by determining the exact flow rates. This can be done after the maintenance person is satisfied that emitter settings are acceptable. The amount of water dispensed to a plant over a given period (a day or week or even a single watering cycle, for example) can be measured by actually catching it in a cup, or other small container, as it leaves the emitter (see Figure 14.2).

   A measuring cup or laboratory graduate is used to determine the number of ounces accumulated. Calculations are made to extrapolate the values to time periods under consideration, for example, ounces per 10 seconds, ounces per day or week, gallons per hour, per day, or per week, etc. This must, of course, be done at each plant and can become quite tedious if many plants are involved, but it is the most accurate way to determine water application rates. The readings will be found useful for keeping the system uniform over long periods. It should be kept in mind, however, that the plants' needs will increase as they grow, and flow readjustments must be made accordingly. New measurements would then be made to recalibrate the system. This type of adjustment is infrequent.

6. Plants in professionally maintained interiorscapes are turned to the light regularly. Care should be taken that system tubing is not disturbed. If necessary, the maintenance technician should remove the emitter assembly (by pulling up the stake) before turning the con-

Figure 14.2 Technique for making flow measurements—a small plastic or glass cup catches the water emitted during one (or more) irrigation cycle(s). (Courtesy of Aqua/Trends.)

tainer, and then replace it. This prevents the tubing lines from getting twisted, tangled, and crimped.

7. The control timing should be checked periodically to be sure all is in order. Accidental or deliberate changes can go unnoticed, until plant health declines. Then it could be too late.

8. As a general rule for any plant maintenance technique, among other things, plant leaves should be closely inspected for a number of reasons. One reason has to do with the fact that leaf tips start to turn yellow, then dry out and turn brown when soil moisture levels are improper. Unfortunately, this can happen when there is too much water as well as when there is not enough, but at least it gives an indication that soil moisture should be carefully checked.

9. During routine inspections, check that the adjustable mini-valve/emitters are above the soil level and not covered by mulch.

10. When built-in planters are involved and control centers are installed within the planter, be sure that mulch does not come near the equipment. The electronics can be damaged by damp mulch (or other materials) around them, and the condition could present a fire hazard as well. A type of retaining wall or moisture-proof enclosure can be installed. It can be as simple as a piece of stiff plastic secured to the wall and bent around the controls. A moisture-proof pump or solenoid valve controller case (as well as timer case) should be used under such circumstances.

11. Mulch should be used to prevent excessive evaporation from the soil surface.

12. A supply of emitters, emitter stabilizers (stakes and clamps), and tubing should be maintained near the installation for occasional repair and replacements.

13. Fertilizer and other plant-care chemicals can be applied with the irri-

gation water. Water-soluble fertilizers, systemic pesticides, and fungicides may be used in dilute solutions when the system is one of the low-pressure versions. The chemical solution is stored in the reservoir. If there is any appearance of sludge or jelly-like accumulations in the reservoir that might be caused by mixing incompatible chemicals, stop the practice and clean out the system. These accumulations can cause tubing and fittings to clog.

14. A similar condition frequently occurs when algae build up in reservoirs and tubing lines, particularly those near windows and other light sources. Clear tubes should be replaced with opaque types where that problem exists.

15. Fertilizer injectors are not recommended for use in high-pressure systems. Problems can be created by the chemical deterioration of fittings and control parts.

16. The periodic inspection and cleaning of input filters and screens is important and should not be overlooked in the routine maintenance schedule.

17. In cold weather climates, the outdoor system should be purged of water before the coming of frost, and the electronics should be shut down for the duration of the cold season.

15 THE FUTURE

Overview The future we face will be filled with technological marvels, for many of the innovations that have come upon the scene in recent years will later become commonplace in our lives. There are also legions of new and yet undiscovered concepts that will further broaden our opportunities for electronic assistance. The buildings we live in, work in, and frequent will be more extensively fitted with technical innovations as they become more available, more acceptable, and more necessary to the homeowner and to the commercial community. Real estate management has become more sophisticated and is seeking new solutions to its problems. There will be less tendency to be overawed by technology itself—we will learn to discriminate and choose systems offering real benefits. Manufacturers will have to listen more carefully to their markets, eliminating meaningless features. The movement toward new power concepts will bring revolutionary changes in the way products are designed and used, as will further application of microprocessor-based control systems. The federal government is now requiring effective solutions, which are soon to be mandated by law, to indoor air pollution. More and more of these everyday problems will be addressed by technical minds, making future life-styles and work styles significantly advanced over the way they are today.

Building architects and interior designers will also be more attentive to our personal and business needs. That movement has gained momentum and will produce structures that are more comfortable, convenient, versatile, cost efficient, business effective and health oriented. The combination of design and technology will become important, and highly functional buildings will become more prevalent.

Live plants have become an important part of the architect's and interior designer's quest for humanistic as well as beautiful interiors. That trend will now be augmented by health-oriented motives and by the new plant-care technologies that make these concepts practical. The following sections will deal with those continuing trends.

Technology in Commercial Buildings The movement toward intelligent, high-tech buildings is experiencing temporary setbacks as commercial construction slows to allow for the absorption of excess inventory. This adjustment will take a few years, but continued growth is imminent. Meanwhile, industry is developing new products to improve on those already available. There are an inordinate number of automated energy management systems, as well as more comprehensive building management systems, in the market or about to enter, and others are being engineered. They run the gamut from remote-control systems to fully integrated systems of varying complexity. The most extensive are tied to the building's security systems, safety and HVAC systems, etc. Aside from the economics involved, one of the dilemmas facing the industry has to do with providing these technologies with so-called *user-friendly* interfaces, that is, operating environments not overly complex, so that building management people can easily understand and operate them and take advantage of their full capabilities without having to be an engineer or computer professional. Complex systems require complex controls, but these early stages of maturity are finding much of the technology outrunning the ability of real estate management to hire, train, and retain capable operating and maintenance personnel. Some feel the systems are too complicated and require engineering or data processor attention rather than lay help. Many claim they confuse people and are not easy-to-use products. That has been a typical problem facing computer-related systems of any type. Developers have had a difficult time "humanizing" their technology. The danger is that the inherent cost efficiencies can be cancelled out by the need for larger and more expensive technically trained system management staffs. Those problems are being addressed.

Other factors have also slowed penetration of technology into new construction. One factor has to do with the rapidly changing concepts being introduced, and the fear of quick obsolescence of the expensive automation packages. So long as the United States and other technically oriented countries are populated with quick and fertile minds, innovations will come in a never-ending stream, so that transcending yesterday's plateau will be a way of life. That should not be a reason to ignore innovation, so long as it provides real benefit to the situations at hand. There must be a common denominator established, however, in order to develop standards that would slow radical obsolescence; in other words, the need to completely scrap a system in favor of the newest and the best. One important area is in the cabling networks that allow control centers to communicate with remote systems scattered throughout a building. They are expensive to install and are frequently unique to the systems they service; however, they needn't be. A standardized, well-designed cabling plan should be able to accommodate many different system configurations and technologies as long as they, too, are designed for the cabling standard. Electrical paths between building "brain centers" and the remote devices they control are fixed. They occupy the same locations, regardless of the communicating or control technology involved. Standardization permits state-of-the-art control equipment to replace the original equipment at some future point in time, without having to rewire the building. Moving remote systems around within the building is also made possible. Additional control channels can be created by using carrier-frequency multiplexing, or other appropriate technologies. Other areas of standardization may not be as apparent but could be just as valid. The point is that related industries must work together in some ways, and the real estate development branch may have to be the catalyst, bringing about cohesive changes. Much of this is already taking place and will speed the adoption of functional systems.

Product development will be branching out into many areas of interest to the property and facility manager; in other words, finding easier and less expensive ways to accomplish common tasks. For example, while we don't foresee robots operating a building, we do envision them being used for many routine service chores. They should be capable of cleaning floors, acting as messengers or product conveyors between tenant areas or building management personnel, helping to load trucks, and doing other tasks that are now labor-intensive. This is not *Star Wars* fantasy. Many of these simpler robotic functions are already commonplace in manufacturing industries, particularly in Japan where many factories use robots extensively.

The impetus for rapid technical change is coming from the federal government, the insurance industry, corporate management, and environmental groups. With the strong concern about indoor air pollution, the industry is being charged with the task of developing better air-handling techniques for energy-efficient buildings. Constant recirculation of air containing asbestos, radon, and a host of other toxic gases and particulates is having serious effects on building management personnel and tenants. Improved air evacuation, as well as filtration techniques, is being developed. The modification of building and decorative materials, as well as office supplies, is also being sought. Laws pending in Congress will mandate change. The high incidence of liability suits against building owners are bringing pressures from the marketplace and insurance companies. The spin-off technology coming from NASA will inevitably change the way we perceive and use ornamental plants. Their value in reducing indoor air pollution will become recognized as an important part of pollution control programs. Buildings will be developed with extensive plantings in mind for tenant as well as common areas. New functional products, such as Aqua/Trends' precision Micro-Irrigation™ Systems, will be installed to work cooperatively with biotechnology related, pollution abatement techniques. Commercial buildings of the future will be healthier and safer, in addition to having other superior attributes.

Those of us that are developing technologies of a lesser complexity will offer relatively inexpensive, simple-to-use systems that meet a wide variety of property and facility management needs. Combined with the large-scale automation of building control systems, they will bring to the building managers of the future an arsenal of automatic tools to make their tasks easier and more cost-effective. Commercial buildings will possess a degree of intelligence far beyond present levels, and the use of technical systems will be much more commonplace. Inefficient and inconsequential products will be eliminated from the marketplace, as building owners and managers become more sophisticated at evaluating technical systems. With training and greater exposure, they are leaving behind their inexperience and awe of state-of-the-art development—as we all are. The cost-effectiveness of intelligent systems will improve, and confidence in building innovation will grow.

Micro-Irrigation Systems are one of the new concepts that promises to have an impact on the management and marketability of commercial buildings, particularly as the interior landscape movement regains momentum. The technology is useful to building management, tenants, and service companies. It is inexpensive enough to incorporate into most buildings and cost-effective enough to pay for itself within capital investment guidelines. Its usefulness in new construction, as well as retrofit situations, will promote automated interior plant care in large portions of the building population. Beauty, design, and functionality, as well as business efficiency and occupant health, are all served by this innovation.

The Outlook for Smart House Technology

Research and development into products and technologies for residential housing is strong and gaining momentum. At stake are the huge markets for new construction and retrofit installations—contract as well as do-it-yourself. Although the present population of homes that could be considered smart is very small, few dispute the inevitability of truly functional houses in the future. Much of this technology exists today. Systems can be purchased that flash the image of a visitor onto your video screen as they await entry at the front door. When identification is assured, a remote-control lock-release unlatches the door. Other devices change computer-controlled energy management programs by simply touching appropriate sections of a video screen, thus putting the house into an "off-to-work" or "vacation" mode, or whatever other programmed conditions are required. Other systems are capable of controlling lights, appliances, sprinkler systems (lawn and fire) or whatever, merely by making a telephone call from work (or from thousands of miles away) and injecting control codes onto the telephone line to be received, decoded, and acted upon at the house. Devices can remotely open windows or drapes, automatically switch on your security lights, time the operation of your VCR, signal police or fire fighters in emergencies, and automatically balance your checkbook. These wonders were unfamiliar only 15 years ago. The next decade will see many more rapid advances in home technology as the wave of the future grows. Momentum is a powerful force, and the home automation industry is just getting started—there is much to be offered in coming years. Coordination and standards are needed here as much as they are in commercial sectors, and that leadership has fallen on the National Association of Home Builders (NAHB). The Electronics Industries Association (EIA) is also taking an active part by establishing wiring and control standards for smart homes. During the 1988 Summer Consumer Electronics show, the EIA demonstrated a range of equipment and appliances utilizing their CEBus concepts.

Although technical development is fragmented, the most concerted effort is being brought to bear in Smart House Development Venture (Project "SMART HOUSE"). With over 100 high-powered industrial companies working in concert with the Gas Research Institute, the electrical power industry, the electronics industry and the home building industry, the project is absorbing millions of dollars in research and development expenditures. This is a limited partnership venture with the hopes of developing proprietary concepts for use by the consortium and licensed to others. The success of Project "SMART HOUSE" will determine in large measure the shape of things to come, in terms of products and services offered the homeowner and in the way devices are controlled. Today's appliances and electronically controlled systems have little compatibility. For the most part, they don't speak the same control language and, therefore, cannot communicate with each other. House wiring consists of a power network and an amorphous mix of conductors for front doorbells, security sensors and lights, phone lines, smoke detectors, intercom systems, central vacuum cleaning systems, audio and video systems, heating/air-conditioning sensors, and whatever other controlled or static devices might be installed. The main concept of Project "SMART HOUSE" is to install a single multiconductor cable throughout the house and to radically change power control concepts to take advantage of the simplified house circuitry. All devices, powered or nonpowered, are to be plugged into special receptacles that transmit and receive electrical power as well as data, appliance control, and entertainment signals. Power is applied under this concept by means of a closed-loop control network. Smart lighting fixtures and appliances having small built-in microprocessors are plugged into the system and communicate with a microprocessor-based control center that permits or restricts the application of power to the remote devices according to the signals received. For example, if a dishwasher calls for power (either switched manually or by means of an automatic timer) and its microprocessor shows a normal condition, then the control center is free to apply electrical power to run its cycle. If, on the other hand, the dishwasher signalled an electrical short condition or overload from a jammed pump, etc., then the control center would remove the electrical power until the abnormal condition was cleared. Electrical safety is inherent in this concept, for one could not get shocked by live sockets or shorted appliances, and electrical fires would be rare. Other channels on the cable would carry signals from room to room and to the control center for processing or display. For example, the stereo amplifier would plug into a receptacle so its musical output could be received and played by remote speakers plugged into local receptacles in other rooms. Or security sensors around the house could be plugged into local receptacles, causing a TV screen at the command center to flash a warning and the location of intrusion and then automatically signalling the police or monitoring service. Appliances can communicate not only with the command center but with each other as well. The kitchen oven can transmit a dinner-ready signal to the intercom stations around the house; for example, 5 minutes before the end of cooking time, a voice-synthesized announcement can call diners. In Project "SMART HOUSE", not only is the power control concept different, but electrical power levels will be reduced and possibly converted to DC (direct current) for safety, economy, and ease of control. As a result, energy requirements will be reduced. If Project "SMART HOUSE" is successful, residences of the future will be efficient and convenient, as well as safer. Although everything is still experimental and in the prototype stages, its wiring standards have been written into the 1987 National Electrical Code (NEC), which lays the guidelines not only for house wiring but for lighting fixtures and appliances as well. The benefits will be particularly meaningful for the elderly and infirm. Many things can be accomplished for them by the technically sophisticated house. The residence of the future will have some of the advanced features of the intelligent commercial building; in fact, many of the concepts of Project "SMART HOUSE" are scaled-down versions of those technologies. Commercially targeted technology will learn from the residential sector as well.

While Project "SMART HOUSE" is being demonstrated and perfected, other more modest concepts will be introduced and will reach varying degrees of acceptance. As already mentioned, most of these concepts have to do with power control, security, and safety systems. The stress will be on digital controllers, remote-control concepts, and proximity and touch switches, as well as sensor-activated systems—and on their electronics and physical interconnections.

Micro-Irrigation automated interior plant-care systems are one of those new breeds of smart devices. It is considered an appliance by most building codes. Other new concepts will be developed to meet diverse human and commercial needs. Many of those will be in response to the federal government's growing effort to reduce indoor air pollution.

Plants in Buildings The trends in interior landscape design are strong. As more architects, interior designers, and real estate developers are keyed into the long-term human factors affecting their building projects, more live plants will grace interiors. As more merchants, hoteliers, and restaurateurs begin to understand the commercial value of natural green environments, more plants will grace interiors. As corporate executives and office managers begin to understand how decorative potted plants can improve employee satisfaction, efficiency, and health, more plants will grace interiors. As homeowners learn of the emotional and physical health benefits of houseplants, as well as how their life-style and well-being can be improved, more plants will be used.

The highly visible work that NASA is doing to combat indoor air pollution with plants can have a greater impact on the movement than most envision. NASA has already demonstrated the ability of plants, through their natural biological processes, to dramatically absorb many toxic gases from the surrounding air. Each house or office is a microenvironment within the earth's ecological system, isolated by the energy efficiency we designed into it. The materials and activities within it are introducing toxics; plants are now being shown to be capable of neutralizing them as harmful agents. Plants are normally taken for granted, but it must be kept in mind that many of nature's higher creations are more complex and performance-effective than anything man can devise. The human body is one example. Higher plants are now known to be another—we should not be surprised. For over a hundred years scientists have recognized a plant's ability to absorb carbon dioxide from surrounding air and assimilate it with physical, chemical and photochemical processes, finally "exhaling" oxygen and water vapor. NASA research has built on that knowledge. They now know that harmful pollutants are also absorbed and digested within these processes, as well as by microbial degradation of pollutants trapped in the soil. Within this wonderful, microscopic cellular structure and planter system (plant/growing medium) is the most highly developed pollution filtration system known to mankind—useful in both air and water pollution control. The technology is more refined than most others mentioned within these pages. It becomes more apparent when we consider fluid and gas transfer patterns within the structure, internal pressure changes, photochemical induced food (sugar) generation, sensor activity, internal electrical patterns, adaptation to environmental changes, part regeneration, waste disposal, photochemical color changes, pore size adjustment to light intensity—these and other processes partially illustrate the complexity and sophistication of these plant systems. NASA's work is just now unlocking their usefulness. The question is no longer whether plants are capable of acting as pollution filters; it is a matter of degree of effectiveness, and much of that has to do with the way they are utilized in home and commercial environments. An extension of this research is the new scientific field of *biospherics*. Pioneered by Soviet V.I. Verdansky, its objectives are the study and improvement of earth's enclosed ecosystems—with eventual space colonies also in mind. The important U.S. project called Biosphere II in Arizona will research ecosystems on a grand scale, with bio-technology playing a major role in controlling environmental purity. As of this writing, NASA has already reached very positive conclusions and has mounted an extensive campaign (in cooperation with the Associated Landscape Contractors of America) to publicize the healthier interior environments possible with live plants. We can be sure, however, even at this stage of discovery, that if we put a lot of live foliage (particularly select varieties) in our homes and commercial places, the air we breathe will be cleaner. Therefore, use is clearly better than nonuse. Plants can become an inexpensive answer to some of our interior air pollution problems. Other avenues of pollution control are being attacked, particularly techniques of air handling and product modifications, but the natural methods have great appeal.

If the Indoor Air Quality Act of 1989 (H.R. 1530) and other related legislation become law, they can be a significant factor in adopting plant biotechnology for future buildings. The bill's stated objective is "To authorize a national program to reduce the threat to human health posed by exposure to contaminants in the air indoors." One of its provisions is to empower the president to establish a national indoor air quality council that would coordinate federal programs in the field. It expresses great concern over the health hazards of our present environments and recommends extensive study and technical development efforts toward indoor air pollution abatement. For example, Section 2(E) states ". . . indoor air pollutants pose serious threats to public health including cancer, respiratory illness, skin and eye irritation, and related effects." And in Section 5(b)(M) ". . . development of control technologies or other measures to reduce the concentrations of contaminants indoors, including control of emissions from internal sources of contamination, improved air exchange and ventilation, and related measures."[1]

NASA's plant biotechnology falls within the purview of this bill. NASA's environmental research with live plants as elements of pollution control will be given serious study in many of the programs generated by this bill. Buildings will be designed and built giving special consideration to indoor horticulture and associated pollution control techniques. Media coverage will tout the advantages of live plants, and greater use will inevitably result in houses and commercial facilities. It is the beginning of *indoor ecology*.

Most people fail to recognize that even if tropical foliage did little to abate pollution, it still provides many other benefits and is worthy of consideration and use. Plants' positive influence on our emotional health is well documented, and their influence on our satisfaction and efficiency as employees is becoming more appreciated, especially as we take some important lessons in this regard from our European cousins. The economic benefits to merchants and commercial ventures cannot be directly measured in dollars, but most businesspeople have a good feel for the need for foliage in their decor. Combined, the reasons are compelling for the use of live plants in our interior spaces.

While we don't envision growth rates as steep as in the past 10 years, future growth will continue to be strong, with short diversions during economic slowdowns. The tropical foliage industry is currently going through one of those adjustments. The demand for its product is high despite a slowdown in commercial construction. The industry has become overproduced, however, with too many chasing the limited market. Shakeouts will result, and the industry should emerge stronger than ever. As mentioned in Chapter 2, the demand for interior plants will take on a new impetus once automation makes significant inroads into interior plant-maintenance practices. There will be a time when lush interior vegetation will be common to most home and commercial environments.

The Outlook for Automated Interior Plant Care

Economic and social factors are leading inevitably toward the use of automated techniques that reduce the time, effort, and cost of interior plant care. These techniques will be oriented toward plant watering, for most other maintenance-related tasks do not lend themselves to automation. It is inconceivable that human plant care can ever be replaced totally, but the new precision irrigation technology will make deep inroads into labor expenditures and promises to radically change the way things are done in the interiorscaping industry. Plant-care techniques are being altered, as are maintenance schedules, in order to find better ways of utilizing available labor. More accounts will be covered by the available work force. The plant-care technicians will learn to work with auto-interigation systems as another tool. They will be trained in the specialized methods and equipment involved. In commercial installations, they will be charged with the responsibility of interfacing with the new systems. Making minor flow adjustments once in a while would be their main concern. The maintenance technicians of the future will be more knowledgeable and a higher caliber of employee. They will be higher paid yet will not work as hard. Those deeply trained in interior horticulture will be able to use their expertise with greater intensity and will have greater satisfaction in their work. There will be much less labor turnover in the industry, leading to a stability that could never be achieved with the current manual practices. That will also inevitably lead to a higher level of client satisfaction.

[1]House Bill #H.R. 3809, "Indoor Air Quality Act of 1987," (Washington, DC: U.S. Congress, December 18, 1987), pp. 1, 2, and 10.

The large-scale use of Micro-Irrigation Systems by the construction and real estate industries will foster better methods of installation, lowering equipment and application costs. Partition studs and baseboard moldings will be factory modified to accommodate irrigation tubing. Furniture and cabinetry will have complete mini-systems built into them at the manufacturing plants, or they will incorporate irrigation tubing harnesses that can later be connected to the building's irrigation system. Future buildings will have greater interaction between energy- or security-related technologies and auto-interigation systems. As buildings become smarter and use more technology, automated plant-care systems will be an important part of the mainstream of peripherals working in concert with digital control centers. If the technology being developed for Project "SMART HOUSE" becomes a future standard, new opportunities will open for Micro-Irrigation Systems. The cabling networks being adopted provide a convenient means of communication between remote areas of a building. These logic channels will be used to send data from individual plant locations to the system controller, which would then react to the conditions there. If things were not normal, sensor data would be transmitted to the controller. Its monitoring circuits would tell the system to alter the irrigation cycle according to plant needs at the time, or it might flash a visual signal or a sound alert identifying the abnormal condition. These are called *closed-loop systems*. They monitor and react to remote situations. Such higher levels of sophistication are possible now. As a matter of fact, soil moisture sensors are commonly used in agricultural irrigation, but in a much different way. Indoors, however, the monitoring needs are more complex—ungainly and expensive interconnections are required. The new cabling methods being proposed would provide built-in channels of data communication, making closed-loop systems highly practical for future Micro-Irrigation System installations.

Other technologies now being introduced to the market will increasingly be interfaced with these precision automatic irrigation systems. Home and office control through television cable networks will provide remote plant-care monitoring as an optional service. Similar control services are also being market tested by telephone companies. The monitoring and control data are transmitted and received over existing telephone lines. This system concept can also be interfaced with automated plant-care systems in a residence or commercial building, and billed services provided.

Integral power control systems are becoming increasingly popular. These built-in automatic switching networks are capable of controlling many things, from security devices to standard lighting and appliances. Previous chapters have mentioned the X-10 Powerhouse System that uses carrier-frequency control methods. Because of its economy, versatility, and ease of installation, it is among the preferred devices for many power control systems. Various manufacturers have adopted it for their own developments under license to X-10 (U.S.A.), Incorporated. As more installations of this kind are made, they will be increasingly linked to Micro-Irrigation Systems, for here, too, convenient interfacing can be accomplished. Aqua/Trends has its own versions in kit form. They are expected to become an important part of that company's line of products.

A Typical Day in the Future

Mr. and Mrs. Smith of the future live in a world wondrous with automated systems. Their home and work environments are safer, healthier, and more endowed with sound and visual esthetics, which are more convenient and less costly to operate. A typical day might go something like this:

The six o'clock waking call is transmitted to the bedroom from a home control center, providing a persistent audible reminder that the world must again be faced and challenged. A sleepy hand lightly touches a contact on the headboard to signal the control center that someone is awake. The waking call stops. Mr. Smith gropes his way to the bathroom where the lights switch on as he passes through the doorway. Meanwhile, the coffee maker and other appliances are prepared for use by the home control center. Coffee beans, premeasured for the family's needs, drop into a grinder. The grinds are then automatically transferred to a brewing section, and the coffee is perked to perfection. Mrs. Smith's hair curlers start to preheat, for her waking call will be coming up at 6:30. Heating and air-conditioning subsystems become active to bring comfort up to preselected levels.

When the Smiths descend for breakfast, they find the toast ready and a precooked meal in the microwave oven. The dog hasn't been forgotten either, for a fresh dollop of food has been dispensed to its dish as water flows slowly into the bowl alongside. The green members of the Smith family are also automatically cared for. They have purchased one of those *healthy* homes, designed for the extensive use of decorative houseplants. Several large greenhouse windows contain dozens of plants growing in a charcoal and soil medium, with a circulation subsystem drawing air through the planter to provide rapid absorption of pollutants. An irrigation system, integrated into the structure and linked to the home control center, gently services those plants and the others scattered around the house with regular doses of water, plant food, and systemic insecticides.

As they dine, a video screen switches on to bring them the TV news, business reports, and personal reminder notes. The security and safety alarm systems are reset, anticipating the family leaving for work. Outside, lights have been turned off and the electrically operated gate latch is opened.

Down at the office, the building's intelligent system starts to adjust the comfort levels. During the night, its automatic interior irrigation system fed small doses of water to the decorative foliage planters (thousands of them around the various lobbies and offices, restaurant, and bank in the building), servicing the plants floor by floor and turning on lights where necessary to supplement life-giving daylight. As Mr. Smith enters the underground parking garage, a key-card slipped into a control box identifies him as an authorized visitor, and the gates swing open. The same card gains him entry to the employees' elevator. As it rises, a pleasant, synthesized voice greets him and calls off the floors as they pass.

Being the first into the office this morning, he again uses his key-card to unlock the doors, disarm the local security systems, and switch on the main lighting circuits. His office lights go on as he passes portal sensors. Office computers and business equipment have been automatically activated to warm and ready them for the busy day ahead. Mr. Smith's computer screen flashes a friendly greeting from the London office—they jokingly ask where he has been for they've been at work for many hours. He is brought up-to-date on business matters, and a teleconference is scheduled for later in the day. He leaves his office to confer with his secretary. Most of the office lights go out behind him. The workstation area is well laid out, with lush tropical plants helping to provide visual as well as acoustical privacy, surrounding the secretaries and sales representatives with natural beauty and air purification. Mr. Smith's secretary reminds him of his daughter's upcoming anniversary. The two scan a video monitor for the gift offerings of a shopping service they subscribe to. A couple of items look appealing, and Mr. Smith makes notes so that he can discuss them later with his wife. A computer disc containing business notes and letters that he entered last night at home is given to his secretary. They are for filing in the data base and transcribing into form messages for wide distribution over the company's electronic mail network.

Meanwhile at home, Herbie the resident robot has left his closet and is busily vacuuming the house. He intimately knows where each piece of furniture is and where the delicate accessories are—they have been burned into his memory. The dog barks while the trusty robot goes about his chores, but Herbie pays no mind. With the Smiths away, the heating and air-conditioning system has automatically been turned down to a more energy-efficient level, and the security and fire safety systems have been turned up to full alert. As the mid-afternoon sun beats down on the west side of the house, a sensor signals the control center, and the vertical window blinds on that side of the building are closed to the bright illumination, protecting the furnishings and helping the house maintain an even inside temperature.

As late afternoon matures, the lawn sprinkling system comes to life and sprays the surrounding landscape with a generous volume of water. Mr. Smith, still at work, remembers that he and Mrs. Smith won't be getting home until late. The dog must be fed, but he knows the control center will feed it on time. The dog needs to be let out though. By dialing his house and tapping a few code numbers into the phone pad, an electronic latch is released, unlocking the dog's exit door to the outside play yard. It will automatically relock when the dog comes back into the house. He then types in a different code, which tells the oven not to begin cooking dinner until a later hour. While on the line, a few more coded entries start a replay of phone messages on the family answering machine. The home computer has also been active while they were away. It has received a signal from their bank that a weekly checking account status report was about to be sent. After activating itself, the computer absorbs the information and checks it against the family's banking data base. A status report is automatically typed out for Mrs. Smith's review later.

As Mr. Smith leaves the office to meet his wife for a shopping trip, the lights go out automatically and the security and fire safety systems are put on full alert. Power outlets are scanned by the building's control center to detect any office machines that might have been inadvertently left on. If so, those outlets are deactivated until the next morning.

At the Smith home, its controller adjusts the heating and air-conditioning to improve the comfort level in anticipation of the family's arrival. The sun has gone down, so outdoor security lights have been switched on, as well as a couple of interior lamps. By this time, the oven has been activated, and bread starts to bake. When the Smiths get home, they turn on the microwave oven while they freshen up. A TV monitor is switched on and the evening news is enjoyed while dinner is put together. The TV goes off, and the Smiths are surrounded by relaxing music from the ubiquitous sound system as they dine. After dinner, relaxation is briefly punctuated by the almost imperceptible sound of water flowing gently to the potted foliage throughout the house, as the automatic irrigation system provides the second daily dose—a reminder that the plants are well taken care of.

The scenario just described is futuristic, but the technology and systems to do these things are available today. Our homes and offices of the future will include this and more. New ideas are spawned by visions of the future—like these. Many new concepts will be introduced as time progresses. The opportunity to meet human needs is almost limitless. There is always a better way of doing something, always a more practical approach to take. Precision, Micro-Irrigation Systems are but one step on the road to a better personal and business environment and to a healthier and more satisfying life-style that provides the freedom to concentrate on more meaningful things.

A INTERIORSCAPING PLANTS AND THEIR PREFERRED CONDITIONS

Mini-encyclopedia of houseplants

COMMON NAME	BOTANICAL NAME	SOIL MIX					LIGHT				WATER				TEMPERATURE			HUMIDITY			PROPAGATION
		All purpose	High humus	Gritty lean	Loose medium	High acid	Sunny	Semi-sunny	Semi-shady	Shady	Keep wet at all times	Keep evenly moist	Approach dryness between waterings	Let dry out between waterings	Cool	Average house	Warm	Very moist	Moist	Average house	
Acalypha	*Acalypha*	•					•	•				•				•				•	Cuttings in fall
Achimenes	*Achimenes*		•				•	•				•				•				•	Rhizome division, seed, stem cuttings
Acorus	*Acorus*	•							•	•		•			•				•		Division in spring or fall
African hemp	*Sparmannia africana*	•						•				•				•				•	Cuttings
African violet	*Saintpaulia*		•				•	•				•				•			•		Seed, leaf cuttings, division
Agapanthus	*Agapanthus*	•					•	•	•			•			•	•				•	Division in early spring
Allamanda	*Allamanda*	•						•				•				•			•		Cuttings of half-ripened stems in spring
Amaryllis	*Hippeastrum*		•				•	•					•			•				•	Remove offsets at potting time; sow spring seed
Anemone	*Anemone*	•					•	•				•			•				•		Seed or offsets
Anigozanthus	*Anigozanthus*	•					•					•			•					•	Seed or root division
Anthurium	*Anthurium*				•				•			•					•	•			Cuttings or offsets
Apostle plant	*Neomarica gracillis*	•					•	•				•				•				•	Division of rhizomes

(*continued*)

App. A / Interiorscaping Plants and Their Preferred Conditions

COMMON NAME	BOTANICAL NAME	SOIL MIX					LIGHT				WATER				TEMPERATURE			HUMIDITY			PROPAGATION
		All purpose	High humus	Gritty lean	Loose medium	High acid	Sunny	Semi-sunny	Semi-shady	Shady	Keep wet at all times	Keep evenly moist	Approach dryness between waterings	Let dry out between waterings	Cool	Average house	Warm	Very moist	Moist	Average house	
Aralia	*Dizygotheca elegantissima*, also see: *Polyscias* and *Fatsia*		●					●	●			●				●				●	Cuttings
Ardisia	*Ardisia*	●						●	●			●			●					●	Seed or cuttings
Asparagus fern	*Asparagus*	●						●	●	●		●				●				●	Seed or clump division
Aspidistra	*Aspidistra*	●						●	●	●		●			●					●	Division of roots in late winter or spring
Aucuba	*Aucuba japonica*	●						●	●			●			●					●	Seed or cuttings
Azalea	*Azalea*	●				●	●	●				●			●				●		Stem cuttings
Baby's tears	*Helxine*		●					●	●		●					●			●		Division of clumps or cuttings
Bamboo	*Bambusa*		●				●	●				●				●				●	Division of large clumps
Bat-wing tree	*Erythrina indica*	●					●	●				●				●				●	Seed
Begonia	*Begonia rex*; *Begonia semperflorens*; Rhizomatous species; Fibrous-rooted, cane species; Tuberous-rooted species; and Fibrous-rooted species		●				●	●	●			●				●			●	●	Cuttings or seed
Bird-of-Paradise	*Strelitzia*	●					●	●				●			●					●	Division of rhizomes, remove suckers in spring
Bougainvillea	*Bougainvillea*	●					●					●				●			●		Seed in spring or cuttings of half-ripe wood
Bromeliad:	*Aechmea*			●			●	●					●			●				●	Remove offsets
Pineapple	*Ananas*				●		●	●					●			●	●				Root top of fruit
	Billbergia				●		●	●				●				●				●	Detach suckers
Earth stars	*Cryptanthus*			●			●	●					●			●				●	Remove offsets
	Neoregelia				●		●	●					●			●				●	Detach suckers
Living vase or Flaming-sword	*Vriesia*			●			●	●				●				●			●		Remove suckers or plantlets
Brassaia (Schefflera)	*Brassaia*	●					●	●	●	●		●				●				●	Cuttings of half-ripened stems
Cactus:	*Aporocactus*; *Astrophytum*; *Cephalocereus*; *Chamaecereus*; *Cleistocactus*; *Echinocactus*; *Echinocereus*; *Echinopsis*; *Gymnocalycium*; *Lobivia*; *Mammillaria*; *Notocactus*;				●		●	●						●	●	●				●	Offsets or cuttings

App. A / Interiorscaping Plants and Their Preferred Conditions

COMMON NAME	BOTANICAL NAME	SOIL MIX					LIGHT				WATER				TEMPERATURE			HUMIDITY			PROPAGATION
		All purpose	High humus	Gritty lean	Loose medium	High acid	Sunny	Semi-sunny	Semi-shady	Shady	Keep wet at all times	Keep evenly moist	Approach dryness between waterings	Let dry out between waterings	Cool	Average house	Warm	Very moist	Moist	Average house	
	Opuntia; Pereskia; Rebutia;																				
	Ephiphyllum; Hylocereus; Selenicereus;	●						●	●			●				●			●		Cuttings in spring or summer
	Schlumbergera; and *Zygocactus*	●						●	●			●				●			●		Cuttings
Caladium	*Caladium*	●						●	●			●					●		●		Divide tubers or clumps in spring
Calceolaria	*Calceolaria*	●							●	●		●			●				●		Seed in April or August
Calla-lily	*Zantedeschia*	●						●			●					●			●		Seed or offsets
Camellia	*Camellia japonica*		●			●	●	●				●			●			●			Cuttings of current season's new wood
Chinese evergreen	*Aglaonema*	●							●	●	●		●				●			●	Root stems
Chlorophytum (spider)	*Chlorophytum*	●						●	●	●		●				●				●	Remove aerial plantlets or division
Chrysanthemum	*Chrysanthemum*	●						●	●			●			●					●	Cuttings
Cineraria	*Senecio cruentus*	●						●	●			●			●				●		Seed in summer
Citrus	*Citrus*	●						●	●			●				●				●	Cuttings of half-ripened wood in spring
Clematis	*Clematis*		●					●				●				●				●	Seed, layering, division, cuttings, or grafting
Clerodendrum	*Clerodendrum*		●					●	●			●				●				●	Cuttings of half-ripened wood or remove suckers
Clivia	*Clivia*	●						●	●			●			●	●				●	Division
Cobra plant	*Darlingtonia californica*		●		●				●	●	●					●			●		Seed or shoots in summer
Coccoloba (Seagrape)	*Coccoloba*	●						●	●			●				●				●	Seed, layering, wood cuttings
Coffee	*Coffea*	●						●	●		●				●	●			●		Seed or wood cuttings
Coleus	*Coleus blumei*	●					●	●				●				●				●	Seed or stem cuttings
Columnea	*Columnea*		●					●	●			●				●		●			Tip cuttings or seed
Creeping Charlie	*Pilea nummularifolia* (also see: *Plectranthus*)		●					●	●			●				●				●	Cuttings

(continued)

App. A / Interiorscaping Plants and Their Preferred Conditions

COMMON NAME	BOTANICAL NAME	SOIL MIX					LIGHT				WATER			TEMPERATURE			HUMIDITY			PROPAGATION
		All purpose	High humus	Gritty lean	Loose medium	High acid	Sunny	Semi-sunny	Semi-shady	Shady	Keep wet at all times	Keep evenly moist	Approach dryness between waterings	Let dry out between waterings	Cool	Average house	Warm	Very moist	Moist	Average house	
Creeping fig	*Ficus pimula* or *Fradicans*		●					●	●			●				●			●		Cuttings
Crossandra	*Crossandra*		●					●				●				●			●		Seed or tip cuttings
Croton	*Codiaeum*		●				●	●				●				●			●		Cuttings
Cup-and-saucer vine	*Cobaea scandens*	●					●	●				●				●				●	Seed
Cyclamen	*Cyclamen*		●					●	●			●			●				●		Seed
Cyperus	*Cyperus*	●						●	●	●						●			●		Division
Cycas fern or palm	*Cycas*	●						●				●				●				●	Seed or dormant suckers
Dieffenbachia (Dumb cane)	*Dieffenbachia*	●						●	●	●			●			●				●	Stem cuttings or layering
Dipladenia	*Dipladenia splendens*	●					●	●				●				●	●				Stem cuttings or seed
Dracaena	*Dracaena*	●						●	●	●		●				●				●	Stem cuttings, layering, root division
	Pleomele	●						●	●	●		●				●				●	Stem cuttings, layering, root division
Easter lily	*Lilium longiflorum*	●						●						●	●					●	Plant bulbs
Euonymus	*Euonymus*	●						●	●			●			●					●	Cuttings of half-ripened wood in fall or winter
Elaeagnus	*Elaeagnus*	●						●	●			●			●	●				●	Cuttings in spring
Fatshedera	*Fatshedera lizel*	●						●	●	●		●			●	●				●	Cuttings
Fatsia	*Fatsia japonica*		●					●				●			●	●				●	Cuttings of branches
Ferns: Bird's-nest	*Asplenium*		●					●	●			●				●				●	Remove offsets or root plantlets
Boston or sword	*Nephrolepis*		●					●	●			●				●			●		Division of clumps
Bear's-paw	*Polypodium*		●					●	●			●				●			●		Division of clumps
Holly	*Cyrtomium*		●					●	●			●			●				●		Rhizome division
Maidenhair	*Adiantum*		●	●					●	●		●				●		●			Division of clumps
Miniature	*Polystichum*		●					●				●				●			●		Division
Rabbit's-foot	*Davallia*		●					●				●			●				●		Rhizome division
Staghorn	*Platycerium*			●				●	●			●				●	●				Remove offsets
Table or brake	*Pteris*		●					●				●			●				●		Division
Ficus or fig	*Ficus*	●						●	●	●		●				●				●	Air layering
Fittonia	*Fittonia*	●						●	●			●			●				●		Tip cuttings
Flame violet	*Episcia*		●					●	●			●					●		●		Root stolens

COMMON NAME	BOTANICAL NAME	All purpose	High humus	Gritty lean	Loose medium	High acid	Sunny	Semi-sunny	Semi-shady	Shady	Keep wet at all times	Keep evenly moist	Approach dryness between waterings	Let dry out between waterings	Cool	Average house	Warm	Very moist	Moist	Average house	PROPAGATION
Flowering maple	*Abutilon*	●					●					●				●				●	Stem cuttings
Flowering tobacco	*Nicotiana*	●						●	●		●					●				●	Seed
Fragrant gladiolus	*Acidanthera*	●					●					●				●				●	New corms in spring
Freesia	*Freesia*	●					●					●			●				●		Seed or offsets
Fuchsia	*Fuchsia*	●						●	●			●			●				●		Cuttings in spring
Gardenia	*Gardenia*	●				●	●	●				●				●			●		Cuttings of half-ripened wood
Gerbera	*Gerbera*	●					●	●				●				●				●	
Geranium	*Pelargonium*	●					●	●					●		●	●			●	●	Cuttings
Ginger:	*Amomum*	●						●	●			●				●			●	●	Clump division
Spiral	*Costus*		●					●	●			●				●			●		Clump division in spring
	Curcuma		●					●	●			●				●		●			Division of tubers in spring
Ginger lilies	*Hedychium*	●					●				●	●				●			●		Division of tubers at rest time
Peacock plant	*Kaempferia*		●					●	●			●				●		●			Seed or clump division
Commercial ginger root	*Zingiber*		●					●	●			●				●		●			Division of clumps in spring
Gloriosa	*Gloriosa rothschildiana*		●				●	●				●				●			●		Seed or tuber division
Gloxinia	*Sinningia*	●						●				●				●			●		Seed, leaf or stem cuttings, tuber division
Gynura	*Gynura*	●					●					●				●				●	Cuttings
Haemanthus	*Haemanthus*	●					●	●				●				●			●		Remove offsets when repotting
Hawaiian ti	*Cordyline terminalis*	●						●	●	●	●					●				●	Stem cutting, layering, root division
Hibiscus	*Hibiscus*	●					●					●				●			●		Stem cuttings
Homalomena	*Homalomena*	●						●	●		●					●			●		Stem cuttings
Hydrangea	*Hydrangea*	●				●	●	●			●	●			●					●	Stem cuttings
Hypoestes	*Hypoestes*		●					●	●			●				●			●		Seed or cuttings
Impatiens	*Impatiens*	●						●	●	●			●			●		●			Cuttings
Iresine	*Iresine*	●					●						●		●	●				●	Cuttings
Ivy, English	*Hedera helix*	●						●	●	●	●	●	●		●				●	●	Cuttings
Ixia	*Ixia*	●					●						●	●	●				●		Bulb offsets

(continued)

App. A / Interiorscaping Plants and Their Preferred Conditions

COMMON NAME	BOTANICAL NAME	SOIL MIX: All purpose	High humus	Gritty lean	Loose medium	High acid	LIGHT: Sunny	Semi-sunny	Semi-shady	Shady	WATER: Keep wet at all times	Keep evenly moist	Approach dryness between waterings	Let dry out between waterings	TEMP: Cool	Average house	Warm	HUMIDITY: Very moist	Moist	Average house	PROPAGATION
Ixora	Ixora	•					•	•					•			•			•		Strong cuttings
Jatropha	Jatropha	•					•	•				•				•			•	•	Seed or cuttings
Jerusaleum cherry	Solanum	•					•	•				•			•				•		Seed
Jessamine, night-blooming	Cestrum nocturnum		•				•	•				•				•				•	Cuttings
Joseph's coat	Alternanthera	•					•	•				•				•				•	Cuttings
King's crown	Jacobina carnea	•					•					•				•			•		Cuttings
Leopard plant	Ligularia tussilaginea	•						•	•			•			•					•	Dividing plants with more than one crown
Liriope or lily turf	Ophiopogon	•						•	•			•			•				•		Division
Lipstick vine	Aeschynanthus		•	•				•	•			•				•		•			Stem or tip cuttings
Miniature rose	Rosa	•					•	•				•				•			•		Seed or cuttings
Montbretia	Crocosmia	•					•					•				•				•	Seed or offsets
Myrtle	Myrtus communis	•					•	•					•		•					•	Cuttings of ripened wood
Nandina	Nandina	•						•	•			•				•				•	Stem cuttings
Norfolk island pine	Araucaria excelsa	•						•	•			•			•	•			•		Seed or root tops of old plants
Oleander	Nerium oleander	•					•	•				•				•				•	Cuttings of firm tip growth in spring or summer
Orchid:	Brassavola				•		•	•						•		•		•			Division in late winter
	Cattleya				•		•	•						•		•		•			Division in late winter
Swan	Cycnoches	•			•			•	•			•			•				•		Division in spring or summer
	Epidendrum				•			•	•				•			•		•			Air layering or division in spring
Tiger	Odontoglossum				•				•				•		•	•			•		Division in spring or summer
Butterfly	Oncidium				•			•	•			•				•			•		Division in spring
Lady slipper	Paphiopedilum				•				•			•				•			•		Division in spring
Dogwood or moth	Phalaenopsis				•			•	•			•				•			•		Division in spring
Oxalis	Oxalis	•					•	•				•				•			•		Offsets or division
Palms: Bamboo	Chamaedorea	•							•	•		•				•				•	Seed or remove suckers
Butterfly	Chrysalidocarpus	•						•	•			•				•				•	Seed or clump division in spring

App. A / Interiorscaping Plants and Their Preferred Conditions

COMMON NAME	BOTANICAL NAME	SOIL MIX					LIGHT				WATER				TEMPERATURE			HUMIDITY			PROPAGATION
		All purpose	High humus	Gritty lean	Loose medium	High acid	Sunny	Semi-sunny	Semi-shady	Shady	Keep wet at all times	Keep evenly moist	Approach dryness between waterings	Let dry out between waterings	Cool	Average house	Warm	Very moist	Moist	Average house	
Date	*Phoenix*	●						●	●		●					●				●	Seed or remove suckers
Fan	*Chamaerops*	●						●	●		●	●			●					●	Seed or suckers
	Licuala	●						●	●		●					●			●		Seed or suckers
	Livistona	●						●	●		●					●			●		Seed or suckers
	Rhapis	●						●	●		●				●					●	Seed or suckers
Fishtail	*Caryota*	●						●	●		●					●			●		Seed in March
Pandanus	*Pandanus*		●				●	●			●					●		●			Remove suckers
Passion flower	*Passiflora*		●				●	●			●					●			●		Seed or cuttings
Pellionia	*Pellionia*	●							●	●	●					●			●		Cuttings
Peperomia	*Peperomia*	●						●	●	●			●			●				●	Stem or leaf cuttings
Philodendron	*Philodendron*	●						●	●	●	●					●			●	●	Stem cuttings or offsets
Pilea (Artillery, Aluminum, Moon Valley)	*Pilea*		●					●	●		●					●			●	●	Cuttings
Pittosporum	*Pittosporum*	●						●	●		●			●					●		Cuttings of half-ripened wood
Plectranthus (Swedish ivy)	*Plectranthus*	●						●	●	●	●					●				●	Seed or stem cuttings
Podocarpus	*Podocarpus*	●							●		●			●					●		Seed or ripened wood cuttings
Polyscias	*Polyscias*	●					●	●	●	●	●				●	●		●			Cuttings
Prayer plant	*Maranta*		●						●		●					●			●		Division
Pregnant onion	*Ornithogalum caudatum*	●						●						●		●				●	Remove offsets
Primrose	*Primula*		●					●	●		●				●			●			Seed
Privet	*Ligustrum*	●					●	●			●					●				●	Stem cuttings
Pyracantha	*Pyracantha*	●					●				●				●					●	Soft wood cuttings in early summer
Radar plant	*Desmondium gyrans*	●					●				●					●				●	Seed in February
Ranunculus	*Ranunculus*	●					●				●				●				●		Seed in March; roots in autumn
Redwood burl	*Sequoia sempervirens*							●	●	●	●					●				●	Plantlets can be rooted
Resurrection plant	*Selaginella lepidophylla*							●	●	●	●					●				●	Purchase in dried form

(continued)

App. A / Interiorscaping Plants and Their Preferred Conditions

COMMON NAME	BOTANICAL NAME	SOIL MIX					LIGHT				WATER				TEMPERATURE			HUMIDITY			PROPAGATION
		All purpose	High humus	Gritty lean	Loose medium	High acid	Sunny	Semi-sunny	Semi-shady	Shady	Keep wet at all times	Keep evenly moist	Approach dryness between waterings	Let dry out between waterings	Cool	Average house	Warm	Very moist	Moist	Average house	
Rhododendron	*Rhododendron*	•				•	•	•				•			•				•		Stem cuttings
Rhoeo (Moses-in-the-Cradle)	*Rhoeo*	•					•	•	•			•				•				•	Remove offset or transplant seedlings from parent
Rosary vine	*Ceropegia woodi*	•					•	•	•				•			•				•	Cuttings or plant bulblets along stems
Sansevieria	*Sansevieria*	•					•	•	•	•			•			•				•	Division of rootstock or leaf cuttings
Saxifraga	*Saxifraga*	•						•	•		•	•			•				•		Remove young plants
Scindapsus	*Scindapsus aureus*	•						•	•	•		•					•			•	Cuttings
Selaginella	*Selaginella*		•						•	•	•					•		•			Cuttings
Sensitive plant	*Mimosa pudica*	•					•	•	•		•	•				•			•		Seed or transplants
Shoo-fly plant	*Nicandra physalodes*	•					•	•				•				•				•	Seed
Shower plant	*Cassia*	•					•	•				•				•				•	Seed
Shrimp plant	*Beloperone guttata*	•					•	•					•			•				•	Cuttings
Spathiphyllum	*Spathiphyllum*		•					•	•			•				•				•	Division of rootstock
Stephanotis	*Stephanotis floribunda*		•				•	•				•				•				•	Cuttings of half-mature stems in spring
Succulents: Century plant	*Agave*	•					•							•		•				•	Remove offsets
	Aloe	•					•	•					•			•				•	Seed or offsets
Ice plant	*Aptenia*	•					•						•			•				•	Stem cuttings
Pony-tail	*Beaucarnea*	•					•	•					•			•				•	Remove offset
	Bowiea	•						•	•				•		•					•	Remove offsets
Propeller, rattlesnake, scarlet paintbrush, etc.	*Crassula*	•					•	•					•	•		•				•	Seed or cuttings
	Eschiveria	•					•	•						•		•				•	Offsets or stem cuttings
Poinsettia, Crown-of-thorns, etc.	*Euphorbia*	•					•	•					•			•				•	Cuttings
Tiger jaws	*Faucaria*			•			•							•		•				•	Seed or cuttings
Baby toes	*Fenestraria*			•			•							•		•				•	Seed or cuttings
Ox-tongue	*Gasteria*	•						•	•				•			•				•	Seed or offsets

App. A / Interiorscaping Plants and Their Preferred Conditions

COMMON NAME	BOTANICAL NAME	SOIL MIX					LIGHT				WATER				TEMPERATURE			HUMIDITY			PROPAGATION
		All purpose	High humus	Gritty lean	Loose medium	High acid	Sunny	Semi-sunny	Semi-shady	Shady	Keep wet at all times	Keep evenly moist	Approach dryness between waterings	Let dry out between waterings	Cool	Average house	Warm	Very moist	Moist	Average house	
Zebra or wart	*Haworthia*	•					•	•						•		•				•	Seed or offsets
	Kalanchoe	•					•	•						•	•	•	•			•	Potting plantlets, seed, or tip cuttings
Living stones, stone face	*Lithops*			•			•						•			•				•	Seed in spring
	Pachyveria	•					•	•						•		•				•	Offsets or stem cuttings
Devil's backbone	*Pedilanthus*	•					•	•					•			•				•	Cuttings in spring
Mistletoe cactus	*Rhipsalis*	•		•				•	•			•					•	•	•		Cuttings
	Scilla	•						•	•			•				•			•		Offsets in autumn
Donkey's tail, coral beads, etc.	*Sedum*	•					•	•						•	•	•	•			•	Seed, cuttings, or division
Starfish flower	*Stapelia*			•			•	•					•			•				•	Division or cuttings in late spring or summer
String-of-pearls	*Senecio rowleyanus*	•					•	•						•		•				•	
Sweet-olive	*Osmanthus fragrans*	•					•	•					•		•					•	Cuttings of half-ripened wood in summer
Tibouchina	*Tibouchina*	•						•	•			•			•					•	Cuttings
Tolmeia (Piggy-back)	*Tolmeia*		•					•	•		•	•			•					•	Pin baby plants into damp soil
Tulbaghia	*Tulbaghia*	•					•	•				•				•				•	Offsets in spring or fall
Voodoo plant	*Hydrosme rivieri*	•					•	•				•				•				•	Remove offsets from tuber
Walking iris	*Neomarica northiana*	•					•	•				•			•				•		From small plants formed by flowers
Wandering Jew	*Gibasis*	•					•	•	•			•				•			•		Cuttings anytime
	Setcreasea	•					•	•	•				•			•				•	Cuttings anytime
	Tradescantia	•					•	•	•				•			•				•	Cuttings anytime
	Zebrina	•					•	•	•			•				•			•		Cuttings anytime
Yesterday, Today, and Tomorrow	*Brunfelsia calycina*	•						•				•				•			•		Cuttings
Zebra plant	*Aphelandra*	•						•				•				•			•		Tip cuttings in spring

Source: "House Plants: Indoors/Outdoors"; Ortho Books, Ortho Div. of Chevron Chemical Co.; San Francisco, 1974.

B TUBING AND PIPE FITTINGS NOMENCLATURE

TEE FITTINGS

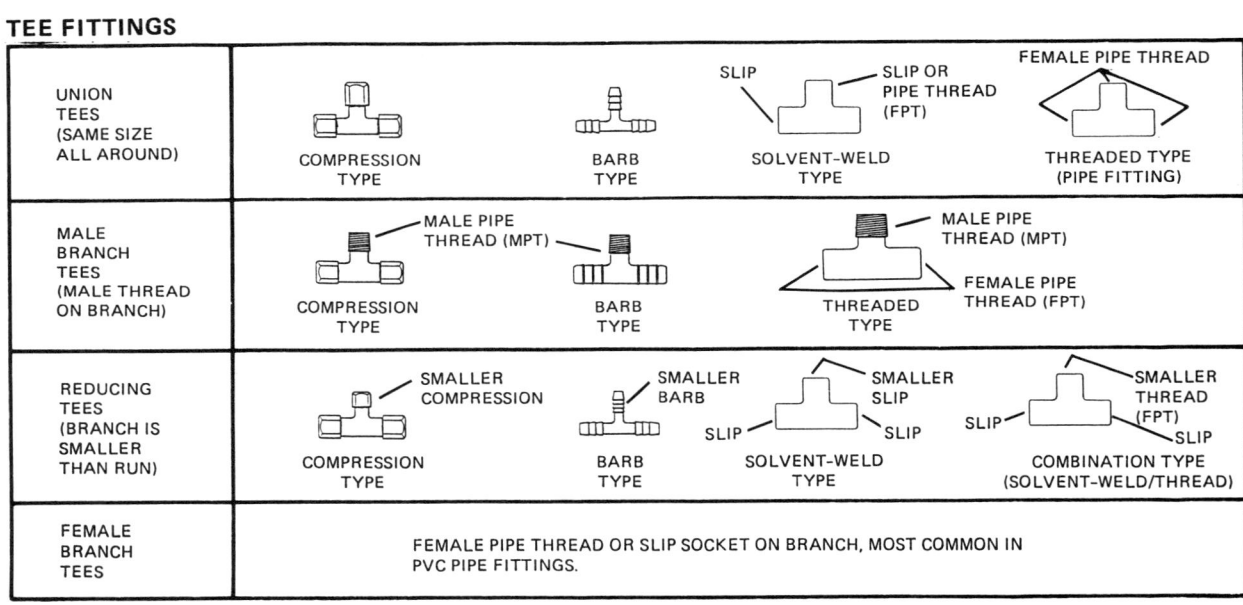

ELBOW FITTINGS (ALSO CALLED "ELLS")

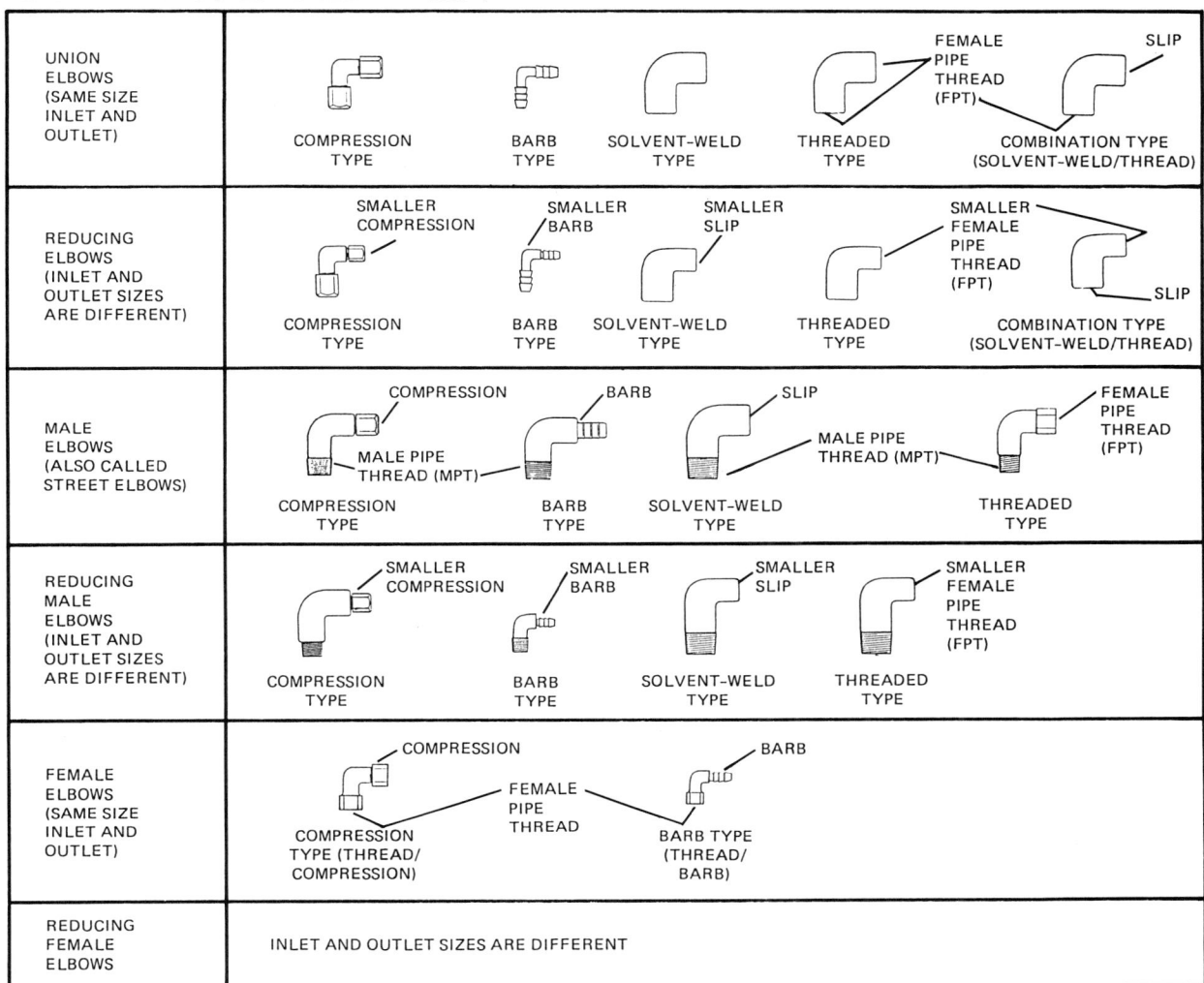

App. B / Tubing and Pipe Fittings Nomenclature

STRAIGHT FITTINGS

MALE CONNECTORS, ADAPTERS OR COUPLINGS (SAME SIZE INLET AND OUTLET)	COMPRESSION TYPE (COMPRESSION, MALE PIPE THREAD (MPT)); BARB TYPE (BARB, MALE PIPE THREAD (MPT)); SOLVENT-WELD TYPE (SLIP, MALE PIPE THREAD (MPT)); THREADED TYPE (FEMALE PIPE THREAD (FPT), MALE PIPE THREAD (MPT))
REDUCING MALE CONNECTORS	INLET AND OUTLET SIZES ARE DIFFERENT
FEMALE CONNECTORS, ADAPTERS OR COUPLINGS (SAME SIZE INLET AND OUTLET)	COMPRESSION TYPE (COMPRESSION, FEMALE PIPE THREAD (FPT)); BARB TYPE (BARB, FEMALE PIPE THREAD (FPT)); SOLVENT-WELD TYPE (SLIP, FEMALE PIPE THREAD (FPT)); THREADED TYPE (FEMALE PIPE THREAD (FPT), FEMALE PIPE THREAD (FPT))
REDUCING FEMALE CONNECTORS	INLET AND OUTLET SIZES ARE DIFFERENT
SLIP COUPLINGS (SAME SIZE INLET AND OUTLET)	SOLVENT-WELD TYPE (SLIP, SLIP)
REDUCER SLIP COUPLINGS	INLET AND OUTLET SIZES ARE DIFFERENT
BUSHINGS (SIMILAR TO REDUCING ADAPTERS BUT MORE COMPACT)	THREADED TYPE (PIPE FITTING) (SMALLER FEMALE PIPE THREAD (FPT), MALE PIPE THREAD (MPT)); SOLVENT-WELD TYPE (SMALLER SLIP, SLIP); COMBINATION TYPE (SMALLER FEMALE PIPE THREAD (FPT), SLIP)
NIPPLES (OR MALE UNIONS)	MALE PIPE THREAD (MPT)
PLUGS AND CAPS	THREADED PLUG (PIPE FITTING) (NO OUTLET, MALE PIPE THREAD (MPT)); SOLVENT-WELD TYPE PLUG (NO OUTLET, SLIP); BARB TYPE PLUG (TUBING FITTING) (NO OUTLET, BARB); THREADED CAP (NO OUTLET, FEMALE PIPE THREAD (FPT)); SOLVENT-WELD CAP (NO OUTLET, SLIP)

HOSE ADAPTERS

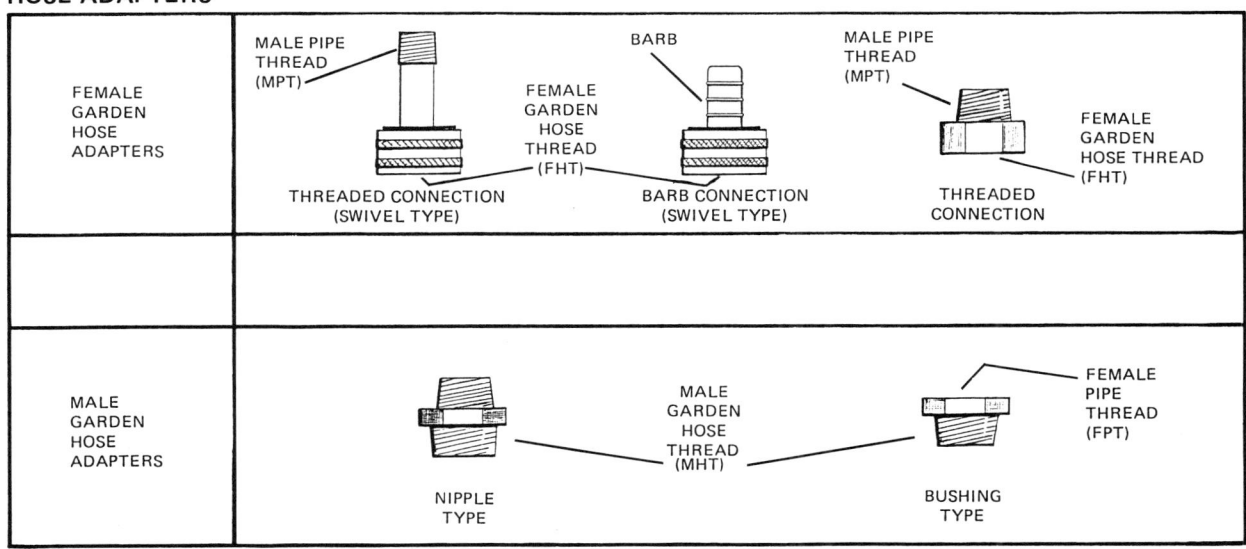

C SOURCES OF SUPPLY

The following lists are by no means complete, but they are representative of the most important supply sources.

MANUAL WATERING MACHINES—Manufacturers
Aquamatic Systems, Inc.
50 Aleppo Street, #5
Providence, Rhode Island 02909

Cascade Designs, Inc.
(The Watering Machine Brand)
157 Clifford Street
Providence, Rhode Island 02903

AUTOMATIC SPRINKLER SYSTEMS—Manufacturers
Champion Brass Manufacturing Co., Inc.
1460 North Naud Street
Los Angeles, California 90012

K-Rain Manufacturing Corp.
1640 Australian Avenue
Riviera Beach, Florida 33404

L. R. Nelson Corp.
7719 North Pioneer Lane
Peoria, Illinois 61615

Rain Bird Sales, Inc.
Turf Division
145 North Grand Avenue
Glendora, California 91740

Rain Master Irrigation
3673 Citronella Street
Simi Valley, California 93063

Richdel, Inc.
1851 South Roop Street
Carson City, Nevada 89701

Roberts Irrigation Products
700 Rancheros Drive
San Marcos, California 92069

App. C / Sources of Supply

Ross Sprinkler Co.
2543 Strozier Avenue
South El Monte, California 91733

Royal Coach/Buckner Sprinklers, Inc.
4381 North Brawley Avenue
Fresno, California 93711

Safe-T-Lawn, Inc.
5350 N.W. 165th Street
Hialeah, Florida 33014

The Toro Company
Irrigation Division
5825 Jasmine Street
Riverside, California 92504

Weathermatic
3301 West Kingsley Road
Garland, Texas 75218

DRIP/TRICKLE IRRIGATION SYSTEMS—Manufacturers

Agrifim Irrigation, Inc.
3081 East Hamilton Street
Fresno, California 93721

Chapin Watermatics, Inc.
740 Water Street
Watertown, New York 13601

Garden America Corporation
(Drip Mist Brand Watering Systems)
Post Office Box A
Carson City, Nevada 89702

Global Irrigation Corp.
8405 Artesia Blvd.
Buena Park, California 90621

Indigo Technologies, Inc.
(Agro-Drip Brand)
2405-650 West Georgia Street
Vancouver, British Columbia
V6B 4N8 Canada

Natafim Irrigation, Inc.
104 South Central Avenue
Valley Stream, New York 11580

Pepco Extruded Products, Inc.
4870 West Jacquelyn Avenue
Fresno, California 93722

Rain Bird Sales, Inc.
Turf Division
145 North Grand Avenue
Glendora, California 91740

Reed Irrigation Systems
Post Office Box X
El Cajon, California 92022

Roberts Irrigation Products
700 Rancheros Drive
San Marcos, California 92069

The Toro Company
Irrigation Division
5825 Jasmine Street
Riverside, California 92504

SUBIRRIGATION SYSTEMS—Manufacturers
Hydrachem Corp.
(Hydropore Brand Pipe)
17400 Dallas Parkway, Suite 105
Dallas, Texas 75252

Pepco Extruded Products, Inc.
4870 West Jacquelyn Avenue
Fresno, California 93722

Rain Bird Sales, Inc.
Turf Division
145 North Grand Avenue
Glendora, California 91740

Reed Irrigation Systems
(Bi-Wall Brand Pipe)
Post Office Box X
El Cajon, California 92022

Leaky Pipe Enterprises, Inc.
Post Office Box 310
Columbus, Texas 78934

Subterrain Irrigation Co.
1534 East Edinger
Santa Ana, California 92705

TPR Products
(MPS Capillary Irrigation/Mona System Brand)
Post Office Box 146
Prairie View, Illinois 60069

SELF-WATERING PLANTERS—Manufacturers
Akvamatik
2312 Central Avenue
Middletown, Ohio 45042

Chatelain Industries
Post Office Box 532
Chester, NY 10918

City Gardens, Inc.
(Water Disc Brand)
455 Watertown Street
Newton, MA 02160

Dr. Jardinier Self-Watering Planter Systems, Inc.
(Dr. Jardinier Brand)
Fresno, California

Hydro-Planter Designs, Inc.
(Genie Brand)
5510 Ambler Drive, #7
Mississauga, Ontario, Canada L4W 2V1

International Concepts, Inc.
(Grossfillex Brand)
11200 Westheimer, #320
Houston, Texas 77042

Planter Technology
(Natural Spring Brand)
999 Independence Avenue
Mountain View, California 94043

Plant Minder, Inc.
(Plant Minder Brand)
22582 Shannon Circle
Lake Forest, California 92630

Plasticom, Inc.
(Watermatic Brand)
Post Office Box 368
Southbridge, MA 01550

Riviera Systems, U.S.A.
(Riviera Brand)
2800 NW 5th Avenue
Miami, Florida 33127

TPR Products
(Mona System Brand)
Post Office Box 146
Prairie View, Illinois 60069

PRECISION INTERIOR IRRIGATION SYSTEMS—Manufacturers
Aqua/Trends, Division of Boca Automation, Inc.
(Micro-Irrigation™ Brand automated interior plant-care systems)
810444 Woodland Station
Boca Raton, Florida 33481

CONSULTING, DESIGN, AND ENGINEERING SERVICES—Automated Interior Plant-Care Systems
Ayers, Snyder and Locke
Post Office Box 164
Boca Raton, Florida 33429

MOISTURE METERS—Manufacturers
AMI Marketing Services
(Instamatic Brand)
Post Office Box 148
Ronkonkoma, New York 11779

Green Thumb Products
(Green Thumb Brand)
Apopka, Florida 32703

Progressive International Corp.
8300 Military Road, South
Seattle, Washington 98108

BIBLIOGRAPHY

"A Look at the Industry—Its Dimensions and Prospects." *Interior Landscape Industry.*

ABRAMSON, A. B., AND CLIFFORD STANLEY. "Intelligent Building Systems." *Forbes* (June 17, 1985).

AMERICAN LUNG ASSOCIATION. "Air Pollution in Your Home." New York. September 1986.

"All About Fertilizers, Soils and Water." Ortho Books [Chevron Chemical Co.]. San Francisco, 1979.

ALLSOPP, BRUCE. *Towards a Humane Architecture.* London: Frederick Muller, Ltd., 1974.

"America's Office Needs—1985 to 1995." Editorial. *Buildings* (January 1987): 46.

AUSTIN, RICHARD L. *Designing With Plants.* New York: Van Nostrand Reinhold Co., 1982.

BARNHART, GARY S. "Landscape Architecture: The Inside Story." *Interiorscape* (September/October 1985): 22.

BENTLEY, MAXWELL. *Hydroponics Plus.* Sioux Falls, SD: O'Connor Printers, 1974.

BRILL, MICHAEL. *The Impact of Office Environment on Productivity and Quality of Working Life.* Buffalo, NY: Buffalo Organization for Social and Technological Innovation, 1982.

BRUNDAGE, JOHN F., M.D.; ROBERT MCN. SCOTT, M.D.; WAYNE M. LEDNAR, M.D.; DAVID W. SMITH, M.S.; and RICHARD N. MILLER, M.D. "Building-Associated Risk of Febrile Acute Respiratory Diseases in Army Trainees." *Journal of the American Medical Association* (April 8, 1988): 2108.

"Business Profile." A survey by *Western Landscape News* (November 1982).

CALOZ, JACK. "Wiring For Intelligence." Administrative Management (January, 1987).

CATHEY, DR. HENRY M., AND LOWELL E. CAMPBELL. "Light Sources for Interior Plants." *Interior Landscape Industry* (April 1984): 34.

CUNNINGHAM, MILES. "Easier Living in Homes That Think." *Insite Magazine—The Washington Times* (June 8, 1987): 42.

DAGOSTINO, FRANK R. *Mechanical And Electrical Systems in Building.* Reston, VA: Reston Publishing/Prentice-Hall, Inc., 1982.

DAWSON, KITTY, AND ANDREW FEINBERG. "Building Intelligent Offices." *Venture* (October 1984): 90.

DIETSCH, DEBORAH. "Intelligent Building—IT INTO IQ." *Interiors* (April 1985): 116.

DIETSCH, DEBORAH. "Picking the Brain of the Intelligent Building." *Interiors* (April 1985): 13.

Dodge Assemblies Cost Data. Princeton, NJ: McGraw-Hill Information Systems Co., 1988.

DOUGLAS, JAMES SHOLTO. *Advanced Guide To Hydroponics.* London: Pelham Books, 1976.

DOWNS, ANTHONY/BROOKINGS INSTITUTION. "How Much Office Space Will We Need in the Future?" *National Real Estate Investor* (June 1984): 20.

DUNLOP, JOHN T., ed. *Automation and Technological Change.* Englewood Cliffs, NJ: Prentice-Hall, Inc., 1962.

ELBERT, V. F., AND G. A. ELBERT. *The Houseplant Decorating Book*. New York: E. P. Dutton, 1977.

FAUST, JOAN LEE. *New York Times Book of Houseplants*. New York: Quadrangle/The New York Times Book Company, 1973.

FREUNDLICH, NAOMI J. "Purify Air the Space-Station Way: With Plants." *Popular Science* (August 1986): 73.

GAINES, RICHARD L. *Interior Plantscaping*, New York: Architectural Record Books, 1977.

GALLAGHER, COLLEEN. "Plants for Better Breathing." *The Miami Herald* (July 10, 1988): 1H.

GILLETTE, BECKY. "Green Machine." *Home Mechanix* (November, 1988): 56.

GILLETTE, BECKY. "Indoor Pollution Solution." *Rodale's Practical Homeowner* (September 1987): 18.

GILLIATT, MARY. *The Complete Book of Home Design*. Boston: Little, Brown and Co., 1984.

GILMORE, V. ELAINE. "Smart House." *Popular Science* (August 1988): 42.

GOODBAN, WILLIAM T., AND JACK J. HAYSLETT. *Architectural Drawing and Planning*. New York: McGraw-Hill Book Co., 1972.

GROOVER, MIKELL P. *Automation, Production Systems, and Computer-Integrated Manufacturing*. Englewood Cliffs, NJ: Prentice-Hall, Inc., 1987.

Habitability Data Handbook. Volume 2. Houston, TX: NASA, Habitability Technology Section, Spacecraft Design Division, July 31, 1971.

HAGUE, WILLIAM E. *The Complete Basic Book of Home Decorating*. Garden City, NY: Doubleday & Co., 1976.

HAMILTON, DAVID L. "Foliage vs. Roots." *Interiorscape* (September/October 1985): 34.

HARVEY, SUZETTE. "Smart Buildings: High Tech Amenities Help Tenants, Leasing Agents Say." *South Florida Business Journal* (December 8, 1986): 82.

HAWKINS, WILLIAM J. "Smarter House—Now." *Popular Science* (August 1988): 56.

HEDBERG, NANCY ANDERSON. "Curing the Sick Building Syndrome." *Buildings* (March 1987): 70.

HEDBERG, NANCY ANDERSON. "What's All the Fuss About the Sick Building Syndrome?" *Buildings* (February 1987): 58.

HENRICH, CRAIG A. "Editorial." *Buildings* (August 1987): 15.

HETTEMA, ROBERT M. *Mechanical and Electrical Building Construction*. Englewood Cliffs, NJ: Prentice-Hall, Inc., 1984.

"Home Automation Turns on More Homeowners and Builders." *The Wall Street Journal* (January 14, 1988).

HORNUNG, WILLIAM J. *Plumber's and Pipefitter's Handbook*. Englewood Cliffs, NJ: Prentice-Hall, Inc., 1984.

"House Plants Filter Out Air Pollution, Says Researcher." *The New York Times* (September 3, 1985): C3.

"House Plants—Indoors/Outdoors." San Francisco: Ortho Books (Chevron Chemical Co.), 1974.

HYLAND, ROBERT. "Modern Technology Solves Personnel Problems: Interior Landscape Irrigation." *Western Landscaping News* (February 1983): 28.

"Indoor Air Pollution: Buildings That Make You Ill." *Air Quality Communique*. Jacksonville, FL: American Lung Association of Florida, Summer 1986.

Indoor Air Pollution—In the Office. New York: American Lung Association, January, 1986.

"Indoor Air Quality Act of 1987." House Bill #H.R. 3809 pending before U.S. Congress. Washington, DC: U.S. House of Representatives, December 18, 1987.

"Intelligent Building Definition Book." Washington, DC: Intelligent Buildings Institute, 1987.

JOHNSTON, JOANN. "Specimen Containers and Specimen Weights. *Interiorscape* (May/June, 1986): 14.

KAPLAN, S., AND KAPLAN, R. *Humanscape: Environments for People.* North Scituate, MA: Duxbury Press, 1978.

KELSEY, PATRICIA A. "NASA Studies Plants To Help Clean Up 'Trapped' Indoor Air." *Air Conditioning, Heating and Refrigeration News* (December 19, 1988): 30.

KINGAARD, JAN. "Artificial Lights and Plant Responses." *Interior Landscape Industry* (March 1982): 54.

KINGAARD, JAN. "The Team Approach to Interior Landscaping." *Western Landscaping News* (October 1982): 32.

KLEEMAN, JR., WALTER B. *The Challenge of Interior Design.* Boston: CBI Publishing Company, 1981.

KLEIN, JUDY GRAF. *The Office Book.* New York: Facts on File, Inc., 1982.

KRAMER, JACK. *Drip System Watering.* New York: W.W. Norton & Co., 1980.

LANG, BURNETT, MOLESKI, AND VACHON. *Designing for Human Behavior: Architecture and the Behavioral Sciences.* Stroudsburg, PA: Dowden, Hutchinson and Ross, Inc., 1974.

LIPMAN, JOANNE. "Smart Doesn't Always Mean Better in Computer-Controlled Buildings." *Wall Street Journal.*

MACFADYEN, DAVID J. "The Home of the Future." *Radio-Electronics* (May 1987): 115.

MANAKER, GEORGE H. "A Planter Primer." *Interior Landscape Industry* (January 1984): 83.

MANAKER, GEORGE H. *Interior Plantscapes.* Englewood Cliffs, NJ: Prentice-Hall, Inc., 1987.

MCDUFFIE, ROBERT F. "The Greening of Interiors." *Interior Landscape Industry* (June 1984): 29.

Means Mechanical Cost Data Book. Kingston, MA: R. S. Means Company, 1988.

MIKELLIDES, BYRON. *Architecture for People.* New York: Holt, Rinehart and Winston, 1980.

MOREY, JEFF. "1986 Interiorscape Contractor 25." Brantwood Horticultural Research Division. *Interiorscape* (September/October 1986): 41.

MOREY, JEFF, AND STUART D. SNYDER. "New Trends for Interiorscapers." *Interiorscape* (May/June 1984): 38.

NAISBITT, JOHN. *Megatrends.* New York: Warner Books, 1982.

National Construction Estimator. Carlsbad, CA: Craftsman Book Co., 1987.

NEWMAN, OSCAR. "Whose Failure Is Modern Architecture?" *Architecture for People.* New York: Holt, Rinehart and Winston, 1980.

NORMAN, JAMES. "Color Your Office a Healthy Green." *Medical Economics* (January 16, 1989): 63.

NUCKOLLS, JAMES L. "Intelligent Building." *Interiors* (April 1985): 101.

NYDELE, ANN. "Are Intelligent Buildings Intelligent Choices?" *Buildings* (May 1987).

NYDELE, ANN. "Controls for Building Management" *Buildings* (March 1988): 74.

NYDELE, ANN. "Future Trends in Hotel Design" *Buildings* March 1988): 60.

"101 Ways to Love, Grow and Care for House Plants." *Woman's Day* Creative Series. *Woman's Day* (March 1981): 80.

ORANS, MURIEL. *Houseplants and Indoor Landscaping.* Barrington, IL: A. B. Morse Countryside Publications, 1973.

OSTER, MAGGIE. *The Green Pages.* New York: Ballantine Books, 1977.

OWENS, DORY. "The Smart Building." *The Miami Herald* (August 12, 1985).

PAREDES, MADELEINE. "Building Designs Often Overlook Maintenance Needs." *South Florida Business Journal.* Miami. (May 11, 1987): 34.

PAZNIK, MEGAN JILL. "Intelligent Buildings Get Smart Enough to Save You a Bundle." *Administrative Management* (January 1987): 25.

"Potted Plants Pose No Mold Problem. . . . And The Right Ones Can Actually Clean Up Your Indoor Air." *Rodale's Allergy Relief Magazine* (December, 1988): 7.

PIERCEALL, GREGORY M. "Interiorscapes: Planning, Graphics, and Design." Englewood Cliffs, NJ: Prentice-Hall, Inc., 1987.

PRECIOUS, TOM. "Housing for the Masses May Soon Be Castles of Comfort." *The Washington Post* (May 24, 1987): E1.

PRINCE, T. A., AND T. L. PRINCE. "How Many Are Saying It with Flowers?" *Interiorscape* (September/October 1985): 46.

PULGRAM, WILLIAM L., AND RICHARD E. STONIS. *Designing the Automated Office.* New York: Whitney Library of Design, 1984.

RADFORD, A., AND G. STEVENS. *Cadd Made Easy.* New York: McGraw-Hill Book Co., 1987.

ROBINSON, NANCY. "Interior Designers: Shattering the Myths." *Decor* (May 1987): 168.

SAINT, ANDREW. "The Image of The Architect." New Haven, CT: Yale University Press, 1983.

SCHNEIDER, CLAUDINE. "The Indoor Pollution Burden." *EPA Journal.* Washington, DC: U.S. Environmental Protection Agency, August 1986.

SCRIVENS, STEPHAN. *Interior Planting in Large Buildings.* New York: Halsted Press/John Wiley and Sons, 1980.

SEABROOK, CHARLES. "Office Maladies Linked to Unclean Air." Science/Medicine Section. *The Atlanta Journal* (September 15, 1987): 48.

SHELDON, L. S., R. W. HANDY, T. D. HARTWELL, R. W. WHITMORE, H. S. ZELON, AND E. D. PELLIZZARI (Volume I); L. S. SHELDON, H. S. ZELON, J. SICKLES, C. EATON, AND T. D. HARTWELL (Volume II). *Indoor Air Quality in Public Buildings.* Volumes I and II. Washington, DC: U.S. Environmental Protection Agency, September, 1988.

SHERMAN, BETH. "Spacecraft Interiors Present Designers with Lofty Challenge." *The Palm Beach Post* (July 19, 1987): 3H.

"Sick Carpets" and "All Things Considered." Radio Program. Washington, DC: National Public Radio Network, May 17, 1988.

SKURKA, NORMA. "The New York Times Book of Interior Design and Decoration." New York: Quadrangle/The New York Times Book Company, 1976.

SMITH, JUDY. "The Friendly Skies Add Greenery at LAX: Drip System to Reduce Maintenance Costs in Airport Interiorscape." *Interiorscape* (July/August 1983): 48.

SMITH, RALPH LEE. "Smart House: The Coming Revolution in Housing." *Electronic House.* (July, 1988): 63.

SNYDER, STUART D. "Popularity of Living Plants in Offices Means Opportunity for Contractors." *Services* (August 1984): 45.

Statistical Abstract of the United States. 1986.

STUBBS, M. STEPHANIE. "Technical Education for Architects." *Architecture.* Washington, DC: The American Institute of Architects, August 1987.

STUSTER, JACK W. "Space Station Habitability Recommendations Based on a Systematic Comparative Analysis of Analogous Conditions." NASA Contractor Report #3943. Moffett Field, CA: NASA, Ames Research Center, 1986.

Sunset House Plants—How to Choose, Grow, Display. Menlo Park, CA: Lane Publishing Co., 1983.

"Survey of the Landscape Business—1982." *Western Landscaping News* (November, 1982).

TANNER, OGDEN. *Garden Rooms; Greenhouse, Sunroom and Solarium Design.* New York: Linden Press; Simon and Schuster, 1986.

THAISZ, DEWEY DE BUTTS. "Maintenance Considerations in Interior Landscape Design." *Interior Landscape Industry* (March 1984): 25.

"The Inside Story—A Guide to Indoor Air Quality." Washington, DC: U.S. Environmental Protection Agency, September 1988.

TRESIDDER, JANE, AND CLIFF STAFFORD. "Living Under Glass." New York: Clarkson N. Potter, Inc., 1986.

TYE, LARRY. "The Menace Within: Indoor Air Pollution, A Crisis in the Making." *The Courier-Journal Magazine.* Louisville, KY. (September 15, 1985).

WALLACE, LANCE A. "The Total Exposure Assessment Methodology (TEAM) Study—Volume I." Office of Research and Development, U.S. Environmental Protection Agency. Washington (June, 1987).

WALLACH, PAUL I., AND DONALD E. HEPLER. *Reading Construction Drawings*. New York: McGraw-Hill Book Co., 1981.

WALSH, MICHAEL. "Towers with Minds of Their Own." *Time* (June 24, 1985): 77.

WOLVERTON, B. C. "Bio-Technology in Housing." Paper presented at the Convention of the National Association of Home Builders. NASA, National Space Technology Laboratory, MS, January 19, 1986.

WOLVERTON, B. C. "Houseplants, Indoor Air Pollutants and Allergic Reactions." Research Report. NASA, National Space Technology Laboratory, MS, December 1986.

WOLVERTON, B. C., R. C. MCDONALD, AND E. A. WATKINS, JR. "Foliage Plants for Removing Indoor Air Pollutants from Energy-Efficient Homes." *Economic Botany* (February 1984): 224.

WOLVERTON, B. C., R. C. MCDONALD, AND HAYNE H. MESICK. "Foliage Plants for Indoor Removal of the Primary Combustion Gases, Carbon Monoxide and Nitrogen Dioxide." *Journal of the Mississippi Academy of Sciences*. Volume XXX. (1985).

WOLVERTON, B. C. "A Study of Interior Landscape Plants for Indoor Air Pollution Abatement." NASA. Stennis Space Center, MS (October, 1988).

WOLVERTON, B. C., ANNE JOHNSON AND KEITH BOUNDS. "Interior Landscape Plants for Indoor Air Pollution Abatement—Final Report." NASA. Stennis Space Center, MS (September, 1989).

WRIGHT, MICHAEL, ed. *The Complete Indoor Gardener*. New York: Random House, 1979.

YLVISAKER, PETER N. "Air Quality: Is It (Wheeze; Cough!) Time To Test?" *Buildings* (May, 1989): 62.

INDEX

Acclimatization of interior plants, 53
Adjustable mini-valve/emitters, 167, 207
Air chamber (anti-water hammer), 231
Air circulation around interiorscapes, 65
Air quality around interiorscapes, 66
American Cancer Society, indoor air pollution and the, 25
American Lung Association, indoor air pollution and the, 25
Architects, 42
Architects, education of, 31
Architects, technology and, 31
Architecture, humanistic approach in, 13
Architecture, team approach in, 14, 31
Asset managers, 41
Auto-interigation, 3, 85
Automated interior plant-care, 2, 28, 84
Automatic drip/trickle systems, 96
Automatic interior irrigation systems, 3, 85
Automatic irrigation in commercial buildings and facilities, 174
Automatic irrigation in office buildings, 176
Automatic sprinkler systems, 55, 84, 89
Automatic sprinkler systems, cycle timing of, 92
Automatic sprinkler systems, emitters for, 90

Backflow preventers (vacuum breakers), installation of, 232
Barb fittings, 159
Bubbler heads, 90, 94
Buffalo Organization for Technological and Social Innovation (BOTSI), 20
Builders/construction companies, 41
Building design and technology, 30
Building developers, 41
Building intelligence, defined, 32, 35
Building managers, 41
Building Owners and Managers Institute, International (BOMI), 41
Building related illness (BRI), 24
Buildings and technology, 31
Buildings as functional structures, 31
Buried reservoir systems, 82, 103
Burolandschaft, 21

CAD systems, 1, 228
Capillary action and moisture diffusion, 80, 83, 129
Carrier, frequency remote control systems, 184, 197
Check-valves, in Micro-Irrigation Systems, 163, 210, 220
Check-valves, installation of, 233
Chlorophyll, 60
Collaborative approach, in architecture and interior design, 14
Compression fittings in Micro-Irrigation Systems, 159
Computer-aided design and costing, 228
Computer control in housing, 38
Computer-controlled power management systems, 232, 256, 258
Computer-regulated structures, 34, 199
Conklin, Everett, 47
Container sizes for decorative plants, 56
Control centers, location of, 177, 196, 207
Cost estimating worksheets, 216, 229

Defining the design problem, 202
Design and costing of Micro-Irrigation System installations, 202, 210
Drip/trickle irrigation systems, 55, 85, 96
    cycle timing used, 98
    emitters used, 98
    operating pressures used, 99
    outdoor applications, 102
    tubing used, 99

Effect of emitters on water flow, 207
Effect of gravity on water flow, 206
Effect of tubing fittings on water flow, 206
Emitter stabilizers in Micro-Irrigation Systems, 173
Emitter tubes, installation of, 250
Emitters used in drip/trickle irrigation systems, 98
Emitters used in Micro-Irrigation Systems, 167
Emitters used in sprinkler irrigation systems, 90
Engineers, 44
Environmental Protection Agency (EPA), indoor air pollution and the, 25

Facility managers, 41
Fitting specifications, 161
Fixture schedules, 215, 229
Floor boxes, installation of, 237
Floriculture, production statistics, 6
Flow characteristics of tubing, 205
Flow reducers, Micro-Irrigation Systems, 166
Footcandles, 63
Fountain plate emitters, Micro-Irrigation Systems, 169
Fungus disease in decorative plants, 94
Future outlooks, technology in commercial buildings, 267
Future outlooks, "Smart House" technology, 269
Future outlooks, automated, interior plant-care, 273
Future outlooks, decorative plants in buildings, 271
Futuristic scene, 274

Grow pots (production containers), 56
Growing medium, 56, 67

Hand irrigation methods, 71
Hand watering methods, 73
High technology in commercial buildings, 31, 267

Index

High technology in housing, 37, 269
High-touch, 1, 8
Home control systems, 38, 197, 256
Houseplants vs. interiorscaping plants, 52
Housing, technology in, 37
Humidity in interior plant environments, 65
Humus in interior plant growing media, 56
Hydroponic irrigation systems, 85, 105

Improved tropical plants, 23
Indoor air pollution, and foliage plants, 24
  and the EPA, 25
  building related illness and, 24
  NASA environmental research and, 1, 66
  physical effects of, 25
  the American Cancer Society and, 25
  the American Lung Assn. and, 25
Installations, layout considerations, 229
Installing,
  control center equipment, 230, 248, 252, 256
  outdoor Micro-Irrigation Systems, 252
  water distribution networks (tubing lines), 232
Intelligent buildings, 3, 31
  Intelligent Buildings Institute (IBI), 35
  building associations and, 35
  the appeal of, 36
  the economics of, 36
Interfacing Micro-Irrigation Systems with other technologies, 256, 258
Interior design,
  humanistic, 13
  team approach in, 14
Interior designers, 31, 43
  and technology, 31
  education of, 31
Interior irrigation system plans, 204, 213
Interior plantscapers, 6, 23, 45
Interiorscape irrigation,
  fully automatic techniques, 89
  practices in, 45, 50
Interiorscape maintenance technicians, 71
Interiorscapers, 6, 23, 45
Interiorscaping- industry profile, 47
Interiorscaping plants,
  atmospheric conditions required by, 65
  basic growth requirements, 60
  biological factors of, 60
  characteristics of, 56
  growth cycle, 60
  light requirements of, 62
  nursery irrigation of, 55
  nutrients required by, 64
  pests and diseases, 69
  sizes, 56
  soil moisture requirements of, 62
  soil used, 56, 67
  varieties, 57
International Intelligent Buildings Association (IIBA), 35
Irrigation contractors, 50
Irrigation, interior, practices in, 45, 50

Irrigation manifolds, installation of, 249
Irrigation receptacles, used in Micro-Irrigation Systems, 170, 175, 211, 217, 223
Irrigation receptacles, installation of, 234, 237, 238

Key switches, Micro-Irrigation Systems, 231

Landscape architects, 44
Laser soaker line emitter, 103, 158, 168
Leaf cleaning and polishing, 74
Loam, 56, 67
Locating control center equipment, 207, 211, 230, 240

Maintenance technicians, plants, 71
Manual fertilization, 74
Manual irrigation practices, 71, 73, 125
Manual plant-care techniques, 71
Marine applications, Micro Irrigation Systems, 200
Micro-climate, 65
Micro-Irrigation Systems, 2, 39, 50, 71, 85, 107, 269
  accessories used in, 169
  adjustable mini-valve/emitters used in, 146
  advantages of, 120
  anti-siphon devices used in, 145
  basics of operation of, 143
  categories of, 143
  check-valve use in, 163, 210, 220
  control centers of, 148, 207, 230, 240
  corporate office installations, 182
  design and costing of, 202
  designing for commercial buildings, 213
  designing for office suites, 217
  designing for single-family homes, 223
  emitters used in, 146, 167
  equipment of, 145
  fittings used in, 158
  flow control devices used in, 146, 162, 210
  high-pressure versions of, 143
  apartment buildings, 194
  apartment suites, 198
  banks, 193
  clubhouses and recreation halls, 194
  hospitals and medical offices, 194
  hotels, 185
  office workstations, 182, 184
  residential buildings, 195
  resort complexes, 186
  restaurants and lounges, 189
  retail establishments, 192
  shopping malls and arcades, 191
  single family homes, 196
  installing, 229
  integrated installations of, 175, 178, 230
  interfacing with other technologies, 256, 258
  irrigation receptacles in, 170, 175, 211, 217, 223
  low-pressure versions of, 146
  marine applications of, 200
  moisture diffusion patterns of, 129
  moisture diffusion studies with, 130
  operating procedures used with, 263

Index

Micro-Irrigation Systems (*Contd.*)
  outdoor applications of, 186, 187, 192, 198, 199
  packaged installation kits, 173
  pressure regulators used in, 166
  Pulse-Flow techniques used with, 125, 127
  pump controllers used in, 146, 152
  pump/reservoir modules used in, 146, 154
  pumps used in, 146, 154
  reservoirs used in, 154
  sequence of operation, 147
  skeleton installations of, 178
  soil composition and, 130
  solenoid valve controllers used in, 146, 149
  start-up procedures, 260
  the X-10 PowerHouse System and, 184, 256
  timers used in, 148
  tubes and pipes used in, 156
  use of mulches with, 139
  uses for, 121
  water distribution networks, 146, 156, 209
  water filters used in, 145, 230
  water pressure levels and, 145, 146
Mist irrigation systems, 55
Moisture meter modifications, 78
Moisture meters, 76
Mulches and moisture retention, 139

NAHB Research Foundation, 39
Naisbitt, John, 1, 8
National Aeronautics and Space Administration (NASA),
  biotechnical studies with plants, 2, 24, 26, 269, 273, 277
  environmental research, 2, 24, 26, 66, 269
  indoor air pollution studies, 24, 26, 66, 269, 273
National Assn. of Home Builders (NAHB), 39
Nursery irrigation, 55
Nursery soil, 56

Osmosis, 60
Overhead watering, 71

Peat moss, 67
Pests and diseases in plants, 69
Photosynthesis, 60
Plant inspection, 74
Plant replacement, 75
Plant rotation, 74
Planter box design, 95
Planter boxes, interior, 42
Planter boxes, interior, permanent (in situ), 42, 95
Planter boxes- installation of Micro-Irrigation Systems in, 251
Planter drainage, 95
Planter pits, interior, 42
Plants, interior,
  acclimatized, 50, 53
  and indoor air pollution, 2, 24, 26, 66
  as a natural element of decor, 10, 12
  as air purifiers, 2, 24, 26
  emotional effects of, 1, 8
  historical summary, 5

Plants, interior (Contd.)
  the workplace, 14
  industry statistics, 6
  leisure time and, 11
  maintenance of, 3
  publicity and, 23
  uses of, 1, 5
Plumbing symbols, 204
Portman, John, architect, 14
Pressure regulators, 166, 210, 230
Project "Smart House", 39
Property managers, 41
Pulse-Flow concept, 125
Pump controllers, 152, 231
Pump/reservoir modules, 231, 240, 253
Pump/reservoir ratings, 204

Quickborner Team, 21

Real estate administrators, 41
Real estate managers, 41
Respiration of plants, 60
Root/shoot ratio of plants, 62
Routing of water distribution networks, 209, 240

Self-watering containers, 80, 103
Self-watering containers, wicking in, 80
Shut-off valves (stop valves), Micro-Irrigation Systems, 145, 162, 230
Sick building syndrome, 2, 25
Skeleton installations, of Micro-Irrigation Systems, 178, 213, 230
Slow-release fertilizers, 74
Smart buildings, 3, 31
Smart homes, 38
"Smart House," project, 39
Soaker hoses, 103
Soil,
  acidity (pH) of, 67
  consistency of, 67
  mixes, 56
  quality of, 67, 73
  -less media, 68
  -less systems (hydroponics), 105
Soldered fittings (sweat fittings), 161
Solenoid valve ratings, 204
Solenoid-valve controllers, used in Micro-Irrigation Systems, 149, 231
Solvent welded fittings, 161
Sprinkler heads (emitters), 90
Sprinkler irrigation systems, 55, 84, 89
Start-up procedures, used with Micro-Irrigation Systems, 260
Sub-irrigation, 80, 83, 85, 103
Subterranean irrigation systems (sub-irrigation), 103
Systemic insecticides, 74

Team approach in building design, 42, 50
Technical considerations in design, 203
Technology used in "intelligent" buildings, 33
Temperature of interiorscape environments, 65
The sample layout in Micro-Irrigation System design, 203
Threaded fittings, 160

Index

Timers used in Micro-Irrigation Systems, 231
Transpiration, of plants, 60, 127
Tropical foliage nurseries, 50
Tropical foliage plants, 5, 53, 57
Tubing/fitting connections, used in Micro-Irrigation Systems, 246
Tubing, installation of, 232, 238, 240, 244, 246, 249, 250, 252, 256

Vacuum breaker (backflow preventer) used in Micro-Irrigation Systems, 145, 162, 165

Water distribution manifolds, used in Micro-Irrigation Systems, 169
Water flow variables, 205
Water stress, on plants, 73, 126
Water-gardening (hydroponics), 105
Watering machines, 71
Wicking, 68
Wolverton, B.C., Dr., 26

X-10 (U.S.A.), Inc., 39
X-10 PowerHouse System, 197, 256, 274